Our Genes,
Our Foods,
Our Choices

Better Health with the Methylation Diet

Triveni P. Shukla

authorHOUSE®

AuthorHouse™
1663 Liberty Drive
Bloomington, IN 47403
www.authorhouse.com
Phone: 1-800-839-8640

Published by AuthorHouse 9/10/2014

ISBN: 978-1-4969-2858-0 (sc)
ISBN: 978-1-4969-2857-3 (hc)
ISBN: 978-1-4969-2856-6 (e)

Library of Congress Control Number: 2014915267

To my wife, Girija, who effortlessly creates from scratch and with love, nutritious, delicious food for the mind and body, and to my children, Rekha, Renu, and Rajan, who live and eat well, inspiring me every day.

Contents

Part Four: Our Choices

List of Tables

Preface

Deoxyribonucleic Acid (DNA), the master molecule behind all human physiology, was already discovered and I heard a lecture by the two Nobel Laureates Drs. Watson and Crick in 1968 at the 100th anniversary of University of Illinois in Urbana-Champaign, IL. This had an everlasting effect on any and all things I did in food chemistry. I had begun believing in intrinsic value of food molecules even during my Master's degree program In India in relations to their daily impact on the DNA dictated operation of our body, the molecular food factory we live by. This belief became a matter of conviction by the time I finished my Ph.D. in Food Sciences at the University of Illinois at Urbana-Champaign. Almost all that happened in new food product development and its processing for mass merchandising and retail since then not only in the United States of America but other 16 major countries where I practiced my art ran counter to my belief. FDA recognized that our children ate too much sugar early in seventies, later it recognized that we consumed a lot of sodium, cholesterol, saturated fat, and trans fat. In recognition FDA came up with laws dealing with nutrition labeling and public education. Not much changed though and chronic diseases spotted early in seventies became well recognized public health problem by late eighties. This is also the prime period of my growth in science and technology. My fundamental knowledge of molecular chemistry and biochemistry of daily foods began to raise a lot of questions in my mind. Questions about the effect of too much sugar and refined carbohydrates, about the sleep deprived American lifestyle, and about a way of life not in sync with solar cycle and our natural circadian rhythms. Our daily food, I began to ponder, was misinterpreted at a large-scale. My conviction turned into an objective of public education, education about what really food molecules are and what they do to and in our body. I chose to prepare an easy to understand narrative on human beings as an electromagnetic being. This book is the narrative that I am devoted to revising more accurately in view of evolving food and nutrition sciences as often as I have to and, given the conflicting daily dose of news on and about our daily foods, I hope this narrative will help my readers learn to choose their molecules of food wisely and live a disease free healthy life..

Our daily life, health, and well-being are determined largely by the air we breathe and glucose (a six-carbon sugar) that is produced in our body by digestion of the foods we eat. The electrons in the bonds of glucose are managed in our cells' mitochondria to reduce oxygen to water. In this process of reducing oxygen to water, proton gradient is created across the inner mitochondrial membrane, and the associated electropotential energy of the proton motive force is used by the mitochondrial apparatus for making ATP, the energy molecule for our lives. Call it the life energy of the orient- well understood Qi in China, Korea, and Japan, Prana in India. Actually, evolution has dictated that our daily life depends on (1) electrons and protons, common to all living creatures on planet earth including plants; (2) our elaborate second brain, associated with our digestive system—the food-processing organ; and (3) our second genome, associated with symbiotic bacteria in our colon—a major determinant of our immune system. Here is how we human beings came to be.

Planet earth formed 4.5 billion years ago; the first living cell with a nucleus appeared on earth 3.5 billion years ago; and plants, with the power of directly harvesting the sun's energy and concomitant production of oxygen by photosynthesis, appeared 475 million years ago. The plants added oxygen to the biosphere by splitting water into hydrogen and oxygen, a process that donates electrons simultaneously. Beyond producing oxygen, plants also convert 10 trillion tons of carbon dioxide each year into foods, of which around half a trillion tons pass through the human digestive system.

Human beings evolved as a creature of oxygen only two hundred thousand years ago, based largely on oxygen, water, and plant foods. The plant foods contained not only carbohydrates, protein, fat, minerals, vitamins, and dietary fiber, but also therapeutic nutrients such as antioxidants, anti-inflammatory (willow bark aspirin) foods, anesthetics like clove oil and opioids, mild laxatives like glycoalkaloids in eggplant, and other psychoactive (coca, coffee, nicotine) food ingredients. The methylation and gene-expression diet that I emphasize in this book thus includes fruits, vegetables, seeds, nuts, herbs, and spices—foods that we humans have been consuming since antiquity. Modern science tells us that a more perfect diet must also be supplemented with nonplant nutrients that the human body can't make. The list includes vitamin B_{12}, vitamin D, essential amino acids, and omega three-type essential fatty acids. The great truth in biology is that we eat foods for stability and routine function of our nuclear genes and more than thirty-seven mitochondrial genes. Furthermore, since we evolved as organisms with the powers of movement, sensing, perception, and cognition, our choices of exercise, yoga and meditation, physical fitness, belief and faith, and social interactions are other necessary complements to the ideal methylation diet, choices that connect our minds and bodies for better health. Oxygen obtained through exercise is much more critical for life processes than food nutrients that we consume after obtaining suffiecient oxygen and water.

In effect, human beings are *beings of oxygen, glucose, and water.* In truth, they are beings of electrons and protons. Oxygen, carbon-based foods, and hydrogen make up 65, 18, and 10 percent of the human body. Oxygen is an electronegative gas that burns (helps oxidize) glucose to carbon dioxide for routine production of energy in our body. This happens with the help of electrons that move along the membrane protein complexes of mitochondria and couple proton force of "proton gradient" across membrane The latter converts into chemical energy for our daily living. We need antioxidants when oxygen goes bad because of electrons that leak out prematurely from the electron-transport system in mitochondria, convert oxygen into free radicals. Whereas our body fights free radicals with its own powerful antioxidant glutathione, we need to get most of our antioxidants from plant foods in order to stabilize DNA, express genes, make proteins, and maintain health.

Saying that "we live by glucose and oxygen" really means "we live by electrons and protons." Foods we eat become chemical energy molecules of ATP in every cell of our body with the aid of oxygen, electrons, and protons. It is proton gradient, a battery with electromotive force, that helps produce ATP, the energy molecule. Oxygen strips electrons off glucose, and thirty-two energy molecules of ATP are produced per glucose, involving ten protons for every two electrons. Water mediates as a solvent in all that happens in the body, and glucose can come from most anything that we eat—carbohydrates, protein, and even fat. A trillion trillion molecules of oxygen (512 g) as a powerful terminal electron acceptor are essential for the daily life of a seventy-kilogram person.

Our cells have 1 billion ATP molecules, like DNA in their core structure, other electron- and proton-managing molecules, and 3,870 proteins as antibodies, neurotransmitters, and receptors. All this depends on our genes. A less-than-optimally perfect diet simply can't meet the challenge of this complexity.

Each cell of our body can generate electricity that underlies all our vital signs—the heart contraction, neuron firing and brain connectivity, the eye's vision, and hearing. There are electron and proton currents across every one of our cells. Our brain and its digital cables would not work without electricity, and neither would our immune system, which often defends us by producing free radicals from oxygen on purpose, a type of molecular electron gun. Our daily life depends on interdependent and sequential execution of at least a trillion chemical and electrical signals each day. Our brain's master clock, our chronobiology, and myriads of metabolic reactions depend on these signals. Actually all temporal events of hormone secretion and release, sleeping, waking, spermatogenesis, female puberty, and signals of hunger and thirst depend on the master clock in our brain. Our genetics, epigenetics, cellular biology, physiology, cerebral response, emotional biology, health, and the state that we call *well-being* all depend on such signals, and the signals depend on what we eat. The dictum is simple: "Eat foods that keep your genes happy." I like to call such foods the *"methylation diet,"* which allows the best of gene stability and gene expression, including the genes of mitochondria, the electron and proton manager in our cells. Also, since what we eat is part of our tradition, culture, ethnicity, and social interaction, the act of eating naturally engages all our senses for maximum pleasure, meaning, and brain-to-body and brain-to-brain (interpersonal) interconnections.

A major objective of this book is to cut through the deception and misinformation perpetuated by commercial diets and to distill the therapeutic values of *common daily foods* against common ailments and chronic diseases. The hypothesis that I propose is simple in that we evolved with foods we ate for thousands of years and those foods by their routine functions are preventives against ailments and diseases, mitochondrial diseases in particular.

I have collected a compendium of research data for each chapter in this book to show that (1) many nutrients (vitamin A, members of the B group vitamins including B_2, B_6, B_9 (folic acid), and B_{12}, vitamins C, D, E, and K, omega-3 fatty acids, choline, antioxidants, and minerals such as cobalt, chromium, copper, magnesium, manganese, selenium, and zinc) are not present in the proper amount either in processed retail foods or in foods consumed away from home at fast-food restaurants, (2) a majority of foods of imbalanced nutrients are regularly consumed by large segments of our population, and (3) a majority of our population consumes foods contaminated with hazardous additives and chemicals. These trends have accelerated during the past sixty years, decade by decade. *Methylation or gene-expression nutrients* in our daily diet, therefore, have been deficient, and the deficiency has been linked to diabetes and many other chronic pains and diseases.

Another component of the methylation diet is probiotic foods (yogurt, cheese, kimchi, and sauerkraut), which help establish our second genome—the gene pool of 100 trillion bacteria in our colon. The methylation diet preserves the stability and integrity of our nuclear and mitochondrial DNA, which is more prone to mutation because of its proximity to the electron-transport chain.

Another objective of this book is to show that *lifestyle choices of exercise, yoga and meditation, stress management, and prayer in faith* complement our daily food by inducing and promoting gene expression, enhancing the immune system, increasing telomerase activity for longer life, turning good genes on and bad genes off, and even inducing nitric oxide production in favor of cardiovascular health. The rule of modern biology is that there is a gene behind whatever happens to and in our body. A lot of it happens with the genes in our mitochondria. Unhindered gene expression, therefore, is a metabolic necessity for our immune and nervous system controls, good health, and our ability to fight pain and disease.

A non-methylation diet is heavily burdened with foods containing traces of at least 646 food additives, around 14,000 chemicals, saturated and trans fats, too much omega-6 fatty acids, too little omega-3 fatty acids, too much salt, too high cholesterol, and too much sugar and sugar substitutes like high-fructose corn syrup. The combined carcinogenic and gene expression-blocking effects of bisphenol A, phthalates, zeranol, bovine growth hormone, styrene, and vinyl chloride have been devastating to our health.

There is no way for evolution to fight such an epigenetic assault. Our air and water supplies are getting more polluted day by day, and chronic stress dictates our life by default. Our precious little nuclear families are sleep-deprived, and our bodies are always in "fight" mode. The net effect of contaminated diets, water, and air is DNA damage, and genomic instability, mutations, and onset of chronic diseases as a consequence. Our digestive system that evolved with plant foods containing complex carbohydrates and dietary fiber sees very little of them anymore. Instead, we literally spike our blood with chemicals foreign to our body: fat, sugar, and alcohol.

The food laws, namely the Food, Drug, and Cosmetics Act of 1938, Nutrition Labeling and Education Act (NLEA) of 1990, Dietary Supplement Health and Education Act of 1994, Food and Drug Modernization Act

(FADMA) of 1997, and Food Safety Modernization (FSMA) Act of January 1, 2011, have in effect become inoperational in view of snack and fast foods that are mass-produced and distributed. The FDA's genuine health claims intended to promote gene expression have not been used with due honesty. Authorities such as the National Academy of Sciences, National Research Council, Food and Nutrition Board, USDA, FDA, USPHS, NIH, and professional organizations have failed to combat aggressive food marketing loaded with half or no truth.

Multiple revision of the food pyramid is a telling story of deficiencies in retail mass-produced foods as they relate to public health. Fortunately, the final version appears close to the methylation diet though. Although folic acid became a mandated dietary supplement rather recently, a significant portion of the US population suffers from deficiencies with respect to minerals such as copper, chromium, iron, manganese, magnesium, zinc, selenium, and molybdenum and with respect to essential amino acids, omega-3 essential fatty acids, B-vitamins, and antioxidants. Deficiencies of choline and methionine are now known to cause genomic instability. Carcinogens from grilled and deep fat-fried fast foods containing DNA-damaging acrylamide are now known to be associated with cancer, diabetes, and hypertension. Emphasis on superconvenience in the United States has now created two generations of chronically stressed overeaters, with total disregard to food portion, food choice, food timing, and daily nutrient balance. Our genome has become unstable and our epigenome has been corrupted in less than seventy years since World War II. Mitochondrial DNA is implicated in many neurological disorders of Alzheimer's, Parkinson's, and Huntington's, diabetes, and hypertension. On top of it all, we have succumbed to physical inactivity, sleep deprivation, and social isolation.

I propose the methylation diet not only for basic nutrients but also for (1) controlling functions of small and big food molecules; (2) managing electron-hopping from one molecule to another during energy production from glucose in our cells' mitochondria; (3) inducing proper photon and electron interaction for our vision; (4) preventing electron leakage, which in turn causes type 2 diabetes and insulin resistance; (5) managing cell-to-cell communication, neurotransmission, proton production for acidity in our stomach, pain management by voltage-gated proton channels, and proton currents in our cells and organs; (6) controlling the glucose level for heart and brain health; and (7) keeping us conscious. To do all this, our daily food must help us manage gene expression.

This book then is about the power of *common foods* to sustain and maintain disease-free health and longevity, a power that manifests by food-dictated functions of (1) our second brain—the enteric nervous system; (2) our second genome—the genetic pool of good probiotic bacteria; and (3) our daily oxygen intake, which traps electrons in a very controlled stepwise process from glucose for routine energy production. This book, I believe, will serve as an educational tool and teach our readers that

1. we are oxygen-dependent creatures;
2. our bodies make their living by transferring electrons and harvesting energy concomitantly by translocating protons;
3. diets based on high-antioxidant foods can prevent and control electrons gone amok under stressful conditions;
4. gene expression, a process that can take from seconds to hours, depends on molecular signals based on foods we eat;
5. food nutrients can control chronic stress, the three-dimensional dance of proteins in our body, production of 600 energy molecules of ATP per second, 600 rounds of ATP recycle, and 2,160 rounds of recycle of other electron-managing molecules per day (actually, nutrients punctuate our evolution day by day);
6. food nutrients, including probiotics, can keep and preserve our first genome that we get from our father and mother, the second genome that we get after the first few weeks after birth from our friendly gut bacteria, the first brain—the evolutionary construct of our first genome, the second brain of the digestive system, and our mind-body connectivity for routine health and well-being; and
7. the methylation diet can thus boost the power of the sun inside us, just as Lord Brown said years ago. The ordered structure and function of our body and mind is produced and kept by foods we eat or the energy of sun used to make our foods.

"Life is a pure flame, and we live by an invisible sun within us."
—Sir Thomas Brown
Triveni P. Shukla (the Author)

Acknowledgments

I began integrating the concept and construct of this book during the most demanding and torturous period in my life. My two daughters, Rekha and Renu, my wife, Girija, my son Rajan's lovely daughter, Asha, and I ran into health problems. I thank each member of my family, especially my wife, Girija, for their extraordinary forbearance of my habit of stealing valuable family time for preparation of this book.

Renu Zaretsky, my middle child, gave me a very valuable piece of advice: to pay attention to the "tone of text" and "tenure of technology." I am indebted for her reading of the roughest first draft. Although she doesn't know it, I sent a rough draft to her on purpose to get the most out of her reading. These two phrases that she came up with shall guide me throughout the balance of my professional undertakings and endeavors.

Dr. Vinod K. Chaudhary deserves my most sincere acknowledgment and appreciation for his reading of the same rough draft. Dr. Chaudhary gave many valuable hints and suggestions as to simplicity that finally got incorporated in the final manuscript.

I owe my deep gratitude to Dr. Anil Dwivedi, who intensely read the third draft with the keen mind of a biophysicist and made suggestions as to the clarity of a total of four fundamentals in human metabolism.

I express my gratitude to my friends Dr. Ila Misra, Dr. Devendra K. Misra, Dr. Umesh Saxena, Mr. B. P. Gupta, and Mr. Udai Bhan Pandey for helping me coin and construct the title for this book— *Our Genes, Our Foods, Our Choices*. The title came from my daughter Renu Zaretsky, but all had their input for its finality.

It is a rare book that is faultless. Mine, I know, will have many errors. Also, I know that many in food science, food technology, and nutrition sciences will argue against and take exception to my interpretation of facts.

I must say to them and to my readers that I have a science-based conviction as to the daily value of what we eat and that I must exercise my conviction. The readers of this book are the final arbiters and judges of what the second edition will look like. For surely, there shall be a second edition.

CHAPTER 1

Introduction

"Foods which promote longevity, intelligence, vigor, health, happiness, and cheerfulness and which are essence of food, substantially bland and naturally agreeable are foods dear to people of pure nature
[The sanskrit Text - Aayuh satva balaarogya sukh priti vivardhana, rasyah snigdhah sthirah hridya aahara satvikapriya, ShrimadBhagawatgeeta 17.08])

Imagine the health values we would be beneficiaries of if we didn't fall prey to the follies of food innovations like myriads of additives, partially or fully hydrogenated artificial trans fats, and chemicals used for mass agricultural production. No doubt we would be better off health-wise if all this didn't happen. We have discovered that diet is connected to chronic diseases and that methylation diets can prevent, cure, and even reverse such diseases. Fast foods imbalanced our daily diet and deficiencies of folic acid, magnesium, vitamin D, and vitamin B_{12} became prevalent in sizable segments of the American population because "variety of foods" went extinct from their daily diet. We would be much better off if we were to consume a methylation diet based on a variety of easily available nonexotic common foods. This is the main story of this book.

Human beings evolved with oxygen, water, and foods of proteins, carbohydrates, fats, minerals, and vitamins. The human body can make many life molecules except a few so-called essential nutrients such as minerals, vitamins, certain amino acids, and certain fatty acids. The evolution of human beings and other animals with oxygen is a mystery by and in itself. The fact, though, is that human beings need twice as much oxygen as all other food solids for energy production and routine maintenance of their metabolism. Combined genes from Mom and Dad at conception steer us for growth and development with foods from maternal blood and milk, and later in life from our daily food. Most anything that happens in our body is due to food we eat. In effect, food sustains us.

Food, suggests folk medicine now receiving scientific support, is a curative; it has therapeutic effects in our daily life. We need foods for immunity, stable DNA, and gene expression. We need foods for protecting our mitochondrial DNA, which sits next to where there is maximum electron activity. We need foods that augment functions of our brain for the best of consciousness and cognition. We need foods that manage and prevent pain. But we need foods with essential nutrients that our body is not evolved to make by itself.

Even the most balanced and ideal foods, however, serve best only when our daily lifestyle includes optimal exercise, mobility, sleep, mind engagement by social interaction, and a purposefully orchestrated mind-body connection for stress-free living. Food therapeutics goes beyond mere food consumption of 2,000 to 2,500 calories per day.

1.01 Routine Therapeutics

Balancing nutrition, sleeping well, exercising, committing to physical activity for oxygen, and drinking enough water constitute the essence of food therapeutics. All inputs need to be balanced, including food, oxygen, and water, for a healthy body and mind connection. We need to sleep for defragging and repairing our brain. We need to concentrate, introspect, and meditate for enhanced food-mediated cognition. We need to force our neural transmissions, and we need to live a stress-free life. We have our genes, we can select our foods, and for optimal function of the two, we need to fashion our choices of daily exercise, yoga and meditation, interpersonal relationship, faith, and spirituality. The lifestyle has a lot to do with the health of our cellular powerhouse, the mitochondria.

Nutrition of the offspring begins right at the time of conception. It is thus initially based on the mother's nutrition. The baby's nutrition during the first year is very critical, including the period of breast-feeding. Three meals a day during the later years of life should provide for routine energy needs, essential needs for fat storage, and energy for all intervening and intermediate physiological needs of the body. We need to eat foods that protect our genome and build our epigenome. Day-to-day therapeutics in essence is embedded in food selection for optimal nutrition for full gene expression.

1.02 Therapy by Calorie Control

Fewer and yet the optimal number of calories should become the norm of menu planning. Consuming small portions, consuming a lot of plant-based natural soluble and insoluble fiber, selecting low-glycemic index foods, selecting nutrient-dense and yet low-calorie foods, selecting high-volume (high air or water) low-calorie fruits and vegetables, and thus restricting total daily calorie intake are rules of common-sense food therapy. Actually, that is exactly how we evolved. Consuming whole grains, nuts, and seeds and getting antioxidant vitamins and key phytochemicals by adding spices and herbs to prepared foods for flavor and therapeutic values is a cost-effective way to live long in good health and with the best of mental aptitude. Mediterranean and Okinawa diets attest to such food values and the values of colorful foods that embody antioxidants and anticarcinogens. Such diets relieve our mitochondria of overwork.

1.03 Therapy by Antioxidants, Minerals, Phytonutrients, and Vitamins

We need to consume a variety of vegetables and fruits like apples, apricots, berries, and peaches. Three percent of the 1.26 liter of daily oxygen that we inhale each day can become 38 milliliters of free radicals in our body as byproducts of the oxygen-mediated process of energy production from glucose. We need to sequester or quench these free radicals by antioxidants in our daily foods. We need to consume required levels of key antioxidant vitamins (A, C, D, E, and K) and minerals like selenium in order to quench the free radicals. Fruits and vegetables simply mean fiber, antioxidants, minerals, vitamin B complex, and anticarcinogenic phytonutrients. Vegetables can supply all such antioxidants, folic acid, and other B vitamins. Salmon, trout, walnut, and flaxseed can add omega-3 fatty acids to our diet that augment the effects of antioxidants. The truth is that antioxidants are necessary in order to prevent or keep mtDNA damage to a minimum and thus protect us from the neurodegenerative diseases of Alzheimer, Huntington, and Parkinson.

We need free radicals to kill invading pathogens, but we don't need a lot of them. Whether superoxide, hydrogen peroxide, or simple but very strong hydroxyl radical, free radicals cause mitochondrial DNA damage and imbalance energy production. They can damage even nuclear DNA and thus cause production of the wrong genes. Our routine physiology is endowed with enzymes (catalase, superoxide dismutase, glutathione peroxidase, and reductase) that kill off free radicals. Plant food-based antioxidants are simply an added necessary defense. They are extremely critical nutrients as we advance in age and lose the power of our mitochondria.

1.04 Anti-Inflammatory Therapy

Herbs and spices are truly "the spice of life" because they are ready sources of antioxidants and antimicrobials. Anti-inflammatory omega-3 fatty acids are a critical part of our brain cells, and this is why they have been classified as essential fatty acids. We have heard of the anti-inflammatory values of vitamin D, cucurmin from turmeric, resveratrol from grapes, and natural anthocyanins and antioxidants from fruits and vegetables. These natural anti-inflammatory foods have been part of the scheme of human evolution. Body of research on diet disease connections todate shows that routine consumption of adaptogenic and anti-inflammatory foods extensively tabulated in this book is therapeutic.

1.05 Color Therapy

Many key nutrients in foods can be selected simply by color: red foods such as apple, cranberry, and tomato for heart health by antioxidants; yellow foods for vitamin C; and dark green vegetables for vitamin B complex and anticarcinogens.

Color is therapeutic. It is used for emotional therapy in ayurvedic medicine in India, a practice that has been common in China, Egypt, and Greece for centuries. Our body tissues vibrate, and so do the color spectra. Our brain recognizes foods by shape, size, sight, and color. The color of fruits and vegetables we eat is not an accident in nature; it is a matter of evolutionary architecture. Our mind is hardwired for food color and appetite. McDonald's red and orange is suggestive of eating and then leaving. On the other hand, one lingers in a blue lounge longer, and yellow tempts one to eat more. Red and yellow stand for enhanced social interactivity. Feeling blue is suggestive of depression or missing imagination and creativity. The color of foods indicates the presence of methyl groups that help turn otherwise-silenced genes on. Color can enhance cognition, and it can successfully treat depression, anxiety, and panic depression (http://www.depressionhelpfiles.com/articles/a20.htm). Visual ergonomics is the key, and we need to choose the color of our food and surroundings wisely.

1.06 Essential Amino Acid Therapy

Essential amino acid phenylalanine is necessary for making tyrosine, which is converted to dopamine, the feel-good and anti-craving neurotransmitter. Our daily diet must contain proteins that supply phenylalanine for controlling addiction to overeating. A clue to this value in our daily diet comes from the fact that obese people have low levels of dopamine.[1] They habitually suffer from food craving and overeating. Also, amino acid tyrosine is therapeutic to those who suffer from phenylketonurea because they are unable to convert phenylalanine to tyrosine. Tyrosine comes from foods like poultry, fava beans, bananas, mustard greens, cottage cheese, and soybeans. It is a curative against depression, attention deficit disorder, and attention deficit hyperactivity disorder.[2,3] Vitamin B_6, which comes from fruits and vegetables, is necessary for making tyrosine. Another amino acid, L-arginine, is necessary for making nitric oxide, a blood-vessel dilator for managing blood pressure. Protein foods like nuts and seeds contain arginine, necessary for good health. Also essential is the amino acid tryptophan. It is necessary for making the neurotransmitter serotonin, which is important for learning, cognition, and sleep. Serotonin converts to melatonin, which induces sleep. Tryptophan therapy can result from seafood diets, including fish and shrimp, and proteins from nuts and seeds. Pea protein is a great source of L-arginine, along with sesame flour and soya protein isolate. Amino acids threonine, serine, tyrosine, and histidine are involved in turning the enzymes that run our body and brain biochemistry "on" or "off."

1.07 Dietary Fiber and Probiotic Food Therapy

Prebiotic dietary fibers, up to 30 grams a day in our diet, add bulk, volume, viscosity, and texture to the stomach juice. The fiber satiates us by filling the stomach volume quickly. Ingredients like chia seed, guar gum, and psyllium can fill up to 20 percent stomach volume for satiety. In effect, this is an indirect portion control. Prebiotic fibers help optimize calorie intake beyond their critical role in peristalsis, immune system maintenance, and overall

---

1 http://www.bnl.gov/bnlweb/pubaf/pr/2001/bnlpr020101.htm.

2 http://www.health.harvard.edu/newsletters/Harvard_Mental_Health_Letter/2009/June/Diet-and-attention-deficit-hyperactivity-disorder.

3 S. Meyers, "Use of Neurotransmitter Precursors for Treatment of Depression," *Altern Med Rev.* 5(1) (2000): 64–71., http://umm.edu/health/medical/altmed/supplement/tyrosine.

digestive health. Probiotics, like India's centuries-old dahi, sauerkraut, yogurt, kefir, and kimchi, have good bacteria that help build our immune system; they talk to our immune system. Chapter 8 deals in more detail with the role of probiotics in building immunity. They have a long history in the human diet.[4,5] Food fermentation began seven thousand years ago. Actually it predates fire by three thousand years.

1.08 Neurotherapies for Appetite and Hunger

Neuroscience closely studies diet and nutrition, and balanced food is necessary for our central nervous system. Carbohydrate, fat, and protein loads get communicated to the brain by a peptide hormone (cholecystokinin) and levels of glucose, insulin, and leptin. Gut hormones—insulin, ghrelin, glucagon—like peptide, are associated with routine cognition. Also involved in this communication are osmomolality (osmotic pressure as a function of concentration) and pH. Both appetite and hunger can be managed routinely by controlled protein intake and by avoiding flavor addiction. Ion diffusion and electrolyte balance and its regulation by kidney is very important in neurotransmission, acid production in our stomach, workings of our sensory systems, and conscious behavior. These are electronic phenomena that can be optimized by diet and lifestyle.

1.09 Lifestyle Therapy

Food intake in terms of quantity, quality, timing, and variety can help maintain our physical, emotional, and intellectual health. Spiritual health, social relationships, faith, belief, and friendship are very closely connected to the health of our mind and body. We are culturally conditioned social creatures, and lifestyle per se has its epigenetic effects. A good and clean environment, a good night's sleep, daily exercise, yoga, meditation, and social and spiritual connectivity can enhance the value of what we eat. Physical activity, exercise, and meditation simply help maintain homeostasis or equilibrium of life processes. Exercise affects mitochondrial function by oxygen supply and maintaining homeostasis. Lifestyle, therefore, can be very therapeutic. Daily foods can prevent aches, pains, and chronic diseases.

---

4 http://www.fao.org/docrep/x0560e/x0560e05.htm.

5 http://www.baofoodanddrink.com/pages/fermented-foods.

They can fix our chronobiology, sleep pattern, and rate of aging. Foods can control the body and mind connection and modify our behavior. Foods, family, friends, and faith are central to the daily workings of the human body and mind. The effects of all combined are now known to lower cholesterol, blood pressure, and blood sugar.

In order to truly understand these therapies, we need to understand our body and its digestive system, upon which depend all other body systems' operations. We need to appreciate that what happens in and to our body is truly molecular, microscopic, and nanoscale. We need to fully understand the role of oxygen and water in relation to our being by mitochondria, what I call the "thread of life" that is responsible for managing electron transfer and proton pumping for making the energy molecule ATP.[6,7] Using data from NASA (National Aeronautics and Space Administration), I find that we use 3.85×10^{25} electrons and 19.25×10^{25} protons per day through our mitochondria, which reduces 840 grams of oxygen into water in our cells in the case of people who live nonsedentary lives.

Let us keep in mind that electrons as a source of energy and protons as an electromotive force and energy can be ten times as much during times of emergency and heavy exercise. Therefore, good health derives from managing not just the composition of foods we eat, but also from managing oxygen intake and the health of mitochondria. I find that commercial dieting and diet programs fail because they rarely focus on the intrinsic value of food molecules as they relate to the health and vigor of mitochondria in our cells—the food we eat, water we drink, and oxygen we breathe. This book attempts to link all such understandings by way of connecting food to the apparent health of the energy-producing systems in our body.

1.10 Our Body

The spatial arrangement of enzyme and protein molecules in our body matters. Let us keep in mind that there are 3×10^{17} atoms of close to sixty elements in our body and twenty-five of them are critical to health. Molecules of carbohydrates, proteins, fats, DNA, RNA, hormones, cell receptors, and neurotransmitters are much bigger than an atom. A majority are composed of carbon, oxygen, nitrogen, phosphorus, and sulfur; the arrangement of these atoms in respective

---

6 http://www.nicholls.edu/biol-ds/biol155/Lectures/Glycolysis%20&%20Respiration%202.pdf.
7 http://users.rcn.com/jkimball.ma.ultranet/BiologyPages/C/CellularRespiration.html.

molecules, surface, and geometry matters in our daily life processes. If atoms were the size of a small pea, a molecule would be the size of a marble and a single cell in our body would be the size of a cruise ship. Our body would be the size of North America. Our daily food preserves the integrity of these essential nano-molecules and their relationship with our body.

1.10.01 The Genome, the Basis of Our Existence: The human body has 10 trillion cells, of which 2.5 trillion are oxygen-carrying red blood cells that have no DNA but are loaded with 95 percent hemoglobin. Ten percent volume of the remaining 7.5 trillion cells is genome- and DNA-containing nucleus, DNA being the master molecule that controls much of our cells' activity. On the other hand, the epigenome is all nonnuclear. Actually it is the membrane of all 7.5 trillion cells. We could call these membranes of 216 types of cells in our body the master software that controls epigenetic effects. The main types of cells that we hear about in a doctor's office are those of heart, lung, muscle, skin, stomach, colon, kidney, and nerve cells of the brain. Cells have their own life span. Stomach cells live only for two days, skin cells for close to a month, red blood cells for up to 120 days, bone cells for up to thirty years, nerve cells and heart cells last throughout life. . All of them have the same DNA, though.

Food can maintain and extend the life of our cells. Our genome in the nucleus (the nuclear genome) has 25,000 to 35,000 genes, which are aided by the gene pool of our gut bacteria. The latter, called microbiome, are critical to our immune system. The mitochondria in the cell have another set of thirty-seven genes (call it mitochondrial genes).[8] Our daily life depends on the expression of all genes, though—bacterial, genomic, and mitochondrial. A total of close to 75,000 enzymes made by the genome run and regulate our body, mind, and metabolic processes. The control of and communication about what goes on in our body depends on sixty-five hormones and one hundred small and large neurotransmitters made under the control of our genes. Things go wrong when our genes go wrong, not only the genes from the nuclear and mitochondrial genome but also the gene effects from our second genome of symbiotic bacteria in our colon.

1.10.02 Optimal Energy Needs: The nano-factory in our body works because of what we eat, drink, and breathe. Our daily food produces enough energy to raise the temperature of 25 liters of water to boiling. The food we consume

---

8 http://en.wikipedia.org/wiki/Mitochondrial_DNA.

needs to be balanced for the right amount of energy production for proper control of the body's nano-factory, or else we begin to boil and succumb to ailments, disease, and pain. I call this boiling a forceful euphemism for inflammation. Keeping this energy in balance means balancing functions of the mitochondria in our vital organs—brain, heart, liver, and lung.

1.10.03 Daily Oxygen Matters: The heart needs to beat at about seventy beats per minute in order to make it possible for food nutrients and oxygen to get to each and every cell over a distance of 60,000 miles of the vascular and capillary system. The lungs need to supply oxygen to the blood. To be functional, the brain needs to be fed enough glucose and oxygen by the blood flowing through it. Actually the brain uses 20 percent of all the oxygen we breathe in.[9] These functions and a myriad of others have to be in control and in equilibrium, called homeostasis.

1.10.04 Homeostasis Is Vital: Physiological control and homeostasis is a very balanced process. Hormones are the chemical messengers in our body for internal communication. The endocrine system (pituitary, thyroid, parathyroid, adrenal, thymus, and pineal glands, plus pancreas, heart, lung, kidneys, intestine, ovaries, and testicles) works in close relationship with the central nervous system. Fat-soluble steroid hormones, derived from cholesterol, can pass through the cell membrane and act on our DNA. On the other hand, protein hormones work at the cell surface because they can't pass through the fatty membrane of the cell and get inside. The thyroid system takes care of metabolism and protein synthesis, the parathyroid controls calcium, the pineal gland controls sleeping and waking states and corresponding metabolism, the adrenal gland controls stress and metabolism, the pancreas is key to digestion, the thymus takes care of of our immune system's T Cells, and the pituitary is the master gland that modulates the entire endocrine system. This control system depends largely on the type and amount of foods we eat that supply the precursors for making the molecular components of the control system.

1.10.05 Electrons and Protons Rule: In essence, our body's nano-factory is a glucose factory wherein the electrons and protons rule. We use electrons to burn glucose (redox processes in mitochondria) and secure energy as

---

9 http://www.disabled-world.com/artman/publish/brain-facts.shtml.

ATP by proton gradient. Our cells breathe by pumping protons.[10] Our body exchanges matter and energy millisecond by millisecond as an open system. It degrades and remakes molecules of life every second of our average 2.5 billion-second life span. We use our own weight equivalent of energy molecule ATP per day for running this nano-factory. The structure of ATP synthase, the enzyme that makes the energy chemical ATP, is conserved in all living systems, and so is the electron-transport chain powered by proton gradient.[11] *Think of the cell membrane as a dam, the ATP-making enzyme as a rotary turbine, and the making of ATP like the making of electricity.* Life exists because of our cells' ability to ump protons across their membranes and because of electrons that hop around from molecule to molecule in the electron-transport chain. These two processes of information transfer by encoded genes and cellular breathing or pumping of protons transcend chemistry. The energy-producing mitochondria, around 10 percent of our cell volumes, matters in our day-to-day health.[12]

1.10.06 The Sun Rules the Life of All on Planet Earth: Chlorophyll and carotenoids in plant leaves are solar energy collectors. Photons from the sun hit the leaves of the plants, and light energy helps knock electrons off water molecules. Electrons, along with proton from water, energize the process of photosynthesis. Proton goes to make ATP, and electron makes NADPH. Glucose made via photosynthesis sustains the lives of most all other living organisms, including humans; it is the source of energy for all living systems. There are twelve electrons and twelve protons used to make one glucose molecule from six carbon dioxide molecules during photosynthesis. Just the reverse happens when glucose is burned in our body. The energy of electrons is used to pump protons throughout the metabolic system, and the electrons are recycled. The sunlight, which can produce electrons from water, thus rules all life on planet earth.

1.11 Our Digestive System and Our Second Brain—the Enteric Nervous System

Our thirty-foot-long digestive system and our daily food that it processes are critical to the sustenance of all other systems and their tissues and organs.

---

10 http://www.nature.com/scitable/topicpage/why-are-cells-powered-by-proton-gradients-14373960.

11 http://users.rcn.com/jkimball.ma.ultranet/BiologyPages/A/ATPsynthase.html.

12 http://www.sciencedaily.com/releases/2013/02/130205123748.htm.

With appropriate signals, checkpoints, and means of control, it is designed to select and extract nutrients from liquids and solids that we eat. The heart beating, respiring, thinking, sleeping, moving around, and perspiring—everything depends on what, when, and how we eat. We can't afford to consume the wrong foods, the wrong amount of foods, or foods that are corrupted or are contaminated with corrupted molecules.

The cell membrane permits movement of nutrients, water, ions, and waste across the cell. This elaborate system of cells and their organization has to recognize and use what we eat and drink and their derivative products and byproducts. There are consequences in terms of aches, ailments, and diseases when cells fail to recognize what we eat and what they need to discard. Different cells have different functions and different nutrient needs. While our body is a busy factory extracting energy from what we eat for daily living, it is also busy building cells and tissues and putting them together as molecular tools.

1.11.01 The Human Body Is a Nanoscale Molecular Factory: The water we drink is only 0.2 nanometer or 0.02 billionth of a meter in size. Flavor molecules are bigger than water. In comparison, casein protein in milk is 100 nanometers or more. A starch granule in wheat flour is much bigger than water, at 20,000 nanometers. So we ingest both large and nanoscale foods and beverage. The big and small—water, oil droplets in salad dressings, and chunky food particles—all get digested in our gastrointestinal tract and go into our bloodstream as nanoscale nutrients (see table 1.01).

Table 1.01 Size of What We Eat

Cellular Materials and Components are received by shape and size.	Size in Billionth of a Meter or 1 nanometer
Diameter of a proton	0.0016
Diameter of an electron	0.0056
Diameter of water molecule	0.1 to 0.175
Distance between DNA base pair	0.34
Diameter of DNA helix	2.0
DNA Helix	10
Water with fats and oils	0.2

Cellular Materials and Components are received by shape and size.	Size in Billionth of a Meter or 1 nanometer
Water with carbohydrates	0.3
Flavor molecules	0.3 and up
Diameter of glucose molecule	1 to 1.5
Molecules of insulin and hemoglobin	5 to 6
Molecules of starch	Up to 1000
Casein units in milk	2
Large Proteins and carbohydrates	1–10
Diameter of insulin	5
Diameter of hemoglobin	6
An average protein molecule	10
A typical virus	75
Smallest bacteria	200
Casein micelle or aggregate	50–100

1.11.02 We Are What We Eat: Food, water, and oxygen that go into our blood make us into what we are: the body structure, our senses, our movement, and our memory, thought, and behavior. Food, no doubt, is our daily medicine. Enzymes in our body work as motors, and the mitochondria in our cells work as a chemical energy (ATP)-generating powerhouse. Energy as ATP is the chemical currency that conducts the business of each and every cell in our body. Enzyme pathways talk to each other and work together, and our cells communicate with each other in order to make us tick. Our mind is connected to our immune cells, it is connected to the rest of the body, and it controls our behavior. Thus our body has a complex communications network. Shape and connectedness in space (topology) matters in routine workings of cellular receptor in our body (see table 1.01). We live by the shape and size of molecules in space and their interactions, and our genes control it all.

1.11.03 Next in Importance to What We Eat Is How We Choose to Live: The foods we eat and our lifestyle can make or break our health. They can kill our cells prematurely, they can make us age faster, and they can cause chronic health problems. This simple fact has been ignored, and the result is poor health and too many interconnected chronic diseases. Diseases are more prevalent today because (1) our DNA is mutating faster today than ever before under environmental pressure and (2) our epigenome is stressed out by poor choices of lifestyle, food, and environment.

1.11.04 Corruption of Food Is Killing Us: Food additives, pesticides, herbicides, and food toxins in our diet, contaminated water, and polluted air are closely linked with autoimmune diseases, cancer, diabetes, lupus, mental disorders, and many genetic disorders. Polybrominated biphenyl ethers (present in furniture and toys) cause brain disorders, carbamates (present in cockroach and mosquito sprays) cause poor brain development, and tetrachloro ethylene (present in dry-cleaning solvents) also causes poor brain development.[13] Time magazine had multiple coverage in May 17, 2010, Dec 12, 2012, and February 14, 2014. Corruption of our environment is getting to be life-threatening. In no more than seventy years since World War II, our government policies on mass production and retailing of food have changed dramatically. What, when, and how often we eat have changed drastically, and so have the stability of our DNA and the epigenetic pathways that operate in our cells. In other words, food additives and our food choices have affected our genome and epigenome. They have changed DNA methylation and caused DNA to mutate. In the case of pregnant mothers, they have even changed the genes for early growth and development of the offspring. They have also changed the timing of puberty in girls.[14]

Processed foods readily available today confuse the hypothalamus in our brain.[15] We end up overeating because modern foods prolong the point of satisfaction. We eat wholesome and minimally processed foods only rarely and do not consume food molecules that offer natural defense. Modern foods have changed the hard wiring in our brain with respect to amount of food, food flavors, and food types that we crave.

The Dietary Supplement Health and Education Act (DSHEA) of 1994 has added more problems than it has solved. We have lately begun to consume exotic food supplements of questionable therapeutic value and unknown side effects. Our cells are unfamiliar with most of them.

---

13 http://content.time.com/time/magazine.

14 http://digitaljournal.com/article/335214.

15 http://www.34-menopause-symptoms.com/hot-flashes/articles/tips-about-sugar-hot-flashes-and-night-sweats.htm.

1.12 Water and Oxygen : Hydration and Breathing

We breathe to inhale oxygen and to exhale carbon dioxide, the waste product of energy production in our cells. The brain is very much involved in when, how much, and at what rate we breathe. Our brain monitors the level of oxygen and carbon dioxide regularly and constantly. Breathing is involuntarily controlled without our willful consent, and it is kept at a steady rate. The medulla in the brain can detect acidity or low pH because of high carbon dioxide and increase breathing and heart rate. We know that alkaline foods (high pH avocado, broccoli, lettuce, pepper) cause slow breathing.[16] Most green vegetables are alkaline foods. Our daily diet and nutrient balance is necessary for this regulation. No oxygen to the body means no energy production as ATP by stepwise transfer of electrons to oxygen and converting it to water. The following relationship is an example of this fundamental.

Glucose + 6O_2 + 36 P^{2-} + 36 ADP + 36 H^+ (protons) = 6 CO_2 + 36 ATP + 42 Water
A 2,000-calorie diet in a day provides a huge number of 216 X 10^{22} molecules of ATP.

Electron and proton movement from molecules of food for energy production is tied to oxygen and reduction and oxidation of molecules of NAD/NADH and FADH and $FADH_2$.

Water is a critical food for us, and our body is a moving and walking sac of water (75%). It moves molecules of life around and even helps make them. The percent of water in our brain is 85 percent, in muscles 75 percent, in the kidney 82 percent, and in bone 22 percent.

Food can get to our stomach even if we stand on our head. All 10 trillion cells work all the time. The heart beats, the lung pumps air, and the kidneys filter our blood. Water is involved in all body structures, their functions, and their uniqueness. The function of glands and various organs in our body depends on water. All molecular traffic in and out of our cells is in water. Water and hydration are critical to health and well-being.

---

16 http://metabolichealing.com/education/articles/acid-and-alkaline-nutrition-shattering-the-myths/.

1.13 Diet Programs Don't Work.

The epidemics of chronic diseases and obesity are the main reason for all the dieting schemes and the more than ninety diet programs. Even the select few listed in table 1.02 under diet food programs suffer from obvious deficiencies and contradictions because none work reliably. The desired effects, if any, are only transient and temporary. None of these diets reflect the (1) food industry's bad innovations of artificially saturated trans-fats and partially hydrogenated trans and omega-6 fatty acids, nor the (2) food industry's fast-pace utilization of foods loaded with unnatural additives, chemicals, flavor enhancers, and industrial chemical contaminants. None of these diet programs transparently revealed the relationship among the variety of foods for genuine taste and sufficiency of antioxidants, vitamins, and minerals for daily consumption and long-term stability of the genome and epigenome. None accorded proper emphasis to the austerity of the body as activity and exercise, austerity of speech in daily social interaction for stress control, and austerity of mind in terms of it connecting to the body and controlling our behavior in regard to appetite, excessive eating, imbalanced eating, and stress reduction. This book on our genes, our food, and our choices embodies due details on all such counts.

Table 1.02 Diet Food Programs

Diet Program	Description
Abs diet	Almond, beans, legumes, spinach, vegetables, oatmeal, protein, & whole grains
Alkaline diet	High protein and low controlled carbohydrate diet
Atkins diet	Limited net carbohydrate, low glycemic foods, high fat foods, and low blood glucose
DASH diet	The diet is approved by the National Heart, Lung, and Blood Institute. It advises eating fruits, vegetables, whole grains, no trans fat, little saturated fat, high fiber and protein, and a good daily intake of calcium, vitamin D, and potassium. A thirty-minute daily exercise and physical activity program complements the daily diet. Controls blood pressure, diabetes, and hypertension. This is close to being the methylation diet.

Diet Program	Description
Dean Ornish diet	Consume 10% polyunsaturated good fat, 70–75% complex carbohydrates, and 15–20% protein. Consume whole grains, fruits, vegetables, and low-cholesterol foods. Avoid dairy foods and caffeine. Exercise and meditate. This is close to the methylation diet but doesn't spell out means to combat nutrient deficiencies.
Digest diet	Claims to shift body in fat-release mode and suggests consumption of fiber, calcium, vitamin C, protein, and poly- and monounsaturated fat. Emphasizes consumption of cocoa, honey, quinoa, and vinegar. Offers a twenty-one-day weight reduction program.
Dopamine diet	Consumption of high-tyrosine foods like fava and garbanzo beans
Dukan's diet	High-protein low-carbohydrate weight-loss diet
Flavor Point diet	Yale University's Dr. David Katz's Sensory Effect Approach for appetite control. Consume fruits and vegetables. (David Katz, Flavor Point Diet, Rodale, Inc.)
Glycemic- Index diet	Whole-grain complex carbohydrate diet that eliminates refined sugar
Jenny Craig diet	Restrict calories, portions, and fat. Recommends exercise.
Mediterranean diet	Whole-grains, fruits, and vegetable diet that permits olive oil and wine. A good diet.
Microbiotic diet	Fruits, vegetables, whole grains, and bean-based diet, close to methylation diet
Okinawa diet	Based on consumption of nutrient-rich and calorie-rich plant-based foods, close to methylation diet
Paleo/Genotype diet	Diet based on plant foods of natural origin
Pritikan diet	A diet program based on straight-from-nature foods like fruits, vegetables, legumes, and lean meat, augmented by daily exercise and physical activity involving weight training and stretching
Slim-Fast diet	Restrict calories and portions of foods
South Beach diet	Dr. Arthur Agston's no unhealthy fat, low sugar, low glycemic, heart-healthy version of Atkins diet

Diet Program	Description
Stillman's diet	High-protein low-carbohydrate diet of that was developed during sixties for weight reduction
TLC diet	Therapeutic Lifestyle Change diet
Traditional Asian diet	High-fiber, high fish-protein diet based on grain, fruits, and vegetables
Vegan diet	Diet that excludes animal products
Vegetarian diet	Diet based on fruits and vegetables, nuts, seeds, sprouts, and milk products
Volumetric diet	Diet of low-density foods with high water and air content
Weight Watchers diet	Create a calorie deficit and consume high-fiber and high-protein foods. Consume fruits and vegetables.
Zone diet	Consume the right dose at the right time for anti-inflammatory effect: 40% calories from carbohydrates, 30% from good fats, and 30% from proteins

1.14 Food Molecules in Our Daily Diet

I select the methylation diet as an answer to our chronic disease problems based on (1) my readings of seven texts in this area, including those by Drs. Dean Ornish's The Spectrum; (2) my critical examination of results of direct trials and meta-analyses of the Adventist Health Study of Loma Linda University in California involving 154,000 participants, the famous Nurses' Study 1976–89 involving 121,000 participants, the EPIC (European Prospective Investigation into Cancer) study of the European Union involving 440,000 participants, the Iowa Women's Health Study involving 41,836 participants and follow-ups involving 51,000 participants; and (3) my review of USDA's new nutritional guidelines of 2005 revised in 2010 and the revised well publisized food pyramid of 2011. All point to a diet that can methylate our DNA and protect it from damage, the diet called the methylation diet. My own research on values of daily foods and their functional constituents in combating ailments and chronic diseases, epigenetic proofing, and promoting general well-being led me to conclude that the methylation diet is common in all cases. To make this important point, tables of data I collected appear in each chapter even at the risk of repetition.

What is referred to as the methylation diet is a combination of proteins, carbohydrates, fat, vitamins, minerals, and other necessary phytonutrients like antioxidants, anticarcinogens, and anti-inflammatory molecules that have been the cornerstone of our evolution over 500 million years. It is a diet that systematically mitigates vitamin and mineral deficiency.

Plants have evolved to consume the sun's energy directly, and human beings evolved later to depend largely on plants for food. As such, a good diet should be largely based on plant foods, including fruits and vegetables of color and variety. Proteins, carbohydrates, and fat should be balanced. Such a diet must provide at least twenty-five to thirty grams of fiber and fifty to sixty-five grams of balanced protein per day. Overall, our daily diet must supply trace minerals, B-vitamins, and all antioxidant vitamins. Also, the meaning of daily food should include clean air and water free of additives and chemicals. A complement of lifestyle punctuated with physical activity, exercise, meditation, social interaction, and living in a clean environment of good air and good water no doubt adds biological meaning to foods we consume. As you will see in subsequent chapters, the methylation diet must provide the recommended daily intakes of nutrients listed in the block below.

Prebiotic dietary fiber
Probiotic foods for good bacteria- our second genome
Choline
Vitamins B_2, B_5, B_6, B_9 (folic acid), B_{12}
Vitamins A, C, D, E, and K
Minerals: copper, cobalt, chromium, iron, magnesium, manganese, selenium, and zinc
Glutathione
Omega-3 fatty acids

A methylation diet is a good and balanced variety of adaptogenic, anti-inflammatory, antioxidant, and antiangiogenic foods including prebiotic dietary fiber and probiotics like yogurt and sauerkraut, fruits, vegetables, whole grains, seeds, nuts, and soy foods, with emphasis on fruits and vegetables.

Note: Chapter 6 on food functions has extensive details on foods for omega-3, water-soluble B vitamins and Vitamin C, fat-soluble vitamins A, D, E, and K, and various minerals, along with their recommended daily allowance.

1.15 Chronic Stress Is Another Killer.

The DNA can't do it all. The "nurture" part, or the environment and lifestyle, must be conducive to good living because stressors from poor environment and bad lifestyle can not only make us sick, they can kill us. We need to design and practice a lifestyle conducive to physical and mental health. Longevity, then, is simply a byproduct of it.

Our visits to the doctor's office include skin disorders—43 percent, osteoarthritis—34 percent, back pain problems—24 percent, cholesterol problems—22 percent, upper respiratory conditions—22 percent, anxiety and depression—20 percent, chronic neurological disorder—20 percent, high blood pressure—18 percent, headaches and migraines—14 percent, and diabetes—14 percent (National Center for Health Statistics, 3311 Toledo Rd, Room 5419, Hyattsville, MD 20782). This can be substantially curtailed simply by balanced nutrition and a planned lifestyle.

Diet and nutrition impact every part of the human body, including bones, bowel, eyes, and muscles. The most important food functions, beyond needed calories, are digestive-system performance, immune-system performance, and functions of the brain and heart. Good nutrition is necessary for preventing colds and flu, weight gain, poor digestive health, loss of energy, fatigue, and poor performance, and it is necessary for skin complexion, good mood, good sleep, good physique, and well-being.

1.16 The Health of Mitochondria

Mitochondria, the power plant for energy production in our cells, are where all electron and proton actions take place. It is the mitochondrial matrix where maximum quantity of electron carrying NADH is produced from carbon to carbon bonds of glucose and where fatty acids from fat we consume are oxidized to acetyl CoA. Actually Citric acid cycle or Kreb cycle operates in mitochondria. I call mitochondria "the thread of life". Unfortunately, they are where free radicals are also produced and unfortunately our execessive calorie consumption based on refined sugar and high glycemic carbohydrates forces them to work over time This organelle in our cells is of probiotic origin, and it shapes, fuses, unfuses, maintains a superb genetic quality control, and multiplies in our cells on demand very much like bacteria.[17,18] Mitochondria have multiple copies of their own genome, whose genes are vulnerable to mutation by free radicals that they produce. The health of mitochondria and the stability of their genes are crucial to our daily health, and mitochondrial health can be enhanced by the methylation diet and our lifestyle including exercise, yoga, meditation, sleep, faith and prayer, and social interaction. Mitochondria in our cells determine our state and quality of living.

As a short summary, the methylation diet, I need to emphasize again, should be based on the common foods in table 1.03, which should be optimized for best nutrition values season by season and region by region depending on what is available.

Let us keep in mind that in order to process our daily two thousand-calorie diet, the turnover of energy-producing enzyme *ATP synthase, considering all recycling,* exceeds 50–75 Kg per day.[19] If the genes that make it and nutrients that run it don't come from our daily food, our body will for sure get sick. It certainly has been getting progressively sicker notwithstanding a number of health-promotion policies and programs during the last fifty years, including creation of low-calorie foods by avoiding refined sugar during the mid-1970s, creation of low-cholesterol foods during the 1970s and early 1980s, the push to remove fat from foods during the early 1990s, the push to remove carbohydrates during the early 2000s, efforts to remove trans fat during the mid-2000s, and the present-day emphasis on gluten-free foods. A variety of often conflicting dieting and weight-reduction programs speak to the poor understanding of our problem by the proponents. The proponents have always misjudged the adverse effects of foods in retail stores and away-from-home eating establishments.

The misjudgments have been in the followings regards.

- Low glycemic index and high glycemic load increases risk of weight gain, type II diabetes, and other chronic diseases[20]. In USA, per capita consumption of high glycemic index carbohydrate increased to 69.10 Kg in 2010 compared to 55.5 Kg in 1970 [21].Consumption

---

17 https://www.google.com/search?q=mitochondria+an+organelle&tbm=isch&tbo=u&source=univ&sa=X&ei=CbzdUo3HCsLYrAH_joDICg&ved=0CCoQsAQ&biw=793&bih=529.

18 http://www.nature.com/scitable/topicpage/mitochondria-14053590.

19 http://jb.oxfordjournals.org/content/early/2011/04/26/jb.mvr049.abstract.

20 http://ajcn.nutrition.org/content/87/3/627.full

21 http://ajcn.nutrition.org/content/81/2/341.full

of omega-6 fatty acids increased, sodium and potassium ratio got imbalanced, micronutrient density decreases, fiber consumption remained low, and deficiencies increased causing evolutionary collision with nutritional quality of modern day foods.

- Excessive use of antibiotics that destroy our second genome in the colon. Very few physicians advise as to use of probiotic foods like low sugar plain yogurt after the course of an antibiotic runs out. We must build our gut microflora week by week even when we feel and act healthy.

- Wholesale lifestyle changes in American population have caused three other problems. First is sleep deprivation[22], [23] second chronic stress, and third is defiance of solar cycle and use of less perfect artificial light[24] as reported by the Economist. Very few of us go to bed at the same time each day and very few us work in sun lit room by the window.

These misjudments have already caused serious problems. Health of Americans has to improve with everyday common foods and not by exotic foods unfamiliar to our body or supplements not present in our food chain and not known to and sometimes incompatible with our physiology.

Table 1.03 Healthful Foods Key to Human Evolution

Adaptogenic foods	Asparagus, ginger, garlic and tomatoes (including ashvagandga and astragalus, rhodiola if available)
Antiangiogenic foods	Berries, dark chocolate, kale, red grapes, tomatoes, and turmeric
Anticarcinogenic foods	Fruits and vegetables; vitamin, minerals, and antioxidants
Anti-inflammatory foods	Berries, cruciferacea vegetables, flaxseed, walnuts, salmon, palm oil
Immune-building foods	Probiotic foods (mother's milk, yogurt, cheese, sauerkraut, kimchi, Indian dahi) Antioxidants (carotenoids, vitamins C and E, minerals selenium and zinc)

---

22 http://www.journalsleep.org/ViewAbstract.aspx?pid=24429

23 http://www.ncbi.nlm.nih.gov/pubmed/20438143

24 http://moreintelligentlife.com/content/features/rosie-blau/light-and-health?page=full

This book emphasizes that the therapies by foods we eat work under a physiological norm punctuated by gene expression, including those of our second genome from the probiotic foods for which exercise, yoga and meditation, sleep, stress-free living, friends, family, and faith are critically important. We have been doing it for centuries under one or the other name and description. The Hindu scripture *Shrimat Bhagavatgeeta* says it beautifully by emphasizing some 2,500 years ago that foods that promote longevity, intelligence, vigor, health, happiness, and cheerfulness and that are the essence of food, substantially bland and naturally agreeable, are foods dear to people of pure nature. Although not validated by scientific experiments originally, purity of human nature in *Shrimat Bhagavatgeeta*, I believe, refers to gene expression for pure conscience. Such foods can help connect body with mind better, control digestion, and modulate a disciplined approach to stress-free living. *They can help create a transcendent person.* They can keep our DNA stable and epigenome uncorrupted. They are good for gene expression and desirable DNA methylation.

Chapters two and three of this book are devoted to genes, the foundation of our life; chapters four through sixteen are devoted to foods and the methylation diet, or gene-expression foods; chapters seventeen through twenty-one are devoted to lifestyle and choice; and chapter twenty-two is exclusively devoted to methylation-diet recipes. Lists of foods containing health-promoting and disease-mitigating nutrients are included in various chapters. Most these common foods are common denominators in maintaining health and preventing chronic diseases. Therefore, master tables in chapter 22 are compiled based on information tabulated in various chapters and included as a summary of the common foods-based methylation diet and its properties. These should serve as a ready reference for daily food design and meal preparation.

The summation in chapter 23 examines and represents foundations of food today and tomorrow in health maintenance and life extension. It introduces the readers to DNA and genomics, gene works on the cellular level or epigenetics, the critical importance of mitochondria (the thread of life) in day-to-day living, our body, our mind, our mind-body connection, the diet-disease connection, and chronic disease control by diet. It emphasizes that while postponement of death for happy years of human longevity is possible by our daily food and choices of living, defeating death is still in the hands of father evolution. The foods we choose to extend our disease-free life have to be as various, common, and natural as possible. Given their importance in our metabolism and physiology, we need to consume foods that metabolize for pharmacological effects like tumor suppression and cancer-cell death in particular.

Citations of credible sources are given chapter by chapter at the end of the book, and topic-specific footnotes are added for topics in need of proof of principles and factual details. Quantitative data and estimates included here are averages from many sources and products of my own calculations.

PART ONE

OUR GENES

Nuclear and mitochondrial genomes help make
gene works possible in our body's cells and
thus perpetuate our day-to-day living.

CHAPTER 2

Nutrition, Genomic Stability, and Epigenetics

"The doctor of the future will no longer treat the human frame with
drugs, but rather will cure and prevent disease with nutrition."
—Thomas Edison

The human genome, both nuclear and mitochondrial, evolved with
paleo foods that human beings have been consuming for centuries and
centuries. Drastic changes in food in a very short time of just the past almost
seventy years can, we now know, bring about drastic changes in our genomes
and the DNA they contain. To this we should add changes in our second
genome because of an improper balance of the gut bacteria. In consequence,
DNA instructions or codes on how our body develops, grows, and works
can also change faster than selection-based evolution accommodates. Our
daily food must help keep the DNA stable and gene expressions uncorrupted.

Modern biology views DNA and its genes as hardware and the
extranuclear epigenome as software responsible for gene expression. For
good health, the hardware must stay robust and stable and the software
for gene expression free of interferences and molecular corruptions.
Since genes affect our physiology, they can directly influence incidence or
prevention of diseases. Our diet, daily calorie intake, prenatal care, smoking,
environmental chemicals, and stressful lifestyle are epigenetic marks that
can alter our health and longevity.[25] The nurture part (epigenome) seems
to dominate over nature (DNA and genome) in our day-to-day living.
DNA is electronegative and prone to oxidative damage. Only our daily

---

25 http://www.nature.com/scitable/topicpage/epigenetic-influences-and-disease-895.

antioxidant-rich food can repair it. A better understanding of DNA's charge-transporting property will no doubt help define foods as specific medicine.[26]

DNA in all our 7.5 trillion cells, excluding 2.5 trillion red blood cells, is the same. Genes by way of proofreading in each of them read the same message except for the epigenetic message that can be different without any alteration in the DNA code. Heart cells are different from brain cells, and they act very differently.

Methylation of DNA is a major epigenetic change affected by foods we eat. A mother's nutrition during pregnancy and a child's nutrition during early life can cause epigenetic changes that persist throughout adulthood. This effect is much more important for daily health because all mitochondrial DNA is inherited from the mother. Such changes may even be passed on to the next generation. Good nutrition, therefore, is about delivering nutrients for DNA stability and gene expression, and good nutrition includes probiotic foods for our second genome via the individual specific personal microbiome discussed in chapter 5.

We live, keep our health, and die by our genome, both nuclear and mitochondrial, and epigenome. Our daily food can protect and keep our genome stable and epigenome uncorrupted. Close to ten thousand changes take place in our DNA per day in every cell, but most of them are repaired by foods we eat.[27] Foods we eat routinely are involved in this repair process of DNA changes or damages. Ailments and diseases come about when the curative effects and repair works fail.

2.01 The Genome, Nucleosome, and Chromosome

All 210 types of cells in our body have the same genome with the same DNA. Nuclear DNA wraps around histone protein, a compact globule of protein, and makes nucleosome, a thirty-nanometer fiber. It is stored in compartments and chromosomes; twenty-three pairs of chromosomes in the case of human beings snake in and out of these compartments as and when needed. On the other hand, mitochondrial DNA is unprotected and vulnerable to attack by free radicals. The human genome has a trillion times more information than any computer microchip known today. It has

---

26 http://www.nimh.nih.gov/health/educational-resources/brain-basics/brain-basics.shtml.

27 http://www.ncbi.nlm.nih.gov/books/NBK26879/.

3 billion base pairs distributed over twenty-three pairs of chromosomes, one copy each that we receive from father and mother. Although evolved bit by bit over 2.5 million years and designed to do and control a very complex yet integrated work of human physiology, these chromosomes need to stay invariant and stable. The chromosomes contain 25,000 to 35,000 genes, and yet this author is different from you, the reader, by no more than 3 million base pairs, or only 0.001 percent of DNA material.

Mitochondrial DNA inside our cells that produce energy from food we eat have a small circular DNA of 16,500 base pairs, making up thirty-seven genes that help make enzymes and RNA (more on this topic in chapter 23). It is well known that we inherit these genes from our mother[28] and that their mutations cause many diseases. Many genetic diseases are traced to mitochondrial DNA susceptible to damage by free radicals because mitochondria, the electron-processing and glucose-burning factory in our cells, produces reactive oxygen species or free radicals in the neighborhood. This oxidative damage can be prevented by foods of high antioxidant values.

2.02 Epigenome, the Software

Epigenetics is not a DNA phenomenon. It deals with changes in gene expression or creation of cellular phenotype or what results via the interaction of genes and the environment. DNA methylation—let us call it a chemical tag on DNA—and modification of histone protein that envelops the DNA are examples of this effect of environment on our genes. Environmental chemicals known to cause transgenerational inheritance are DDT (dichlorodiphenyltrichloroethane), BPA (bisphenol A) like food and beverage toxins, and fungicides like vinclowzolin.[29] The methyl groups as chemical tags become part of the cells' memory and part of cellular differentiation or making of body cells of various types from stem cells. The part that is associated with egg or sperm cells is inheritable and as such can influence evolution in a short time. Much of it is established very early in childhood, during the first nine to twelve years.[30] We now know that

---

28 http://ghr.nlm.nih.gov/mitochondrial-dna.

29 http://www.smithsonianmag.com/innovation/the-toxins-that-affected-your-great-grandparents-could-be-in-your-genes-180947644/.

30 http://www.cam.ac.uk/research/news/scientists-discover-how-epigenetic-information-could-be-inherited.

the concept of genetic determinism by DNA alone is an oversimplification because patterns of methylation can pull back a gene's (DNA) activity.

Diet can have effects on mitochondrial function, growth, and development and, therefore, epigenetic changes. Epigenetic marks can turn a gene "on" or "off." They can make us sick or healthy and well.[31] Actually epigenetics is the bridge between nutrition and health. The membrane of our cells and the receptors on them are the biochip responsible for epigenetic effects. Epigenome, the software, tells our genes to express or not to express, and the effects are preserved as our cells continue to divide and differentiate throughout our lifetime. Disease results when heart, kidney, and brain cells do so under dietary overload, smoking, and other environmental stressors. The end points could be cancer, autoimmune diseases, mental disorders, and diabetes.

Prenatal diet and environmental factors can affect DNA methylation and chromatin (the histone proteins) modification, and the effects can pass on to the next generation. Obesity, for example, is a case of overexpression of the Agouti gene. On the other hand, overexpression of the SIRT6 gene increases longevity. Royal jelly that is exclusively consumed by the queen bee is a classic example of the epigenetic effect. Gene that encodes DNA methyltransferase is silenced, and the epigenetic effect is a longer-lived fertile queen bee. Other examples are hypermethylation of cancer-suppression genes involved in (1) aberrant methylation of the VHL (Von Hipple Linday) gene, which causes kidney, pancreas, eye, brain, and spinal-cord cancers; (2) HLTF-gene silencing in human colon cancer; and (3) folic acid deficiency and incidence of head and neck cancer.

There are a few historical examples related to diet and eating: grandparents who were underfed during World War II in Europe had grandchildren of enhanced longevity, compared to those with abundant food during World War II, who begot grandchildren of short life spans because of diabetes and heart disease. Eating low-calorie foods, it is clear, has a powerful impact on our health. Human studies have shown that even autism and bipolar disorders are due to environmental stressors.[32]

---

31 http://advances.nutrition.org/content/1/1/8.full.

32 http://www.alternativementalhealth.com/articles/walsh.htm.

2.03 Genomic Stability

Central to genomic stability is preventing damage to DNA or changes in our chromosomes. DNA damage can result from (1) cellular metabolism or (2) routine errors in DNA replication and recombination. Common exogenous genotoxic agents are ultraviolet light, oxidative stress, and chemical mutagens. Our cells have means of repairing damaged DNA, but only with proper diet or else many defects prevail. We know that defective enzymes due to poor gene expression cause fifty genetic diseases. One hundred seventy-five genes are known to be involved in cancer.[33] The integrity of DNA is crucial for tumor suppression, both for current and future generations, and DNA is personal and individual in relation to chronic diseases because of a potential 3 million common gene variants among human beings. Table 2.01 has a summary of common diseases caused by gene defects.

Table 2.01 Common Diseases Due to Changes in DNA

Mutation	Chromosomes	Disease
Mutations	Chromosomes	
Gene APOE4	Chromosomes 1, 14, and PS1, PS2, APP, andChromosome 9	Alzheimer's
Deletion/duplication	Chromosomes 15 and 16	Autism, 55 genes involved
Variants of gene $SLC_{16}A_{11}$	Chromosome 17 (17p13)	Very common in Mexico
Obesity gene MTHFR		Blood pressure
BRCA1 and BRCA2	Chromosomes 17 and 13 respectively on their long arms (q)	Breast cancer
DNA methylation gene		Cancer
Single and regulatory genes	Single gene disorder	Celiac
Autosomal recessive gene	CFTR, gene therapy possibly	Cystic fibrosis
Tumor suppressor	Chromosome 5, gene mutation	Colon cancer

---

33 http://www.ncbi.nlm.nih.gov/pmc/articles/PMC1373651/.

Mutation	Chromosomes	Disease
Mutations of at least onesusceptible gene	Chromosome 6, not well understood	Type 1 diabetes
	Chromosome 21	Down syndrome
Gene APOE (E2, E3, E4), multiple involvement	Single-gene disorder, E3, is most common; E4 causes very high cholesterol	High cholesterol
Apo-1 gene	Chromosome 10	Heart disease & cholesterol
	Huntington	Huntington's disease
	Gene for enzyme lactase "off"	Lactose intolerance
	Dystrophin	Muscular dystrophy
Gene COMT	Chromosome 22	Inactivates green-tea effect
Vitamin D receptor genes	Chromosome 12	Poor calcium absorption
Fat mass and obesity associated FTO genes, there are 50 gene effects involved in obesity	Long arm (q) of Chromosome 16	Weight variation

South East Indians are lactose-intolerant, Alaskans suffer from increased risk of heart disease, and the Masai group in South Africa have many health problems when on a corn-and-bean diet. Mutations in the gene PAH that codes for enzyme phenylalanine hydroxylase cause phenylketonurea.[34] Serotonin receptor gene (HTR2A) variants are the cause of panic disorder.[35] The variant angiotensin gene that codes amino acid threonine for methionine is the cause of hypertension in African American populations. The sickle cell anemia gene on chromosome 11 is because of mutation of the HBB gene,

---

34 http://ghr.nlm.nih.gov/gene/PAH.

35 http://www.ncbi.nlm.nih.gov/gene/3356.

causing coding of valine in place of glutamic acid. This single amino acid change causes a beta chain of hemoglobin to become rigid and thus a poor carrier of oxygen.[36]

The majority of chronic diseases are tied to mutations in mitochondrial DNA, which is very prone to free radical damage on the one hand and slow to repair on the other.

2.04 Epigenetic Stability

A methylation diet can stabilize the epigenome. The healthiest food ingredient choline (found in eggs, peanuts, lettuce,), B$_5$ (pantothenic acid), B$_6$ or pyridoxine (found in beans, fish, meats, poultry), B$_{12}$ (found in dairy products, fish, cheese, and beer), B$_9$ or folic acid (found in vegetables, beans, spinach, sunflower seed), methionine-rich protein (found in sesame seeds, sunflower seeds), and zinc (found in oysters, other high-protein foods) are known to be necessary for DNA methylation. Our daily foods should be rich in methyl-donating groups like amino acid methionine, S-adenosyl methionine, choline, and betaine. For epigenetic stability, we need to prevent suppression and under-expression of good genes and overexpression of bad genes. To be able to do so, we need to enhance folic acid utilization necessary for DNA methylation. Positive effects of green tea and leafy green cruciferacea vegetables on reducing the risk of stomach cancer are now well-known.[37] We know that a stress-free lifestyle can help manage the epigenome.[38] This involves routine exercise, relaxation, and meditation for neural transmission and memory, and avoiding smoking and alcohol. Avoiding chemicals like bisphenol A from plastics (bottled water) and canned foods is an absolute necessity.

The FDA is now approving drugs that counteract epigenetic effects. Good examples are Azacitidine by Celgene Corp that treats MDS (blood malignancy) and others that stimulate tumor-suppressing genes. Genes can be silenced or made dormant in the case of cancer, schizophrenia, autism, Alzheimer's, and diabetes.

---

36 http://ghr.nlm.nih.gov/condition/sickle-cell-disease.

37 http://www.aicr.org/foods-that-fight-cancer/foodsthatfightcancer_leafy_vegetables.html.

38 http://www.nimh.nih.gov/health/educational-resources/brain-basics/brain-basics.shtml.

Table 2.02 Effect of Dietary Constituents on Epigenetic Stability

Diet Constituent	Common Foods	Effect
Methionine	Protein foods	S adenosyl methionine synthesis
Folic acid (B_9)	Baker's yeast, sunflower seeds	Methionine synthesis for fighting allergies
Vitamin B_{12}	Beer, cheese, yogurt	Methionine synthesis for fighting allergies
Vitamin B_6	Fruits and vegetables	Methionine synthesis for fighting allergies
Choline	Egg yolk	Methane group donor to SAM
Betaine	Sugar beets and wheat	Breakdown of toxic products from SAM synthesis
Resveratrol	Red wine	Histone deacylation and longevity
Genistein	Soy foods	Methylation and cancer prevention
Sulphoraphane	Broccoli	Histone acylation and cancer prevention
Butyrate	Dietary fiber in colon	Histone deacylation and longevity
Diallyl sulfide	Garlic	Histone acylation

Note: Diet constituents listed above are key to the methylation diet.

2.05 Nutritional Genomics

Nutrigenomics deals with the effect of diet on the epigenome; one could call it the response of gene expression to diet. Human beings have 99 percent identical genes. Only 1 percent difference accounts for all the genetic diseases. Some chronic diseases are due to effects of more than one gene. Although the response may vary from individual to individual, disease prevention can be attained by diets that preserve cellular function. There are close to ten thousand disease-causing genes. Many diseases are monogenetic, but those controlled by more than one gene are complex and often difficult to treat. Insulin resistance is a good example of a multigene effect involving forty-seven different gene products for its signaling. Even in the case of this complexity, current research shows that diet can control diabetes.[39] Hope comes from the diet dependence of physically expressed phenotype attributes of color, weight, inflammation, and chronic diseases in general.

Examples of multigene control relate to cholesterol, homocysteine, and obesity. The number of genes involved, common genetic variants, diet, environmental factors, behaviors, and socioeconomic factors are part of the total epigenetic effects. Diet, it appears, plays a significant role, but epigenetic marks also depend on lifestyle and genotype. Dietary chemicals can and do change gene expression.[40,41] Diets can have bad effects also when they cause uncontrolled methylation. A case in point is the bitter-taste sensitivity experienced by diabetics that happens due to a minor change in DNA sequence.[42] Much more important in this respect is the stability of mitochondrial DNA.[43]

2.05.01 Foods for Stable Genes and Better Health: Consumption according to the methylation diet of fruits, vegetables, beans, spinach, peanuts, Brazil nuts, tofu, fish, cheese, and yogurt beginning early in childhood can bring about positive epigenetic health effects. This is very easy to do. What we eat as a family matters in managing the health of our next generation. The following examples, which will be amplified in subsequent chapters, demonstrate many nutrient values of daily foods.

1. Choline must be part of the daily diet (lecithin in egg yolks).
2. Consumption of vitamins B_2, B_6, and folic acid helps manage homocysteine level and blood pressure (variety of vegetables).
3. Antistress vitamins B_{12} (from dairy products and fish) and B_5 (pantothenic acid from eggs, peanuts, green vegetables) should be part of the daily diet.
4. Asparagus is an adaptogen. It is anti-inflammatory and a telomerase activator for longevity. It is good for liver health and prevents DNA damage. Ginseng, Indian ginseng (ashvagandha), licorice, and

---

39 http://www.hsph.harvard.edu/obesity-prevention-source/obesity-causes/diet-and-weight/.
40 http://www.ncbi.nlm.nih.gov/pubmed/16787199.
41 http://web.udl.es/usuaris/e4650869/Morella06/BB/Diet_disease%20gene%20interactions.pdf.
42 http://www.plosone.org/article/info%3Adoi%2F10.1371%2Fjournal.pone.0003974.
43 http://onlinelibrary.wiley.com/doi/10.1002/dmrr.1203/abstract.

shiitake and mitake mushrooms are also adaptogens for stable homeostasis.

5. Antioxidants like vitamin K, eugenol, limonene, flavonoids, orientin, and vicerin control DNA damage.

6. Genistein, a soy protein component, regulates genes and affects receptors in our body. Soy protein polypeptides modify chromatin structure and regulate tumor-suppression genes, cell death, cell-cycle control, and DNA repair.[44]

7. Green and black tea antioxidants prevent coronary heart disease and cancer.

8. Methionine-rich sesame and sunflower seeds and oysters (high in zinc also) are good for DNA methylation and DNA repair. Oranges with vitamin C and hesperidins prevent DNA damage, prevent cancer and varicose veins, and improve digestion.

9. Pomegranate prevents DNA damage.

10. Wolfberries with zeaxanthin, lutein, and polyphenolic antioxidants (a Chinese fruit) improves vision.[45]

Diet and environment can solve chronic disease problems. Since eating habits, lifestyle, environmental exposure, stress, cultural interactions, and food preferences are individual, the effect of diet is individual also. Exercise, physical activity, and practice of yoga and meditation vary in their effects from people to people. Effects of diet on control of LDL cholesterol, triglycerides, and blood pressure are also individual. This is why the propensity to chronic diseases varies from person to person. This chapter has specific citations on personalized medicine, epigenetic reprogramming, transgenerational response to nutrition, the effects of diet on the bisphenol A chemical in plastics, and the effect of genesteine from soybeans that can bring about a phenotypic change in coat color in mice.

2.05.02 Molecular Diagnostics and Personalized Therapy: Human beings have individual genetic profiles. Table 2.03 provides a general picture. A cancer cell or a virus can take more than one route to bring about disorder and bad effects. Gene variations and epigenetic effects do not allow two different people to respond to the same medication identically. Levels of specific

44 http://jvi.asm.org/content/73/2/1245.full.
45 http://www.sciencedaily.com/releases/2010/03/100330102835.htm.

proteins, presence of specific genes and mutations, and risk factor analysis apply to individual patients. Personalized DNA kits are now available, and diet and individual genetic profiles can now be cross-referenced (Consumer Genomics Laboratory, CA) for diet-gene relationships. This will soon allow us to evaluate genome-wide diet related interactions of all our 25,000 genes . Even more important, it appears, is the development of an i-Pod-size gene-radar test device that diagnoses diseases (HIV, malaria, tuberculosis, and even cancer) with genetic footprints of DNA and RNA markers in less than an hour (Anita Goel, Nanobiosym, Cambridge, MA) for less than $20. A saliva or blood sample on a chip is compared against the library in the gene-radar device.

Table 2.03 People Have Different Genetic Profiles.

100% have 23 pairs of chromosomes	56% have male baldness	19% have Alzheimer's
100% can smell asparagus in urine	44% have attached earlobes	17% have multiple sclerosis
88% have wet earwax	41% have migraines	17% are left-handed
67% are prone to cancer	29% have diabetes	14% have perfect pitch
	22% are lactose intolerant	13% have restless leg syndrome

Today predisposition to a chronic disease can be detected, screened, and diagnosed. Current technology allows better drug efficiency and management of side effects also. Although none of this is easy yet in the case of polygenetic diseases, there are possibilities of microscopic vials (capsules) that can monitor our digestive system's function routinely as the technology matures.[46] We may have the benefit of nanobots, nanomotors, and nanocopters powered by ATP, and motorized nanomotors for day-to-day therapy in the near future.

There is success in cancer therapy already, and genome- and proteome- (enzyme proteins and receptors in particular) based therapies may become

---

46 http://www.hopkinsmedicine.org/healthlibrary/conditions/digestive_disorders/digestive_diagnostic_procedures_85,P00364/.

common in a few decades.[47,48] Repeat performance of life molecules may be used to enhance brainpower, health, and longevity via genetic profiling, monitoring protein profiles, and assessing metabolism. We may be able to do so for specific tissues and probe into blood and infected tissues. The probe shall reveal genetic variation and find out missing sequences or even an entire gene responsible for a disease. This will end trial-and-error health care, with new individualized treatments for cancer, HIV, and depression.

While we need to manage epigenetic effects responsible for temporal and spatial control of gene expression, quick and reliable noninvasive diagnostics is already possible by blood testing for diseases such as Alzheimer's by detecting the amyloid protein, autism by a panel of proteins, cancer by tumor proteins, Parkinson's by antibodies that signal dying nerve cells, and depression by a panel of eleven proteins.

Today we have the technology to silence a bad gene, modify and activate genes including the X-linked chromosomal gene, bookmark and mutate genes, and manage DNA methylation by managing *choline and folate* in our diet. Molecular diagnostics shall in the near future help manage chronic diseases and cancer by automating best choices for nutrients in foods and necessary food supplements. A day will be here soon when we get our genome analyzed for personalized medicine at Walgreen and CVS pharmacies. The same day, food selection by phone with a PGA (personal genomic assistant) will become downright practical.[49] Such selections in effect will be made for foods that facilitate gene expression and keep DNA stable.

Abbott Nutrition, GlaxoSmithKline, Kellogg, Kraft, Mead Johnson Nutritionals, Nestle, Pfizer, and Unilever are working toward such a goal. Nestle Health Sciences already emphasizes personalized nutrition in order to treat diabetes, cardiovascular disease, obesity, and Alzheimer's. *23andMe* genetic testing provides routine testing for over one hundred traits and diseases as well as DNA ancestry. Anne Wojcicki, CEO, says, "Our mission is to be the world's trusted source of personal genetics." The company's website www.23andme.com/about/ has much more information. Another company, Pathway Genomics, offers nutrition and exercise recommendations based on an individual's genes. "We have speed gene (a variant of $AcTN_3$ of fast-twitch muscle fibers), strength-training gene $INSIG_2$, and there is a gene

47 http://www.ncbi.nlm.nih.gov/pubmed/19702439.

48 http://www.cancer.net/publications-and-resources/asco-care-and-treatment-recommendations-patients/hormonal-therapy-hormone-receptor-positive-breast-cancer.

49 http://scienceroll.com/2010/01/26/smartphone-as-a-personal-genome-assistant/.

for overeating by diabetics," according to Pathway Genomics and other sources (http://www.ncbi.nlm.nih.gov/gene/51141; http://www.medscape.com/viewarticle/720690_3; http://www.wellnessgene.com/images/PDF/WellnessGeneSample-Test.pdf) . The company claims to individualize nutrition. The use of such genetic testing is not approved by the FDA yet, and the practice may be contrary to bioethical standards.

Personalized medicine has a few common hurdles in its way, though, namely, needs of physician training, special requirements, genome sequencing cost, regulatory compliance, transitioning with traditional medical practice, payment system, and resistance by the pharmaceutical industry. But the future looks good. In the meantime, we should cure ourselves with diet, oxygen, good water, a stress-free lifestyle, exercise, and planned physical activity. As to food nutrients, we should optimize intake of choline, B vitamins B_2, B_5, B_6, folate (B_9), and B_{12}, vitamin C, and vitamin K—key nutrients of the methylation diet. Much of it can routinely come from the common foods listed in Table 2.02. Planned nutrition thus can prevent many of our ailments and chronic diseases, cancer in particular.

As things stand today in clinical practice, no physician willfully educates us to diagnose our health by examining our daily stool and urine; paying attention to our abilities of taste, smell, texture, and color perceptions for controlled consumption of foods; and meeting our daily needs of oxygen and water. My own physician has never said a word about the effects of overeating on routine physiology. This book, I hope, will teach about the values of such precautions and managing health by the foods we eat and choices of exercise and socially interactive ways of living.

2.06 Major Reference Institutions

The institutions listed below are leaders in research on nutrition, genomic stability, and epigenetics.

The Salk Institute, La Jolla, California; University of California, Davis; Tufts University Research Center on Aging; Penn State University; Netherlands Nutrigenomics Center; The J. Craig Venter Institute; and Cornell Institute of Nutritional Genomics.

CHAPTER 3

Our Body, Our Mind, and the Mind-Body Connection

"Our thoughts and imagination connect mind and
body for enhanced health and well-being."
—Triveni P. Shukla

All that happens in and to our body, mind, and body-mind connection is controlled by the DNA in our cells' nucleus, the mitochondrial DNA, the genome associated with our individual microbiome, and the epigenome on the membrane of our cells. Interacting genes control our mind and behavior.[50] A gene identified as SLITRK1 suppresses anxiety and depression, and another identified as SLITRK6 is involved with our neural circuits, say scientists at the Riken's Brain Science Institute (http://www.gensat.org/GeneProgressTracker.jsp?entrez=15552). Close to one hundred membrane proteins with amino acid leucine repeat units and genes encoding these proteins like software may have a lot to do with our mind and behavior. Also, the methylation diet as a major determinant of nutritional genomics shouldn't be ignored.

Our Body

Our body is created by foods our mother ate and then foods we eat later in life day by day. By composition, 99 percent of our body is two thirds hydrogen, one quarter oxygen, and one tenth carbon. Oxygen, critical to our life, is the majority by mass. Water stands first in line with all its hydrogen as the medium for life processes and source of electrons and protons. Water helps

---

50 http://phys.org/news/2010-11-genes-mind-behavior.html.

hemoglobin to carry oxygen on its back. Next in line after water are proteins. The remainder is minerals, small but essential for our daily metabolism. Electrons and protons rule in terms of our bodily functions of genes, enzymes, and neurotransmitters. Although an energy equivalent of a mere one-hundred-watt bulb, our body is not a random factory. It is loop controlled and maximally organized for efficiency with no more than 200–500 mV electricity. All that happens in our body is fast and electrochemical in nature. The ions, electrons, and protons are in play all the time in controlling human body operations as receptors for vision, hearing, taste, and smell; they help convert and converge signals to the brain at the rate of ten to fifty impulses per second. No neuron circuit and no impulse simply means death. Food affects the circuits and, therefore, our body's function. Furthermore, our body is a self-sustaining nanoscale molecular factory. All its ten functional systems are under the control of the nervous, endocrine, and immune systems. A 70-kilogram or 154-pound person has 25,000 to 35,000 genes in twenty-three pairs of chromosomes, 2 million protein molecules, 10 trillion cells, and 5 trillion molecules. On a macroscopic level, our body has 45 percent protein devoted to repair, water balance, muscle contraction, energy production, skin health, hair health, nail health, and regulation and protection and upkeep via enzymes, hormones, and molecules for immunity. We are a kind of protein house. Next to DNA, proteins are the most important molecules for our body, a molecular factory that runs on what it is fed, that is, the food we eat, the liquids we drink, and the air we breathe. All affect our energy level, mood, food cravings, stress level, sleep, daily behavior, and even genetics. There is a gene behind everything that happens in our body.

Our Mind

We are conscious because of our mind and because of our genes. Mind analyzes, thinks, perceives, imagines, commits to memory, and solves problems. The mind is aware of past, present, and future. The mind can be trained and self-regulated by invoking an internal state of relaxation, energy, and single-pointed concentration. As a child I knew it as single-mindedness or *akaagrata* in Sanskrit, but today the fashionable descriptor is "mindfulness." Single living cells with a nucleus evolved 2.0 billion years ago, multicellular organisms evolved a billion years ago, and the human body and mind evolved 500,000 years ago. Given the evolutionary timescale, the human brain evolved rather recently. We exist as complex and conscious beings because of our brain and mind. We need our brain for sensory perceptions and physical movements, wherefrom emerges the mind with quality of consciousness and transcendence. The mind can shape the brain, and the brain, communicating to itself, can control and drive the mind. Our religious, social, and cultural perceptions continuously inform and train the brain, which is a physical device for transmission, coordination, and communication. Communication is all electrical, with neurons acting as batteries. Inside of a nerve cell is negative, and outside is positive, and this polarity is reversible. The underlying action potential is the basis for propagation by neurotransmitters and communication. Life continues because action potentials travel down the axons in the neural network. The multidirectional consciousness appears to be the firing pattern of neurons. Each neuron affects a thousand of its neighbors connected by synapses; it thus affects 100 trillion synapses. The totality of it all is the adaptable neural network that we understand very little about. We do know, however, that energy production is the basis of brain function and its communications network. The brain depends on glucose and oxygen, and thus it depends on exercise and movement. This is why the brain consumes 20 percent of all the oxygen we inhale. A lack of oxygen supply to brain cells even for ten minutes due to poor blood circulation can kill them. An enormous number of well-functioning mitochondria, the power plant in the brain cells, is necessary for its health and speed of work. To allow consciousness, the blood must flow through all 60,000 miles of blood vessels throughout the body. We know that Parkinson's disease results when brain cells abandon their mitochondria, thus limiting energy production. Everything our body does involves electrical signals, including energy production, beating of the heart, movement of our hands and legs, and even greeting of a friend by seeing, hearing, and speaking. All this depends on our daily diet, and the diet effects can be augmented by exercise, a good night's sleep, and meditative relaxation.

Our Mind-Body Connection

Mind and body are always connected. Nothing happens in the body without intervention by the mind and the brain, the physical basis of mind, but the mind can't exist by itself without assistance from the rest of the body system. Body and mind, therefore, are to coexist. The brain dictates behavior, which is shaped by perception and emotions. We should continuously use emotions to train our mind. It is now believed that consciousness is a multidimensional *information* system composed of its own space-time system. The seat of consciousness is the brain, though. We need to connect matters of the mind with the body in order to stay conscious.

Homeostasis—call it equilibrium—is the web of life, and the psychosomatic network is mind-body connectivity. This is the network that facilitates conversation with our nervous, endocrine, and immune systems. The conversation is ionic and electronic, and its connections depend on foods we eat and the lifestyle we choose to maintain. Diet, exercise, and meditation can help keep this connection in balance for better health and longevity. Conditions of high cortisol, stress, physical tension, aches, and pain can be reduced by mind control. Meditation or contemplation, in particular, is used to train and self-regulate the mind for relaxation, emotional balance, and internal energy.

Common in the United Kingdom and Germany, psychosomatic therapy or manipulation of the mind-body connection for chronic disease treatment is now a growing new trend. The foundation of therapy, for example in the case of guided imagery, is mind-body interconnectivity by chemicals and receptors. For psychosomatic therapy to be effective, though, the mind must be connected to the body. The nervous and immune systems must be constantly in communication, and the hormone cholecystokinin must talk to the brain all the time.

3.01 Our Body

The real power for keeping us alive rests with the DNA in our cells. All DNA combined measures around 3.00 meters. Unravel your DNA and it would stretch from here to moon. We have a total of forty-six chromosomes (twenty-three pairs). Each pair is composed of paternal and maternal DNA on a 50:50 basis. This is where our 25,000 to 35,000 genes reside. Our body makes 75,000 enzymes in each of our 7.5 trillion cells (excluded are 2.5 trillion red blood cells). The enzymes make more than 173,000 different proteins under the direction of these genes. In addition, there are fifty hormones, one hundred neuropeptides, many other steroids and proteins, antibodies, and antenna-like receptors on each cell involved in cell-to-cell and organ-to-organ communication. This complexity is unfathomable.

Whereas carbon is central to the organic structure of the human body, nitrogen and oxygen are critical to the life processes of reproduction and daily living. The mystery of the molecules of carbon, oxygen, nitrogen, phosphorus, and sulfur atoms is augmented by minerals and vitamins in our daily life processes. Electrons and protons rule and even keep our health.

Antioxidant vitamins C, E, and K and other antioxidants like flavonoids, beta-carotene, and selenium are electron managers. Therefore, minerals, at around 1 percent in our total daily food, are very important.

3.01.01 Cells in Our Body: A cell in our body is the smallest *macromolecular* factory that we know of. Everything in our body happens at the cellular level, including repair of broken bone, reproduction, and fighting of infections. Cells, with their plasma membrane, the interactive and protective barrier around them as a sheath, have the power to communicate and exchange nutrients. They have channels and pores, and receptors to translocate chemical molecules in and out. Sugars, lipids, glycolipids, and phospholipids are the membrane's core constituents. Actually each cell is an organism by itself. Table 3.01 lists the size and type of various cells of the human body.

Table 3.01 Size of Our Cells

Cell Type	Nanometers	Description
Average human cell	50,000	10 trillion cells in our body
Size of nucleus in the cell	5,000	This is where DNA resides in the cell.
Mitochondria	3,000	Energy factory for ATP
Smallest cell	4,000	Granule cells of the cerebellum, the brain
Red blood cell	9,000	Carry oxygen on hemoglobin to cells of various organelles
Sperm cell	25,000	With flagella
Egg cell	100,000	Unfertilized ovum
Largest cell	135,000	Cells in spinal cord
Skin cells	1,000,000	Divide and reproduce quickly
Glandular cells	15,000	Produce hormones and enzymes
Nerve cells	100–600	Do not divide and reproduce
Lung cells	7,000	Produce mucous
Muscle cells	100,000	Attached to one another, contract and relax
Mouth cells	60,000	Produce saliva

Cells make tissues, tissues make organs, and organs make an organ system. There are ten such systems in our body, all constructed and operating around and in close conjunction with our open digestive system. Human cells vary in size (see table 3.01), and so does the weight of the human body and its height. Table 3.02 contains daily input and rate information on body systems.

Table 3.02 Our Body's Capacity Data

Average weight	48 Kg + 2.3 Kg for each extra inch over 5 feet
Average waist	35–38 inches; two inches less for women
Daily intake	6 liters of air, 2 liters of water, and at least 0.5 lb food solids
Daily output	2 lb urine and 0.75 lb feces per day
Heart and blood	Half–pound heart acts as a pump and circulates 2,000 gallons of blood per day (3 oz per beat) through 60,000 miles of vascular system
Lung and breathing	12–17 times per minute; inhale 660 CC air per breath; oxygen and carbon dioxide exchange and air sterilization by microphages
Blood filtration	25% of blood pumped per beat goes through kidneys, and 99% of filtrate is reabsorbed.
Resting heart rate (morning)	Average 70 beats per minute
Body mass index	A measure of body fat based on weight and height;
Total calcium in our body	2.0–2.2 lb; 99% of it is in the skeleton

To put things in perspective, an average consumption of 2,000 nutritional kilocalories (2 million calories) is equivalent to 7,920 BTU per day or 330 BTU per hour of energy. This equates to 96.7 watts per hour or 0.13 HP per hour. The human body is not random, but maximally organized and loop controlled.

3.01.02 Molecules of Our Life in Our Body: As the molecule of heredity, deoxyribonucleic acid (DNA) is the master molecule of our evolution. It is a semiconductor, with electrical properties. DNA is composed of five-carbon sugar ribose, phosphate, and pairs of nucleotide bases—adenine, thymine, guanine, and cytosine.

Whereas glucose is the fuel for all cells in our body, it is not involved in the structures of DNA, RNA, the energy molecule adenosine triphosphate (ATP), and electron transfer molecules like NAD/NADH and FAD.FADH$_2$.

Proteins and peptides, including enzymes, hormones, and neuropeptides, are the next most important molecules of life. In our cells, proteins are molecular motors that tote DNA-laden chromosomes, all kind of protein cargo, and DNA-copying enzymes as a "sliding clamp."

Gas molecules such as oxygen and nitric oxide perform very critical functions. Everything in our body happens in water, the liquid of life. Actually, all that lives is water.

The human body has its major systems housed in the skeleton, made of bones, joints, and muscles. There are 206 living bones, making 10 percent of body weight. Bones are made of the mineral calcium and marrow. There are thirty-three bones in the spinal column alone. A majority of bones are in our hands and legs. There are joints balanced by tendons, ligaments, and muscles at critical points for flexibility and motion, and that is how we move under voluntary control by the cerebrum of our brain. The muscles, all six hundred of them, have specific and critical functions, including those of the intestine, bladder, heart, blood vessels, and skeleton.

3.01.03 Major Body Systems: There are ten major functional systems in our body.

1. The digestive system begins with the mouth and continues through the esophagus, stomach, and small and large intestines, ending in the rectum. Its function is to grind and break down foods we eat into small molecules, sterilize them and kill bacteria, and then help absorb nutrients to the circulating blood. More than 50 percent of the immune system resides in our digestive system. The truth is that the digestive system supports all other body systems, and its health depends on what we eat.[51]

2. The respiratory system includes the nose, trachea, and lungs. It is responsible for the exchange of oxygen and carbon dioxide

---

51 http://preventioncare.org/index.php?option=com_content&task=view&id=30&Itemid=2.

and circulation of nitrous oxide. Our lung capacity is six liters. A 150-pound male needs around 443 liters of oxygen per day or 4.5 ml/Kg/min. The need can be as high as 1.2 liters per day for a person who does extensive aerobic exercises. We need three square meters of grass or one acre of Christmas trees to produce enough oxygen for us. Our respiratory system exchanges by number as many air molecules in a single breath as the number of stars in all the galaxies of our expanding universe. The air we inhale has 0.04 percent carbon dioxide, and what we exhale has 4 percent carbon dioxide. *Carbon dioxide is just as critical in maintaining blood acidity as oxygen is for producing energy.* The air that we breathe in must meet the requirement of 20.8 percent oxygen, and it must be sterilized by macrophages (a kind of virus in our nasal system) that engulf bacteria, chemicals, and dust that accompany the air. The air should be clean and unpolluted.

3. The skeletal system is made of 206 bones and 600 muscles; it supports our body with bones, cartilage, muscles, and ligaments. Our hands and feet have most of the bones. Cartilage lines the bones of joints and ligaments and confers bone-to-bone connection.

4. The muscular system, with skeletal and smooth muscles working in pairs, provides for body movement per se and for control of movement of material through organs.

5. The circulatory system, with heart, blood vessels, and blood, helps transport nutrients, gases, hormones, neurotransmitters, and waste. The heart pumps about 2,000 gallons of blood per day that circulate through 60,000 miles of blood vessels carrying oxygen and nutrients to each of 7.5 trillion cells in our body. Nutrients and gases reach our cells through this system.

6. The nervous system includes brain, spinal cord, and peripheral nerves. It controls our body's functions and directs our emotions, behavior, consciousness, and movement. It is our communications network. The system relays signals chemically and electronically in close cooperation with hormones of the endocrine system.

7. The endocrine system, with glands such as the hypothalamus, pituitary, thyroid, pancreas, and adrenal, helps relay chemical messages. It works in cooperation with the nervous system for the control of our physiological processes.

8. The reproductive system, with ovaries, oviducts, uterus, vagina, and mammary glands in the case of women, and testes, seminal vesicles, and penis in the case of men, manufactures and distributes cells that allow reproduction. A sperm from the male and an egg from the female become a zygote, which becomes an embryo, which finally differentiates into a male or female human being.

9. The lymphatic system includes structures of the lymph nodes, vessels, spleen, thymus, lymphoid follicles of the digestive system and tonsils, and bone marrow dedicated to carrying interstitial fluid (lymph) containing white blood cells and T and B cells to cells of tissues in various organs of our body. Our immune system fights infection due to invading bacteria and fungi. It is critically important for the control of (1) fat and chyle absorption by lacteals of the intestinal tract, (2) transportation of cancerous cells, and (3) fighting food allergies. The digestive tract is the largest immune organ in our body. *Blood is never in contact with the parenchymal cells and tissues of our body; it is the interstitial fluid that contacts such tissues.* It too is supported and maintained by what we eat.

10. The excretory system, with kidneys, uterus, bladder, and urethra, helps filter wastes, toxins, and excess water or nutrients from the circulatory system. Our kidneys filter 1.5 gallons of total blood once every forty-five minutes or Forty-five to forty-eight gallons per day. They filter 25 percent of blood per heartbeat under pressure. Our body, the nano-factory, reabsorbs 99 percent of the filtrate. Also, the kidneys produce erythropoitin, a hormone that helps mediate blood-cell production. Close to 1.5 liter of urine, the waste, per day helps remove drugs and byproducts of our daily metabolism. Breathing puts out roughly 1 Kg carbon dioxide per day. Daily human excreta weighs 0.75 lb, more than half of which is moisture. Of 50 percent solids, almost half is bacterial mass.

The biology of our being is highly organized, and the organization is maintained by what we eat and drink. All these ten systems represent no more than a one-hundred-watt bulb in terms of energy from 1,888 food calories. This efficiency in the human body simply means efficiency of intricate organization, feedback, and recycling. The effects of balanced food for building a good and healthy body can be further enhanced by exercise and meditation. Thirty minutes of exercise and fifteen minutes of meditation per day is known to boost the efficiency and performance of our body.

3.02 Our Mind

The brain is a physical organ capable of making us balance, listen, move, perceive, see, and speak. It regulates and coordinates all that happens in the body. It communicates by transmission of electrical impulses. Most important of all, although it communicates to itself, it is not an independent organ. It is connected to the vascular, immune, and nervous systems. Autonomous functions like heartbeat and repair and growth of cells are regulated by the brain.[52] The low-voltage current in the brain defines the range of its possibilities, say speech and speech decoding. The brain is fed, nourished, and repaired all the time with nutrients from our food while it is communicating with the body. It is always connected to the body.

The mind, on the other hand, as an electromechanical pathway and its connectivity, is a data phenomenon that exists and manifests in the substratum. It is dynamic, and it makes us conscious of the body and the world around us. *Consciousness, however, cannot be reproduced.* Meditation via single-point thinking or mindfulness can control the mind. Is biological consciousness the same as electronic consciousness? Is there something other than the physical brain responsible for consciousness and personal identity? While we do not have all the answers to such questions, it is clear that the brain controls the body and the body and mind are connected. It is the synapses and their 10^{15} connections that constitute mind. It appears that the mind can be controlled for better health and freedom from diseases that result from mitochondrial inefficiency and partial failure.[53]

3.02.01 Brain and Mind: The brain is a physical structure; it is an organ. It is the basis of the mind, about which we understand very little because manipulating and dealing with the individual nerve cell is not yet possible. The proposed project by the White House under President Obama's leadership on the "Brain Activity Map" may help improve our understanding of the mind.[54] With the advantages of nanotechnology

---

52 http://www.cse.msu.edu/~weng/research/CIM-FNB-AMD.pdf.

53 http://www.ncbi.nlm.nih.gov/pubmed/19664343.

54 http://arep.med.harvard.edu/pdf/Alivisatos_BAM_12.pdf.

and the power of data computing and analysis, we may soon begin to understand nerve-cell connections and collaboration among them. We do follow brain activity via fuel consumption (by functional magnetic resonance imaging) on a centimeter scale. But on a micron scale, we can barely follow the working of an individual nerve cell only. All in between is dark and not understood at all. The brain activity map will surely help pin down the effects of food components on cognition, daily behavior, mood, and longevity.

The mind relates to awareness and consciousness; it is free will and the capacity to understand, imagine, think, and process present, past, and future. It solves problems like making tools to see through ultraviolet light, although human beings can't see through ultraviolet light. The mind is a collection of thoughts, patterns, perceptions, beliefs, memories, and attitudes. *We should use our mind to control the brain and not vice versa.* Culture, religion, faith, and interpersonal relationships do influence the mind. We can reduce stress by mind control, diet, meditation, and exercise.

The same electrons, the same protons, and similar organic and inorganic molecules are in our brain as in the rest of the universe. But what a difference! The brain is much more than a simple tissue composed of cells and half a million cortical columns, each with 60,000 neurons. Emerson was right in saying that the integrity of our brain, as commander in chief of the body, is sacred and a superbly encoded marvel of nature. He said, "Guard your integrity as a sacred thing. Nothing is at last sacred but the integrity of your own mind." At the present, it appears to be beyond our 25,000 to 30,000 genes. It directs all voluntary and involuntary activities and controls daily metabolism and behavior.

On an average. our three-pound brain, less than 2 percent of our body weight, along with the spinal cord, controls our body, the molecular nano-factory. One neuron can produce 200 mVolts of bioelectric potential, which is measured in electroencephalogram (EEG) testing for electrical impulse as brain activity. *Neurotransmitters create a message, neurons power the message, and receptors receive the message.* Such messages can travel at a speed of 150 miles an hour. Actually the psychosomatic network and connectivity by chemicals and receptors is the foundation of mind-body interconnectivity.[55] Our central nervous system and immune system function in cooperation in response to

---

55 http://www.ncbi.nlm.nih.gov/pubmed/9656499.

the environment and psychosocial factors, and our emotions and general behavior are related to neuropeptides that regulate our immune system.

3.02.02 The Communication System: Higher intelligence comes to us from the molecules we eat and drink. Our brain is built from molecules, operates with molecules, uses molecules for its complex information network, and keeps body functions in checks and balances with molecules. About 100 billion nerve cells or neurons in the brain and 1.3 percent of the 7.5 trillion body cells are involved in communicating with the rest of cell-based organelles by an amazing synergy and coordination. It is this synergy we understand very little about. The brain receives billions of electrical and chemical impulses as information about inside and outside the body. It analyzes information and instructs body organs, glands, and muscles to respond in time and in an intricate and speedy coordination. Feeling is all chemical and electrical, and communications occur across synapses via specific receptor molecules. Since neurotransmitters can be destroyed by enzymes, so can feelings.

The brain and spinal cord make the cable of nerves. Beyond the brain there is the peripheral nervous system, consisting of the peripheral spinal cord, the cranial nervous system in the head and neck, and the autonomic nervous system (ANS) of involuntary functions of breathing, sweating, blood circulation, digestion, gland activities, and heartbeat.[56] Composed of parasympathetic, sympathetic, and enteric nervous systems, the ANS controls stress, rest, digestion, muscles, and glands.

Heart rate and the blood vessels are regulated by the sympathetic autonomic nervous system (ANS), and digestion is regulated by the parasympathetic ANS. Also, the body's connection with the brain is via the endocrine system, which produces hormones and adds them to the bloodstream directly. It is an information system that acts slowly but lasts longer than the fast and short-lived nervous system.

3.02.03 Endocrine Glands

3.02.03.01 Adrenal glands atop the kidneys produce glucocorticoids for controlling metabolism, aldosterone for potassium and sodium control, and epinephrine and norepinephrine for stress management.

---

56 http://faculty.washington.edu/chudler/auto.html.

3.02.03.02 The pancreas, in addition to enzymes and bicarbonate as digestive aids, produces insulin for glucose metabolism and even control of brain activities.

3.02.03.03 The pituitary gland of our brain is the master gland that controls the endocrine system at large. It produces the growth hormone, the adrenocorticotropic hormone (ACTH), and the antidiuretic hormone gonadotropins.

3.02.03.04 The pineal gland in the brain produces serotonin, the precursor of melatonin for controlling sleep and wake states. Serotonin and melatonin production depends critically on consumption of high-tryptophan proteins.

3.02.03.05 The thyroid gland produces thyroxin, calcitronin, and triodothyroxin, critical to metabolism and protein synthesis.

3.02.03.06 The thymus produces thymosin, for immune system control and maintenance.

3.02.03.07 Testes in men produce testosterone, and ovaries in women produce estrogen and progesterone.

Hormones are chemical messengers. Traveling through blood, they work with the nervous system in internal communication with the brain for control and coordination of growth and repair of tissue, metabolism, blood pressure control, and the stress response. Hormones help us cope with the internal and external environments. Whereas the steroidal hormones work with the genetic materials and DNA, the protein and peptide hormones work with receptors on cell membranes. Our senses—eyes, ears, nose, taste, and skin—all connect to our brain via electrical or chemical signals and through receptor activity.

3.02.04 Structure of the Nervous System: There are groups of nerve fibers that travel together in the central nervous system (CNS). They are called pathway or tract or commissure. They are present in both the right and left half of the brain. Information in the CNS passes along two types of pathways:

1. Long neural pathways, in which neurons with long axons carry information directly between the brain and spinal cord or between

different regions of the brain. Thus, there is no alteration in the transmitted information.

2. Multineuronal or multisynaptic pathways that are made up of many neurons or synapses. This pathway can integrate new information into the transmitted information.

Cell bodies of neurons having similar functions cluster together, and such clusters are called ganglia in the peripheral nervous system and nuclei in the CNS.

3.02.05 Components of the Central Nervous System and the Brain

3.02.05.01 The brainstem controls the motor, cardiovascular, and respiratory systems.

3.02.05.02 The cerebellum coordinates movements and controls balance and posture.

3.02.05.03 The forebrain: The larger component of the forebrain, the cerebrum, consists of the right and left cerebral hemispheres, which have an outer shell of gray matter, the cerebral cortex. Each hemisphere is divided into four lobes: frontal, parietal, occipital, and temporal. *The cortex is the most complex and integrating area of the brain.* The central core of the brain is formed by the diencephalon, consisting of the thalamus—a collection of several large nuclei—and the hypothalamus, the master command center for neural and endocrine coordination.

3.02.06 Major Parts of the Brain and Their Functions

3.02.06.01 The frontal lobe: Functions of planning, organization, problem solving, memory, impulse control, decision making, selective attention, behavior, and emotion are controlled by the frontal lobe.

3.02.06.02 The parietal lobe: Sensory integration, sensation (hot, cold, pain), direction control, spiritualism, and faith are controlled by the parietal lobe.

3.02.06.03 The temporal lobe controls sound processing, speech, and memory.

3.02.06.04 The occipital lobe controls processing visual information, and perception of shape and color. The latter is quite complex.

3.02.06.05 The cerebellum (the back of the brain) controls balance, movement, and coordination.

3.02.07 The Central Nervous System and Spinal Cord: The spinal cord lies within the vertebral column. The central gray matter is composed of interneurons, cell bodies, dendrites, and glial cells. This is surrounded by white matter, composed of myelinated axons of interneurons. The fiber tracts either descend to relay information from the brain or ascend to relay information to the brain or transmit information across different levels of the spinal cord.

Afferent fibers from the peripheral system enter on the dorsal (posterior or back) side of the cord via dorsal roots and form the dorsal-root ganglia. Efferent fibers leave the cord on the ventral side via ventral roots. Dorsal and ventral roots from the same level combine to form a spinal nerve outside the cord, one on each side. There are thirty-one pairs of nerves, designated by four levels of exit: cervical (8), thoracic (12), lumbar (5), and sacral (5).

Since the entire human body is connected with neurons, the preceding description of the components of the central nervous system should aid in understanding the mind-body connection as influenced by the powers of daily meditation, exercise, and routine perception. We need to pay special attention to a balance between the sympathetic and parasympathetic peripheral systems: sympathetic for stress control and parasympathetic for control of exercise, digestion, and functions related to digestion. A simple monitoring of what we eat and our lifestyle of exercise and meditation can tell a lot about mind-body connections.

3.02.08 The Peripheral Nervous System: The peripheral nervous system transmits signals between the central nervous system and receptors. It consists of twelve pairs of cranial nerves that connect with the brain and thirty-one pairs of spinal nerves that connect with the spinal cord. The efferent system is further divided into a somatic and an autonomic system. Table 3.03 presents a schematic description of the human nervous system.

Table 3.03 Our Nervous System

Nervous System	Central nervous system	Brain: Heart, blood pressure, breathing, temperature, emotion, hunger, thirst, circulation, rhythm		
		Brain and Spinal cord		
	Peripheral nervous system: The system of receptors	By Direction	Sensory/Afferent sends information to Central Nervous System	
			Motor/Afferent sends information from central nervous system to muscles and glands	
		By function	**Somatic**: Information from central nervous system to skeletal muscles under conscious control, speech, respiration, willful and habitual muscle movements	
			Autonomic) nervous system (ANS) (**Unconscious**) CNS to smooth muscles, heart, glands; can adapt to changes in the environment. It is strongly related to breathing.	**Sympathetic:** Thoracolumbar part of spine is responsible for fight-or-flight response, including energy, arousal, digestion; it can be sedated by breathing and relaxation. The effects are global, and it gears up for emergencies, including inhibition of salivation, acceleration of heartbeat, dilation of bronchia, inhibition of peristalsis/ secretion, inhibition of bladder contraction, conversion of glycogen to glucose, dilation of pupils, and secretion of adrenalin.
				Parasympathetic: Craniosacral portion with vagus nerve controls heart, lung, liver, stomach, and other organs … all this is unconscious. Stimulates salivation, slows down heartbeat, constricts bronchia, stimulates peristalsis and secretions, stimulates bile acid, and contracts bladder.
				Enteric brain that controls digestion

3.02.09 Somatic Nervous System: The somatic nervous system innervates skeletal muscles and consists of myelinated axons without any synapses. Activity of these neurons leads to excitation (contraction) of skeletal muscles, and therefore, they are called motor neurons. They are never inhibitory.

3.02.10 The autonomic Nervous System: *Most autonomic responses usually occur without conscious control.*

The autonomic nervous system innervates smooth and cardiac muscles. It consists of parallel chains, each with two neurons that connect the central nervous system and effector cells. The synapse between these two neurons is called the autonomic ganglion. The nerve fibers between the central nervous system and the ganglion are called preganglionic fibers, and those between the ganglion and the effector cells are called postganglionic fibers.

The *sympathetic peripheral system* is the fight-or-flight control system. It relates to stress control. The *parasympathetic peripheral system* is the rest-and-relax component of the system. Sympathetic ganglia lie close to the spinal cord, while parasympathetic ganglia lie close to the organs. *The power of meditation resides in the parasympathetic nervous system.*

The sympathetic system is arranged to act as a single unit, while the parasympathetic system is arranged such that the parts can act independently. Many organs and glands receive dual innervations from both the sympathetic and parasympathetic fibers. The two systems generally have opposite effects and work together to regulate a response.

3.02.11 Blood Supply, Blood-Brain Barrier, and Cerebrospinal Fluid: The brain is highly dependent on a continuous supply of glucose and oxygen via blood. Although the glucose level in the cerebrospinal fluid is less than in blood, in a range of 50–75 mg/100 ml, it is very critical for brain function. Oxygen consumption by our three-pound brain is 46 cm^3/min or 3.3 ml/100 g/min. The brain alone uses 0.07 gram of oxygen every minute.

The exchange of substances between blood and the extracellular fluid in the central nervous system is highly restricted via a complex group of blood-brain barrier mechanisms. The central nervous system and the extracellular fluid in the brain are in equilibrium with each other but maintain a difference with respect to blood. Oxygen, water, and glucose move freely across the blood-brain barrier. The neural tissue of the CNS is covered by three membranes called meninges—the outermost thick *dura mater*, the middle spiderweb-like *arachnoid mater,* and the inner *pia mater membrane.* The space between the pia and the arachnoid, the subarachnoid space, is filled with shock-absorbing *cerebrospinal fluid (CSF).*

3.02.12 Nutrition and Mind: Mental health depends on the brain and mind. *Sixty percent of the brain's dry weight is fat, and 20 percent of the fat is essential fatty acids.* Too much saturated and trans fat is bad, and a 1:2 ratio of omega-3 and omega-6 fat is good in our daily diet. Saturated fat is known to cause dementia.[57] Balanced protein nutrition is essential for neurotransmission and mental health. Folic acid, omega-3 fatty acids, tryptophan, and the mineral selenium can cure depression.[58] Antioxidants can prevent schizophrenia.[59] ADHD (Attention Deficit Hyperactivity Disorder) can be cured by diets rich in omega-3 fatty acids and the antioxidant mineral selenium.[60] Another popular URL is http://www.drweil.com/drw/u/QAA400126/OPC3-An-Antioxidant-Supplement-for-ADHD.html

Our daily food and drink is linked to proper brain function and mental health. As a matter of fact, diet can modify the brain because nutrition and exercise are related to brain function, mood, and behavior.[61] Exercise is known to elevate brain-derived neurotrophic factors. Decision making and routine executions depend on neurotransmitters that come from our daily food. We can transcend our limitations with commonly available food, and we can improve more by exercise, yoga, meditation, and sleep. Foods and diet, it appears, do have healing effects.

The intake of B vitamins, including 40 mcg B_6, 800 mcg B_{12}, and 400 mcg folic acid, vitamin E, carotenoids, and flavonoids (found in kale, mustard greens, and spinach) is essential for a healthy mind, and so are antioxidants from nuts, seeds, and whole grains. Selenium from Brazil nuts is a powerful dietary antioxidant. Such nutrients are key components in a methylation diet.

Amino acids tyrosine and phenylalanine for making dopamine and adrenaline and tryptophan for making serotonin improve mental health, and therefore, a high-protein diet is a must for routine physical, emotional, and cognitive ability. Table 3.04 on diet for brain health is included below

---

57 http://www.neurology.org/content/59/12/1915.short.

58 http://www.foodforthebrain.org/nutrition-solutions/depression/about-depression.aspx.

59 http://www.schizophrenia.com/prevention/antioxidant.html.

60 https://totallyadd.com/relationship-between-epa-and-dha-on-adhd/.

61 http://mindfull.spc.org/vaughan/Vaughan_MPH_SleepNutritionExercise.pdf.

for reference. Brain health is discussed in detail in chapter 15, "Foods for Long Life."

Table 3.04 Brain-Healthy Diet
This table is based on multiple resources including
(http://www.ncbi.nlm.nih.gov/pubmed/22359306)
(http://www.rodalenews.com/brain-food)

Common Food	Cognitive Function	Authority
Apple	Quircetin promotes blood flow and helps prevent dementia and Alzheimer's.	Cornell University (http://foodpsychology.cornell.edu/)
Avocado	Excellent methylation food, with olive oil, fat-soluble beta-carotene and lycopene, fiber, folic acid, vitamins K, C, B$_5$, and B$_6$.	http://www.avocadosoy.net/clinical-trials.html; multiple studies
Beets	Increase blood flow	Multiple studies
Brazil nuts	Selenium as antioxidant	http://www.ncbi.nlm.nih.gov/pmc/articles/PMC2698273/
Broccoli	Fights free radicals	Multiple studies on genomic stability. http://www.huffingtonpost.com/2012/02/29/broccoli-cancer-sulforaphane_n_1310634.html
Chocolate	Polyphenols prevent DNA damage	Multiple studies (http://www.ncbi.nlm.nih.gov/pmc/articles/PMC2684512/)
Cinnamon	Proanthocyanins and aldehydes prevent stress	UC, Santa Barbara (http://www.mdlinx.com/psychiatry/news-article.cfm/4640446/)

Common Food	Cognitive Function	Authority
Coffee and tea	Promote memory, reduce dementia	Finnish study (http://ije.oxfordjournals.org/content/33/3/616.long)
Concord grape juice	Polyphenols prevent oxidation	University of Cincinnati (http://healthnews.uc.edu/news/?/7001/)
Curry	Turmeric and ginger inhibit Alzheimer's	Multiple studies on turmeric. http://www.huffingtonpost.com/2012/02/29/broccoli-cancer-sulforaphane_n_1310634.html
Eggs	Choline for neurotransmitter acetylcholine	Well-established
Flaxseed	Omega-3 fatty acids	Well-established
High-protein foods	Amino acids tyrosine, tryptophan, and phenylalanine for dopamine and serotonin	Well-established
Lentils	Provide B vitamins and folate	Well-established
Nuts and seeds	Promote mental health and improve HDL	Widely approved
Oats	Digestive health and cholesterol lowering	Widely approved. FDA approved under health claim for reduction of cholesterol.
Sage oil extract	Improves cognition and promotes acetylcholine	http://www.ncbi.nlm.nih.gov/pubmed/20937617 and http://www.sciencedaily.com/releases/2003/09/030901091846.htm
Salmon and sardines	Omega-3 fatty acids and vitamin D	UC, Los Angeles. Anti-inflammatory; Omega-3 fatty acids are part of brain cell structure.

Common Food	Cognitive Function	Authority
Spinach containingfolate, Vitamins K, E, luteolin	Prevents dementia. Luteolin calms immune cells. Swiss chard and kale good source of magnesium	http://www.naturalnews.com/026273_spinach_cancer_brain.html; http://www.caring.com/articles/super-healing-foods-spinach; http://olwomen.com/foods-to-boost-your-brain-power/
Virgin olive oil	Oleocanthal improves mental health	Monell Chemical Senses Center, Philadelphia, PA
Walnuts	Antioxidants and omega-3 fatty acids	Well-established
Whole grains	Fiber and antioxidants	Well-established

Dielectric myelin, the protective sheath around neurons, is 30 percent protein and 70 percent galactose-lipid complex. It is critical to the speed of impulse conduction. Phospholipid-type fats build our brain. Omega-3 fatty acids support synaptic plasticity.[62] DHA, docosahexanoic acid, as a major omega-3 fatty acid is present in brain cells; it is a big contributor to brain growth. *Without fat, phospholipids, and essential fatty acids (omega-3 and omega-6), brain cells cannot think and feel.* The work of each of 100 billion neurons and 1,000 to 10,0000 synapses depends largely on protein and fat.

Vitamin B$_6$ coverts tryptophan to the neurotransmitter serotonin, which becomes dopamine for alertness. Vitamin B$_9$ or folic acid is key to alertness, focus, and memory. Vitamin B$_{12}$, cyanocobalamine, is necessary for a healthy nervous system. B$_{12}$ helps make myelin, helps build DNA, and is critical to production of red blood cells. *This energy vitamin boosts metabolism and as such can help reduce weight and prevent obesity.* It is produced by gut bacteria, so we should really call it a probiotic vitamin.[63] Egg, fish, meat, milk, and cheese are good sources of B$_{12}$. Vitamin E, alpha-tocopherol, protects brain cells from oxidation and free radicals.

Magnesium and manganese protect brain health also. *Fifty-six milligrams*

---

62 http://www.sciencedaily.com/releases/2008/07/080709161922.htm.

63 http://www.understanding-your-irritable-bowel-syndrome.com/probiotics-basics.html.

of magnesium are required to process a single glucose molecule. Minerals, although in trace amounts, are critical to brain and cognitive health. Cytochrome C oxidase, necessary for electron transport and energy production, is copper-dependent. *Zinc is essential for thinking and cell-to-cell communication.* It is involved in the functioning of some two hundred enzymes.

Lithium delays aging, prevents neurological disorders and Alzheimer's disease, improves brain-cell function, improves mood and behavior, fights environmental stressors, and cures headaches and muscle pain.[64] It affects through serotonin and norepinephrine neurotransmitters.

Hydration with clean and sufficient daily water is good for cognition. Diet controls of brain chemistry and behavior are related, and personality types go with brain chemistry via hormonal control and neurotransmission. For example, dopamine B monooxygenase is necessary for production of norepinephrine (from dopamine) for neurotransmission. However, too much dopamine is bad news for health. Parkinson's disease is due to an imbalance of the neurotransmitter dopamine. Gamma amino butyric acid (GABA) in tomatoes and chickpea hummus inhibits neurons for reducing pain.

We can tweak our behavior by responding to ghrelin, the hunger hormone, and leptin, a hormone that commands us to stop eating. Low leptin from fat cells is bad news. Only glucose can supply energy to the brain, and a good supply is a must for brain function. Our brain is hardwired for sugar.[65]

Low serotonin leads to depression. Most important of all for brain structure and function is omega-3 fatty acid. The upcoming chapters will reveal that foods such as pumpkin seeds, sunflower seeds, almonds, pistachios, and cashew nuts deliver methylation nutrients for brain health and for gene expression.

3.02.13 The Unknowns: We know very well that mind controls temperature,[66] the network of information for perceptions, our daily movement, thinking, and reasoning, and our daily behavior. All these functions depend on foods we eat.

The human mind is still a mystery. We know very little about how information is coded, how memory is stored and retrieved, what the baseline

---

64 http://www.ncbi.nlm.nih.gov/pubmed/6240662.

65 http://www.ncbi.nlm.nih.gov/pubmed/12828186.

66 http://www.sciencedaily.com/releases/2013/04/130408084858.htm.

activity of the mind is, how the mind deciphers time and the future, what emotion is, how it controls sleep and dreams, how brain parts interconnect, and finally, what conscience is. My personal advice is to be mindful, meditate, explore, and do all that is possible to maintain mitochondrial health.

3.03 The Mind-Body Connection

The human brain is an enormously complex cellular constellation. To repeat, 10 trillion cells in our body are connected by more than 100 billion brain cells or neurons that transmit information. This connectivity runs our body systems, keeps us conscious, and gives us the power of free will and decision making. We can recognize millions of patterns, each of which may involve more than one hundred neurons. Our neocortex, the center of all that we sense and perceive, folded over the top of the brain, makes up 80 percent of the brain's mass. It has half a million cortical columns, each with 60,000 neurons. The brain receives messages from sense organs of speech, hearing and listening, smell, and touch. It controls circulation, respiration, muscle coordination and balance, and glandular activities, and it helps do complex calculations. It interprets images from the eye.

The brain gives us consciousness, an individual subjective state of sentience including creativity, judgment, memory, perceptions, reason, and thoughts. *It appears that the firing of our neurons and the rate of firing underlie our being conscious.* Anxiety and nervousness are part of consciousness, but not digestion, growth, mitosis, and meiosis.

The human brain exemplifies the timescale of interconnected events that take place in our body. Eighty-eight peptide hormones, including insulin and secretin, move around in the pancreas alone. Other hormones, like gonadotropin-releasing hormone from the hypothalamus, angiotensin from the kidney, and peptide P, which lowers blood pressure, are clear-cut examples of mind-body interconnectivity. The amygdella, hippocampus, and cortex have 85 percent of all neurotransmitting peptides. The immune system, which includes the spleen, bone marrow, lymph nodes, and white blood cells, works because of instant and effective communication. The immune system can learn, and it has memory. *The greatest news is that the immune system can be consciously controlled. The spleen is the brain of the immune system because it not only manufactures immune cells and antibodies but also communicates with the brain parasympathetically using the vagus nerve.* It has a receptor for cholecystokinin (CCK), a peptide hormone that talks to the brain all the time. Our capacity to reason, self-control, think, and plan is attributable to the brain, which is hardwired for thinking, emotions, and behavior via billions of cells interlinked by trillions of synaptic connections.

3.03.01 The Mind: The mind has exclusive characteristics beyond the physical brain. *The mind is nonphysical and nonmaterial.* Although it changes all the time, it is always energy and intelligence. Thought, imagination, consciousness, memory, awareness, and free will depend on the hardwiring of the brain. Thought, it must be emphasized, is a creative agent that directs our lives. "We are what our thoughts are," said Buddha some three thousand years ago.

Although very complex, our body is designed to autoregulate its growth and maintenance through the brain. Mind, thought, and consciousness form the basis of human behavior, and *consciousness is not limited to just three dimensions; instead, it envelops seven dimensions.*[67] We could call it the third eye. The images that are imprinted on our brain by the mind is what we see as reality. It acts as a transistor of ions and electrons that travel at a relatively slow speed of only 125 kilometers per second. The autoregulation by the mind depends to a great extent on diet, the methylation diet in particular.

We need to have answers to many questions in order to understand the mind better than what little we know today. *Can mind affect the behavior of subatomic particles and matter? What is reality? Is living mind a defined and orchestrated vibration for its characteristics of consciousness? Is it pure energy?* We need answers to all such questions. We need to eat well for now, though.

3.03.02 The Connections: The mind-body connection is really hundreds of terabytes of information that travel from and to the brain in our body every second. Neurons control our metabolism and behavior, and therefore, we do need to understand the mind-body connection for maintaining and improving our health and well-being—the end product of our daily metabolism and physiology. At around only three pounds, the human brain is the largest in terms of body weight among all mammalian brains. This extra size is largely devoted to the cerebral neocortex and its frontal lobes. The four components of the cortex (frontal, parietal, temporal, and occipital) are involved in functions of execution.

The mind and body are inseparable, and the workings of the body are

---

67 http://therealjeffhall.blogspot.com/.

influenced by the mind. Disease is what happens to the body, its cells, and its organs, and the mind can intervene and mediate. Full understanding of health can come only from a combined understanding of all biological, psychological (thoughts, emotions, and behavior), and social factors. All such factors play a significant role in body functions and program the body's physical existence. All illness or disease stem from malfunctioning of the body. This malfunction can be traced back to self-control, thoughts, emotional turmoil, and negative thinking. Even social interaction, culture, and religion influence health because they influence the mind.

Our cells can have up to six receptors—the buttons of the cellular engine. They are memory's molecular dimensions in the form of scanners or antennae. Emotion is a neuropeptides-ligand duo, where the ligand binds to the receptor for message transfer. The shape of a ligand acts as the finger for pushing receptor buttons. There is intra- and intercellular communication. *One neuron may have a million receptors acting as invisible sense organs*, and the body and brain are molecularly connected at the neuropeptide-ligand junction, where shapes are the key. When viruses, our enemies, share the same receptors, they succeed in creating diseases.

I choose to call the mind-body connection a routine physiological transformation, whereby thoughts convert to a physical response. The physical, emotional, mental, and spiritual must be in balance and in synergy for the best of brain chemistry and the best of body response.

3.03.03 Multiple Connections: There is no doubt that hyperthyroidism, migraines, high blood pressure, arthritis, peptic ulcers, skin problems, and aches and pains are connected to the state of mind. The main risk factors underlying it all appear to be stress and the shortening of the telomere. The telomere in each of our cells is a kind of clock, denoting life span based on the number of times a cell has divided. The telomere shortens every time the cell divides. Short telomeres are known to be associated with HIV, osteoporosis, heart disease, and of course aging.[68] The enzyme *telomerase* helps keep telomeres, the end cap of chromosomes, intact, and thus genes unaltered and cells young. Foods can improve telomerase activity.[69]

The stress hormone *cortisol* in blood suppresses the enzyme telomerase in immune cells, making the telomeres shorten and immune systems

68 http://learn.genetics.utah.edu/content/chromosomes/telomeres/.
69 http://press.thelancet.com/lifestyletelomeres.pdf.

weaken. When there is damage of immune cells, chronic inflammation and diseases ensue. *Stress is our enemy number one because it can even make us irresponsive to vaccines.* Good food can control stress, which is responsible for 90 percent of all diseases, including memory loss.[70]

Increased cholesterol and fatty acids in blood decrease protein synthesis and immune response and increase allergies. High blood pressure increases metabolism and respiration. Localized inflammation causes fast blood clotting. Increased production of blood sugar leads to stomach acidity. Many of these problems can be managed by methylation foods. Food choice by appropriate selection for minerals like magnesium, selenium, and zinc, B vitamins including vitamin B_{12}, and adaptogens can prevent aging and diseases such as HIV, cardiovascular problems, and osteoporosis.

3.03.04 A Good Mind for Good Health: The food and the environment matter in maintaining body-mind connectivity. Regeneration of skin, hair, and nails depends on our diet. The mind affects the body, and the body affects the brain. Mind and body are part of the same whole, and there is a conduit between mental and physical well-being that can be balanced and equilibrated by

1. Exercising and breathing control for relaxation of muscles.
2. Avoiding alcohol.
3. Decreasing anxiety and stress by mind relaxation and meditation.
4. Practicing guided imagery, music, and art for peace and calm.
5. Practicing mental focus and yoga.
6. Enhancing and optimizing sleep.
7. Increasing the sense of control.
8. Building a strong immune system.
9. Pursuing spiritual practices.
10. Building a network of relationships.

All ten approaches are means to better the mind-body connection. We can easily find that chronic pain, change in appetite, chest pain, high blood pressure, weak immune system, constipation, dry mouth, and tiredness stem from stress, which is a matter of the mind. Thoughts and emotions can be controlled by exercise and yoga because the body responds to how and what we think. Meditation is a great tool to manage how and what we think. The

---

70 http://www-group.slac.stanford.edu/esh/medical/wellness/stressfoods.html.

daily food and exercise when complemented by mind control can cure many of our ailments and chronic diseases. Rightfully the United Kingdom and Germany are practicing the mind-body connection approach for effective health-care delivery by psychosomatic therapy.

Protein food is of primary importance because it helps the brain with (1) production of the sleep chemical serotonin from tryptophan, (2) production of norepinephrine and dopamine from proteins for alertness and energy, and (3) endomorphins, like endogenous drugs that kill pain and maintain well-being. We need to select our food wisely for routine mind-body connection and stress avoidance. The brain requires the best of nutrients for unhindered gene expression.

3.04 Our Genes

Neurons or brain cells are very specialized cells in our body with their receiving dendrites and transmitting axons.[71] Somatic retrotransposition, a DNA sequence that can change its position and cause mutation, alters the genetic landscape of the human brain.[72,73] Since chromosomes hop around and brain genes are not really fixed, relocated and mutated genes can affect neuronal synapses and thus the very act of neurotransmission. There can be more than a thousand such mutations affecting our brain's cortex.

The almighty ApoE gene, which codes a fatty protein that protects the brain cells and the control tower in our brain, namely N-methyl-D-aspartate receptor (NMDA), is controlled by three different genes—GRM5, PPEP2, and LRP1B, which determine the strength of our nerve cells. There are genes for Huntington's, Parkinson's, and Alzheimer's diseases. There are genes that control electrical flow via the most common transmitter, glutamate. There are genes behind the neurotransmitter serotonin, which controls mood, sleep, and appetite. Our brain and body run by some five hundred protein (kinases) molecules coded by almost 2 percent of all genes in our genome, and the phosphate common to DNA, RNA, and these proteins

---

71 http://www.nimh.nih.gov/health/educational-resources/brain-basics/brain-basics.shtml.

72 http://www.ncbi.nlm.nih.gov/pubmed/22037309.

73 http://www.research.ed.ac.uk/portal/en/publications/somatic-retrotransposition-alters-the-genetic-landscape-of-the-human-brain(a1a51986-f159-4df5-bbad-295371d4961e)/export.html.

is the heart of our biology.[74] The genes determine the structure-function relationship of enzymes that control everything that goes on in our body and mind; they control body-mind interaction.[75] Genes per se and their relocation and mutation have been the basis of our evolution.[76] Genes control our consciousness.[77] Our genes, in interaction with the environment, regulate our nervous system, and our daily diet and lifestyle are definitely part of the environment. As a matter of fact, Harvard researcher Dr. John Rinn reports that even junk DNA, the long intergenic noncoding RNA transcript, matters in building our brain—the layers of cortex.[78] The genetic science behind behavior and education is on the march for a clearer understanding and solution.[79] We have to pay much closer attention to prone-to-mutate mitochondrial DNA, the DNA responsible for energy production and for better mental health.

3.05 Major Reference Institutions

The following research institutions conduct cutting edge research on brain foods and connection between food and mind.

Mind Research Network, New Mexico; The Franklin Institute, University of California, Los Angeles; National Center for Complementary and Alternative Medicine; Duke University; University of Rochester; Harvard Medical School; Center for Mind Body Medicine.

---

74 http://www.ocf.berkeley.edu/~aathavan/libraire/whynaturechosephophates.pdf.

75 http://www.sciamdigital.com/index.cfm?fa=Products.ViewIssuePreview&ISSUEID_CHAR=4E3B6963-3358-4EBE-9524-A7CFD3F328A&ARTICLEID_CHAR=65D4DBF3-00DB-485A-A2A3-D6F9CC71E40.

76 http://epress.anu.edu.au/austronesians/austronesians/mobile_devices/ch09s04.html.

77 http://www.ncbi.nlm.nih.gov/pubmed/22742996.

78 http://www.economist.com/news/science-and-technology/21590878-newly-recognised-class-genes-really-does-matter-junking-idea-junk.

79 http://people.virginia.edu/~ent3c/papers2/three_laws.pdf.

PART TWO

OUR FOODS

Human beings are gene products, and the genes have evolved with foods they ate over thousands of years, including mitochondrial DNA, which evolved maybe 175,000 years ago.

CHAPTER 4

Our Second Brain and Food Processing in Our Body

"Works of digestion relate to all functions of our body systems,
exerting direct effects on our mental and physical state."
—Triveni P. Shukla

Our digestive system is a true mechanical, chemical, and biochemical food-processing factory designed to convert chunky foods we eat into small molecules that can diffuse into the blood. This musculo-membranous tube connected to gall bladder and pancrease has evolved for working slowly in a time-phased manner and under complex control by our second brain (the enteric nervous system) in order to maintain the rest of the body.

Food is chewed and broken down in the mouth, and ground and churned in the stomach, where proteins in it are broken down to amino acids, and starch into glucose. What goes into the blood is invisible to the naked eye. The digestive system has our second brain, called the enteric nervous system (ENS), developed from the neural crest that works with both the sympathetic and parasympathetic nervous systems with about 100 million neurons embedded in the lining of the gastrointestinal tract; it works with the central nervous system, coordinates autonomous reflexes, deals with mechanical and chemical conditions, controls motor neurons for peristalsis, and operates with thirty neurotransmitters, including acetylcholine, dopamine, serotonin, glutamate, and norepinephrine. The nutrient composition of what we eat and drink affects the response of the enteric brain. The ENS controls mixing, churning, grinding, absorption, and extraction of nutrients and vitamins in the entire gastrointestinal tract. It controls our mood, behavior, and appetite.

Blood circulates from the heart to every cell of the body. Glucose becomes available to the cells for energy production and production of proteins, hormones, neurotransmitters, and other molecules that build and maintain our body and immune system.

On an average, we consume approximately 1.2 liters of oxygen, two liters of water, and one pound of food solids every day. Our digestive system outputs one liter of saliva, one liter of stomach juice, 0.5 liter bile acid, and 30 ml of pancreatic juice to supply enzymes and hormones in order to break down foods into small molecules of nutrients for absorption in the bloodstream. Foods that we eat include molecules of carbohydrates, proteins, and fat, often embedded in a particulate matrix. The variety of foods that reaches the stomach supplies minerals, vitamins, prebiotic dietary fiber, and even probiotic bacteria—the source of our second genome. Glucose from carbohydrates and even from proteins and fat must travel 60,000 miles to get to all 10 trillion cells of each and every body organ for the production of energy and other vital molecules. Although proteins, fat, and carbohydrates are processed at different rates, the digestive system converts starch into glucose, proteins into amino acids, and fats and oils into fatty acids and glycerol. Fat takes the longest time for complete digestion and absorption. The digested nanoscale molecules can be absorbed in the bloodstream along with vitamins, minerals, antioxidants, and anticarcinogens. Dietary fiber and probiotic bacteria do their job further downstream in the colon, although prebiotic dietary fiber assists in peristalsis all along the digestive tract.

The digestive process consumes up to 25 percent of total daily calorie intake. Proteins take the maximum energy for digestion, followed by fruits and vegetables and complex carbohydrates. Fat requires the least energy for digestion. Thus, high-protein diets, in complement with fruits and vegetable for fiber, deliver lower net energy than carbohydrates. Dietary fats deliver maximum net energy.

Our daily food influences homeostasis, the immune system, and gene expression. The colon should have the right blend of bacteria for the right ratio of anti- and pro-inflammatory cytokines. The wrong bacteria can cause weight gain, autoimmune diseases, anxiety, and even depression. Good bacteria promote digestive health, confer immunity, and protect the intestine.[80]

When we are under stress and under an overload of the hormone cortisol, the brain instructs specialized cells in our stomach lining to produce

---

80 http://www.ilsi.org/Europe/Publications/Prebiotics-Probiotics.pdf.

inflammatory chemicals. In the absence of a threat of infection, cortisol causes muscle contraction, leading to bloating, cramping, and irritation. Thus, aerobic exercise and stress control are physiological needs for perfect and predictable digestive health.

Dietary needs seem to be individual and hereditary. Biology, geographical locations on the planet earth, and cultures have had a history of interaction and even of influencing the genes. Just visualize the carnivorous diet of an Eskimo in the cold climate of North America and the fruit- and vegetable-based diet of people by the equator for obvious differences.

We have PillCam technology today to see through this food-processing plant. One can swallow a pill-size camera along with food, record step-by-step throughout the gastrointestinal system, and see its workings. *A fist-size stomach expands to two liters after a full meal, and 0.33 liter of blood, is diverted to the digestive tract after every breakfast, lunch, and dinner.*[81], [82] Digestion of the methylation diet is good for digestion itself and then for the entire body, including the mind it houses.

Evolution has put certain limitations on our being. Although we are more dependent on the plant foods, we depend on the animal kingdom for a number of key essential nutrients. We need our optimal blend of vitamins, minerals, antioxidants, and phytonutrients from plants or animal sources each day. Plant sources have primary importance because that is where dietary fiber, minerals, vitamins, antioxidants, and phytonutrients come from. The role of dietary fiber in the functioning of our digestive system truly defines our dependence on plants. Called prebiotics, dietary fiber is essential for digestive health, immunity, and peristalsis in our gastrointestinal tract, and it comes from plants only.

Digestion converts big molecules into small molecules that can be absorbed by the blood circulating across the small intestine. It converts carbohydrates to glucose, proteins to amino acids, and fat to fatty acids and glycerol. It lets anticarcinogens, antioxidants, and many phytonutrients pass to our bloodstream unchanged. What is absorbed is circulated via blood over 60,000 miles to all 7.5 trillion cells in our body for energy, for building other molecules of life, and for routine communication and control. A typical adult circulates seven liters of blood through his or her body every day many times

81 http://pillcamcrohns.com/how-pillcam-works?s_mcid=ps-ppc-google-branded-capsule-endoscopy-4c&ef_id=t9NOJa1MJXIAAEGQ:20131205205721:s.

82 http://www.iffgd.org/site/gi-disorders/digestive-system

over. What is not absorbed from the intestine becomes waste or feces past the colon. A good deal of water is recycled from the colon.

4.01 Our Genes and Digestion

Although 98.80 percent of the genome of human beings is the same, our genetic profile is very individual. Our diet, therefore, should be fine-tuned to individual needs, including the bacteria that inhabit our colon. People of the same age, sex, and weight can have different metabolic rates. They can have differences in food tastes, calorie needs, and even propensity to chronic diseases. Digestion is under the control of genes, which get activated the moment we put something in our mouth.[83,84,85]

There are more than ten enzymes (see table 4.03) and ten hormones (see table 4.04), including insulin and glucagon, involved in nutrient production by digestion and then absorption in the blood and transport to cells. The timing and chronobiology are controlled by many neuropeptides associated with routine control of the digestive system. Each neuropeptide is under control by one or more genes. Any defect in these genes can create digestive malfunction or diseases.[86]

Phenylketonurea is a disorder of a single gene that produces an enzyme to convert phenylalanine to tyrosine. A defective copy of the SGLT1 gene on chromosome 22 causes lactose intolerance. IBS or inflammatory bowel syndrome is due to genes CD19 and CD11 on chromosome 16, and pancreatic cancer results from partial loss of chromosome 18, causing autoimmune assault on beta cells of the pancreas and stopping insulin production. Inheritable galactosemia is caused by genes GALE, GALK1, and GALT. This short list of genes is included just to point out that digestion is under strict genetic control.

4.02 Our Food

Whether prepared at home, made at restaurants, or bought at retail stores, food functions go beyond nutrition. Food and food ingredients add structure,

---

83 http://www.ncbi.nlm.nih.gov/pubmed/7683841.
84 http://www.ncbi.nlm.nih.gov/pubmed/12794182.
85 http://www.ncbi.nlm.nih.gov/pubmed/10448517.
86 http://www.ssi.dk/English/News/R%20and%20D%20News/2012/2012_2_Genetic%20 variants_digestive%20system%20disorder.aspx.

flavor, texture, color, water, and visual appeal to our dishes. Ingredients can have similar and synergistic functions. An average American consumes between 2,000 and 2,500 calories per day, and consumption beyond this average causes weight gain, obesity, and even early female puberty.[87]

The energy that we need for muscular and neural activity, making enzymes and proteins, repairing DNA, digestion, excreting waste, and maintaining body temperature is substantial; it can raise the temperature of twenty-five liters of water to boiling. For good health, there is an optimum for this daily need.

Beyond energy needs, our daily food should supply nutrients that prevent inflammation, regulate metabolism, keep cholesterol in balance, help maintain blood pressure, prevent heart disease and cancer, and promote digestion. Such foods must include dietary fiber, essential fatty acids, essential amino acids, minerals, vitamins, and stress-controlling adaptogens for preventing chronic diseases, improving cognition, and conferring longevity. Such foods should make up our daily methylation diet.

The USDA's food pyramid is a good guide for our daily eating, but its message is lost because of too many contradictory quick-fix diet programs. All programs, including the Mediterranean, Atkins, South Beach, Okinawa, and many other diets, represent selection criteria for foods we should eat. However, none of these programs work equally well for all of us. The patchwork of low-calorie diets, low-sodium diets, low-cholesterol diets, low-fat diets, low-saturated fat and low-trans fat diets, low-carbohydrate diets, dopamine diet, DASH diet, and low glycemic index diet often miss the key message of *optimum nutrient delivery for a rather self-regulating human body*. The government laws, policies, and guidelines are generally miscommunicated, creating confusion and contradiction. That we have genetically dictated food requirements and that these requirements can be fine-tuned by the appropriate choice of exercise, meditation, and overall lifestyle must be clearly understood.

Common sense tells us to eat protein for satiety; fiber for digestive health, immunity, and satiety; and antioxidants, omega-3 fatty acids, magnesium, vitamin B complex, and vitamin D_3 for fighting inflammation and stress control. Common sense also tells us that we should eat on time and pay attention to our chronobiology. We can select foods that fight ailments and diseases and foods that confer an enhanced immune system, normal blood flow, and stability of

---

87 http://www.foodproductdesign.com/news/2013/03/mindless-eating-causes-43-of-overconsumption.aspx.

our genome and epigenome. We should pay attention to the fact that a regimen of exercise, mobility and physical activity, meditation, and stress-free lifestyle is a *physiological need* for a more-perfect healthy digestive function.

What we eat and drink is directly linked to problems of obesity, high cholesterol, blood-sugar imbalance, diabetes, heart disease, irregularity, lack of energy, high blood pressure, propensity to cancer, and a weak immune system. High-fat diets cause electron leakage off the electron transport system, causing production of free radicals that cause DNA damage, cancer, and stress. Therefore, our diet must have antioxidants in order to quench and sequester free radicals. Uncontrolled astray electrons are linked to obesity, type 2 diabetes, and insulin resistance.[88,89]

On an average, we live by 11,000 liters of air containing about 20% oxygen 2 to 2.5 liters of water, and a variety of foods that can provide 2,000 to 2,500 calories per day in the form of major nutrients such as carbohydrates, proteins, and fat. Table 4.01 provides details on daily input to and output from our body, and Table 4.02 summarizes an ideal daily food consumption.

Table 4.01 Major Input and Output to and from Our Body

Input	
Air, liters per day	11,000.00
Oxygen, liters per day	0.55–0.60
Water, liters per day	2.50
Bacteria in the large intestine, trillion (buildup from probiotics like yogurt)	100.00
Output	
Water, liter/day (urine, breath, and sweat); cells will draw water from blood if intake is not enough.	2.50
Stool: Organic and inorganic waste, % of calories consumed	50.00 organic, 75.00 inorganic
Stool moisture, %	30.0
Food Residues, %	30.00
Dead bacteria and GI tract secretions, %	10–20
Cholesterol and fat, %	2–3

---

88 http://essays.biochemistry.org/bsessays/047/0053/0470053.pdf.
89 http://www.ncbi.nlm.nih.gov/pmc/articles/PMC3122475/.

In Metabolic Process	
ATP/Energy turnover, Kg per day for a 75-Kg person	75.00
Source: John R. Cameron, James G. Skoforonick, and Roderick M. Grant, *Physics of Body* (Madison, WI: Medical Physics Publishing, 1999), p. 182; William D. McArtlle et al, *Exercise Physiology* (city of publication: Lippincott Williams & Wilkins, February 24, 2014. .	

4.02.01 The Oxygen: We are creatures of oxygen and inhale up to ten liters of air per minute, and a brisk walk even doubles this amount. *Eleven thousand liters of air per day provides 3.209 lb (2,246 liters) of pure oxygen.*[90,91] This amount could be as high as three liters of oxygen per day, depending on physical activity, gender, and age. This is almost three times the amount of food solids we consume each day for a 2,000-calorie-per-day diet. *High calorie consumption requires high oxygen input, which can come only from exercise and sustained physical activity.* Energy production in all living systems is directly proportional to oxygen consumed.[92] Just like food and water, oxygen is part of our daily business of food utilization. Oxygen is required for brain growth and repair,[93] and the need is ten times greater than the rest of the body.[94]

Table 4.02 Ideal Food Consumption

Food Component	Daily Value	Calories
Total fat	65 grams (g)	< 600
Saturated fat	20 g	180
Cholesterol	300 mg http://www.heart.org/HEARTORG/ Conditions/Cholesterol/ PreventionTreatmentofHighCholesterol/Know-Your- Fats_UCM_305628_Article.jsp	
Sodium	2,400 mg	
Potassium	3,500 mg	

90 http://highered.mcgraw-hill.com/sites/dl/free/0073048763/232418/chapter01.pdf.
91 http://health.howstuffworks.com/human-body/systems/respiratory/question98.htm.
92 http://cnx.org/content/m42153/latest/?collection=col11406/latest.
93 http://nacd.org/journal/riggs_my_brain_needs_oxygen.php.
94 http://www.fi.edu/learn/brain/micro.html.

Food Component	Daily Value	Calories
Total carbohydrates	300 g	1000–1200
Dietary fiber	25 g	
Protein	50 g	200–260

4.02.02 The Water: An adult human body is comprised of 60 percent water, which moves freely along the osmotic gradient throughout the small and large intestines. It is needed for the breakdown of all major nutrients (proteins, fat, and carbohydrates), absorption and transport of nutrients and oxygen to the cells, mediation of metabolic reactions, protection of tissues and organs, lubrication of joints, moisturization of eyes, mouth, and nose, and waste removal from kidney, liver, and digestive tract. Water is the source of protons, and its structure is a vehicle to carry protons in our body.[95] *Our daily water need is at least two liters.* Optimum hydration and sufficient water intake in the case of children and senior citizens is very critical. Water, as a major solvent, is critical for metabolic reactions as a source of electrons and protons, proton transfer, and a medium for electrolyte balance. Positive (sodium, potassium, and calcium) and negative (chloride and bicarbonate) electrolytes are present in water in a critically balanced ratio. They are present as ions, and as such are critical to electrical signaling and electrical impulses of nerve cells, muscle contraction, oxygen delivery, fluid balance, homeostasis, neurological functions, and acid-base balance. The two kidneys maintain this balance.

4.03 The Digestive System and Process of Digestion

The human body is an extension of the digestive system—from the mouth to the point of elimination of food waste at the end of the rectum. Everything in the body depends on digestion. What, when, and how much we eat and how well we eliminate food waste determine the conditions of our health and onset of a disease. Digestive glands, nasal passages, trachea, lung, kidney tubules, collecting ducts, bladder, and reproductive-system linings are continuous with the surface of our body. Endocrine glands deposit their secretions directly in the blood. Secretions of exocrine glands, around 9.2

---

95 http://www.eduplace.com/science/hmxs/ls/modf2/cricket/sect1cc.shtml.

liters from the salivary gland, stomach, duodenum, pancreas, liver, small intestine, and large intestine and all indigestible material are truly outside the body.[96]

Food can reside in our twenty- to thirty-foot-long gastrointestinal tract for almost two days. Although half of the stomach content is emptied in less than three hours, total emptying may take up to five hours. Half of the small intestine content empties in three hours, and transit through the colon can take up to forty hours. All nutrients are outside our body until absorbed, digestion occurs outside our cells, and the entire intestinal lumen is kind of external to our body. The transport of nutrients to the bloodstream is a very selective extracellular mechanical and chemical process under hormonal control and concentration-dependent active transport (not simple diffusion).[97] Digestion begins in the mouth with salivary amylase. Chewing foods for enzyme release and enzyme activity is not just a good idea but a need. Plant enzymes that we ingest as sprouts, green vegetables, and fruits are stable under highly acidic conditions of the stomach. They are active for more than an hour before food reaches the small intestine. Other enzymes listed in table 4.03 are delivered to the digestive tract at different stages. The enzymes break down carbohydrates, proteins, nucleic acids, and fat to a nanoscale, making easy-to-absorb, small, soluble molecules.

Digestion is controlled by the brain, the autonomic nervous system.[98] The digestive system is the largest hormone-producing organ, with a total of twenty peptide hormones. Molecules of flavor work through our nose, and we begin to crave food. The taste-bud receptors on the tongue can detect foods we like or dislike. We begin to salivate the moment we think of eating or smell a flavor of fast food while driving on the interstate highway. The brain thus becomes busy with digestion earlier than we even get the sight of food. Saliva is ready with a slurry of enzyme even before we sit at the table to eat. Other digestive enzymes come from glands (exocrine) lining the inner surface of our gastrointestinal tract and other organs attached to it.

Digestion, absorption, and transport of food nutrients use 5 to 10 percent of our total basal metabolic energy requirement. High-protein foods require more energy for digestion, about 30 percent of calorie intake. Next are fibrous

---

96 http://en.wikibooks.org/wiki/Medical_Physiology/Gastrointestinal_Physiology/Secretions.

97 http://www.biology-online.org/dictionary/Active_transport.

98 http://www.kidport.com/reflib/science/HumanBody/DigestiveSystem/NervesDigestSystem.htm.

fruits and vegetables, which can use up to 20 percent of total calorie intake. High-fat foods require the least energy for digestion. Energy consumption during digestion also depends on body composition. *Lean and well-muscled people use more energy for digestion than overweight and obese people.* Clearly, net energy gain is low for muscular people who are on diets of high protein, fruits, and vegetables.

4.03.01 Mechanical Digestion

4.03.01.01 The teeth: Foods must be chewed down in size and well-buffered by saliva for further processing in the stomach. Enzymes from saliva and from food itself begin to work right after ingestion. Saliva has traces of hydrogen peroxide as a disinfectant. During the oral phase of digestion, teeth and gums tear, grind, and chop the food before it passes down the throat. *A 20 sq cm surface of muscles in our mouth can produce 200 Kg force or 142.23 pounds per square inch for mastication.*

Table 4.03 Various Digestive-System Enzymes

Enzymes	Function	Site/Origin
External enzymes	Many protein food molecules break down what we eat.	Raw fruits and vegetables
Salivary amylase	Breaks down starch	Begins in mouth
Pepsin	Breaks down protein	Stomach
Lactase	Breaks down lactose in milk	Small intestine
DPP IV (dipeptidyl peptidase)	Breaks down milk protein	Small intestine
Maltase and sucrase	Break down sugars and starches	Small intestine
Pancreatic amylase	Breaks down starch and sugars	Pancreas
Lipase	Breaks down fat/oil	Pancreas
Trypsin	Breaks down protein	Pancreas
Chymotrypsin	Breaks down protein	Pancreas
Nuclease	Breaks down DNA and RNA	Pancreas
Bile	Emulsification of fats and lipids	Liver

4.03.01.02 The tongue: Stated in terms of lb force, a bite can have a strength of 975 pounds for two seconds. The tongue, a structure with sixteen muscles, is a scraping device designed to move, shift around, dislocate, and help swallow food. It acts with ease and flexibility against the palate and shoves food back of the mouth to the pharynx. Also, it functions as a taste and acceptance device. Taste and mastication signal the rest of digestion to begin. The bolus moves beyond the pharynx to the stomach through the esophagus by peristalsis.

4.03.01.03 Digestive sphincters: Sphincters are muscle valves in our digestive tract for directing and controlling the quantity of food that passes down the tract and inhibit the backward movement of undigested food. This mechanism begins with the esophagus.

4.03.01.04 Peristalsis: Liquid and solid foods move along the esophagus, stomach, and intestine by mechanical peristalsis, which counteracts gravitational pull. This is why we can swallow food lying down or standing up. Peristalsis, a locomotive force, also kneads, agitates, and pounds the food solids.

4.03.02 Chemical Digestion: Chemical digestion converts whatever we eat to nanoscale molecules that can be absorbed and then transported through the blood. While our body is designed to produce glucose for energy more readily from carbohydrates, it can convert protein and fat into glucose and energy also. Food is further broken down and sterilized in highly acidic juices in the stomach for about one hour. Hydrochloric acid produced in the stomach unfolds and unravels ingested proteins so the enzyme pepsin can break them down to amino acids at pH around 2.00. The gastric hydrogen potassium ATPase, a kind of molecular proton pump, is a key enzyme, encoded by two genes for production of 0.5 percent hydrochloric acid in our stomach.[99]

4.03.02.01 Stomach: The stomach is a J-shaped organ immediately below the diaphragm. It is both a chemical and mechanical food processor. It blends, churns, and mixes its contents. A bolus is a pulpy mass when it reaches the stomach about ten inches below the esophagus. Like a deflated balloon

---

99 http://www.ncbi.nlm.nih.gov/pubmed/18536934.

when empty, the stomach is a food-processing cistern. When full (one foot long and six inches wide), it can hold two liters of food and drink. The stomach wall has millions of gastric glands (parietal cells, chief cells, mucous-secreting cells, and hormone-secreting endocrine cells) that secrete up to 800 ml of gastric juice (hydrochloric acid) per day in order to digest our meal. The juice has hydrochloric acid and an intrinsic factor that carries vitamin B_{12}. Chief cells secrete pepsinogen, the precursor of the enzyme pepsin, which breaks down protein. High acidity in the stomach converts pepsinogen to the active pepsin enzyme and at the same time denatures proteins that need to be further digested. Food from the stomach transfers to the duodenum in a very controlled manner. Only alcohol, water, certain ions, and aspirin-like drugs are absorbed into the bloodstream directly from the stomach.

4.03.02.02 Liver and gall bladder: The liver is involved in regulation of the digestive process and the metabolism of carbohydrates, proteins, lipids or fat, and vitamins. It produces bile, an alkaline fluid that helps emulsify fats during digestion and further their absorption in the small intestine. The bile salts combine with fat globules and break them down into small droplets for absorption. Bile acid accumulates in the gall bladder, a mini-hopper, for delivery to the duodenum. Also, the liver is a central metabolic clearinghouse, an organ for glycogen storage and detoxification. It is the highest energy user as part of the metabolic rate.

Blood draining from the intestine goes to the liver, which is the gatekeeper between the intestine and general circulation. It screens the blood by removing absorbed materials like glucose, other sugars, and amino acids and detoxifies drugs and other harmful chemicals. The liver can convert glucose to glycogen, glycogen to glucose, and certain amino acids to glucose. Cholecystokinin induces gall-bladder contraction for delivery of 500 ml bile acid per day to the duodenum.

4.03.02.03 Pancreas: The pancreas has many endocrine and exocrine cells whose secretions (sodium bicarbonate, enzyme amylase for breaking down starch, enzyme lipase for breaking down fat, and other enzymes like trypsin, trypsinogen, chymotrypsinogen, elastase, carboxypeptidase, nuclease, and pancreatic lipase) go to the duodenum. It also produces hormones secretin and cholecystokinin (CCK). The latter controls its secretions.

The pancreas secretes 0.3 milliliter of juice per minute to the duodenum even between meals. This goes up to 3.0 milliliters in response to a meal. We can call it a mini-factory for producing enzyme proteins and hormone proteins of insulin and glucagon. Pancreatic mitochondria must remain fully functional for optimal digestive functions. [100] Pancreatic juice can have up to 7 mg/milliliter enzyme protein (0.7%) plus added insulin and glucagon during stimulation by secretin and CCK.

4.03.02.04 Duodenum. The duodenum is the first segment of the intestine, where the stomach ends. It is ten inches long. Mechanical destruction of food is completed in the duodenum. Pancreatic juice (enzymes, bicarbonate, and hormones) and bile acid from the gall bladder (connected to the liver) enter the duodenum. The emptying rate depends on duodenal pressure, particle size of food, and protein and fat intake. *The duodenum accepts food from the stomach at a constant caloric rate.* High calcium foods empty slowly. Fat in liquefied foods stimulates secretion of the hormone cholecystokinin (CCK), which induces delivery of fat-emulsifying bile acid. Acidity of chyme is neutralized by carbonate coming from the pancreas. Lipase from the pancreas produces fatty acids from fat. In the duodenum, the ingested food becomes nanosized molecules of sugar, amino acids, fatty acids, and glycerol. Protein and starch have already been converted into amino acids and glucose respectively. The duodenum produces lactase in order to break down lactose. Lactose-intolerant people don't have the gene to make this enzyme.

4.03.02.05 Small intestine: The small intestine is twenty-two feet long. It is divided into three main segments: duodenum, jejunum, and ileum. Most of the nutrients are absorbed in the small intestine by its villi, kind of long, slender, hairlike structures. *The bases of villi have stem cells that divide continuously by mitosis. Stem cells become epithelial cells, goblet cells (which produce mucous), and endocrine cells (which produce a variety of hormones).* These cells die and are created continuously. Villi and microvilli provide a surface area of 200 square meters for effective absorption. *This is one hundred times the exterior surface of our body, or roughly the area of a tennis court.* One would die of starvation if there were no microvilli.

---

100 http://www.ludc.med.lu.se/research-units/molecular-metabolism/research-areas/mitochondrial-metabolism-in-pancreatic-beta-cells/.

The plasma membrane of villi contain additional enzymes like aminopeptidase and enzymes for sugar breakdown (maltase, sucrase, lactase). Fatty material is emulsified for drainage as droplets to lymph vessels. Absorption in the intestine varies with nutrients' intensity and profile. Also, the small intestine produces hormones and lysozyme, which kill any bacteria that survive the low pH of the stomach. Zinc is necessary for the production of lysozyme.

4.03.02.06 The jejunum: Food enters the jejunum from the duodenum. The villi of the jejunum absorb amino acids, sugars, fatty acids, and glycerol and send them to the bloodstream. Also, vitamins and minerals, including sodium and calcium, are absorbed in the jejunum.

4.03.02.07 The ilium: This is the main part and a third of the small intestine, where maximum absorption takes place through its six million villi, which are constantly oscillating, pulsating, narrowing and widening, and shortening and lengthening for complete recovery of nutrients. Nutrient absorption is complete at this point. Vitamin B_{12} is absorbed at the end of the ileum as a complex of intrinsic factor.

4.03.02.08 The large intestine: The large intestine is only four to five feet long, and it is so named because of its increased diameter. It receives one pint of residual chyme post nutrient absorption. Divided into four major areas of ascending, transverse, descending, and sigmoid colon, the functions of the large intestine include fermentation of prebiotic dietary fiber, reabsorption of close to one liter of water each day, and absorption of fat-soluble enzymes vitamin K, B_{12}, and biotin and minerals potassium and chloride.

The colon has 100 trillion bacteria, ten for each cell in our body. These bacteria live in symbiosis, synthesizing vitamins for us, digesting nonstarch carbohydrates for which we have no enzymes, and producing short-chain fatty acids good for our health and good for an extra 10 percent energy. They contribute to 50 percent of our immune system. Colon bacteria make up 50 percent of dry stool weight. No more than 3 to 5percent of protein escapes digestion. The rest of the stool is water, cellulose, bacteria, and other indigestible residues. The immune systems come from the second genome, which humans have because of the bacteria in their colon.

4.04 Food and Beverages as Part of Our Daily Diet

We need to maintain a lean body and consume low-glycemic starchy foods, high protein, high fiber, and low-fat foods in order to minimize daily net energy gain. We need to consume foods of high air and water content, which in effect have low caloric density. We do not need to consume high amounts of refined sugar or glucose because there is a physiological limit to our body's glucose-managing capacity. High refined sugar and highly processed carbohydrate consumption have been the most serious problem related to our health during the post-World War II period.

Food variety can help. We evolved with a variety of foods containing antioxidants, minerals, and vitamins. Our paleo diet for centuries has included fruits, vegetables, roots, herbs, and spices. This diet afforded our evolution a variety of proteins and *natural* fats, a fact that most diet programs fail to emphasize. Most diet programs also fail to emphasize two most important aspects of human existence: (1) humans exist because of their second enteric brain, and (2) humans exist because of their second genome from their personal microbiome. The paleo diet in effect is the methylation diet.

No doubt sugar, high-acid foods and soft drinks, hydrogenated trans fat, high-acrylamid fried foods, additives and chemical residues, and other contaminants should be avoided because they collectively affect our digestive health adversely.

4.05 Control of the Digestive Process

We eat intermittently, causing excesses and deficits of nutrients in the bloodstream. Therefore, ingestion needs to be regulated and controlled in regard to nutrient availability to the liver and to the rest of our body cells. The enteric brain exercises multiple control in this regard including (1) parasympathetic nervous system control via the vagus nerve of 25 percent of pancreatic activity stimulated by salivation, sight, smell, taste, and eating of food; (2) control of stomach distension; and (3) hormone control of the remaining 50 to 75 percent. Pancreatic polypeptide hormones are secreted for inhibiting secretion. The endocrine system produces many hormones listed in table 4.04 that get transported through the cardiovascular system.

Table 4.04 Hormones for Control of Digestive Processes

Gastrin: Production of peptide hormone gastrin is triggered by protein and peptide foods. It causes the stomach to produce hydrochloric acid.
Ghrelin: This hormone is produced in the stomach also. It controls appetite and feeding.
Hormone peptide YY: It inhibits appetite.
Cholecystokinin: Production of this hormone is triggered by hydrochloric acid and fat in chyme. It induces emptying of the gall bladder and stimulates secretion of water, enzymes, and bicarbonate in the pancreas.
Gastrin-inhibitor peptide (GIP): Triggered by fat and glucose in the small intestine, this hormone decreases stomach churning and stimulates insulin production. It inhibits gastric secretion.
GLP(glucagon-Like-Peptide): A glucagon-like peptide hormone that stimulates insulin production.
Gastric-releasing peptide (GRP): Hormones GRP and ghrelin control food and energy intake. Ghrelin is a hunger hormone and it also directs energy distribution in rather indirect involvement.
Vascular intestinal peptide (VIP): VIP is a vasodilator. It prevents gastric secretion and water in the pancreas. It helps contract smooth muscles.
Insulin and glucagon: These hormones are produced in the pancreas for regulating glucose in our body and in our cells.
Cytokines: These hormones modulate the immune system, energy production, and energy expense, and coordinate leptin and insulin functions.
Motilin: This hormone is produced in the duodenum, and it helps housekeeping in terms of general functions, pepsin production, and motility.
Hormones involved in nutrient transport

There are sensory neurons for acid, amino acid, glucose, stretch, and tension. While the sympathetic system deals with secretion, sphincters, and blood flow, the parasympathetic system is involved in stimulation.

Diet itself controls pancreatic secretion.[101] Fatty acids and monoglycerides in the duodenum are more powerful stimulants than proteins and carbohydrates. Nutrient contact with mucosa determines stimulation. Glucose is high after a meal. The hormone insulin helps lower blood glucose.

---

101 http://www.ncbi.nlm.nih.gov/pubmed/20439184.

Production of glycogen from glucose happens in the liver under control of another hormone, glucagon, produced by the pancreas. The liver monitors glucose and fatty acids continuously. Satiation is due to distension of the stomach and the hormone cholecystokinin.

No doubt what, when, and how much we eat matters to our health and longevity. A customized diet, I believe, can reverse even unhealthy and defective gene expressions. Foods can control cardiovascular diseases, blood pressure, cholesterol, and onset of cancer and diabetes. They can control mood, memory, and cognition.

Beyond food, our problems today are loneliness, chronic stress, unmotivated living, loss of control of our sensory organs, and loss of control of our own mood and behavior. The daily food along with exercise, yoga, meditation, a good night's sleep, skillful social interaction, a chat at the dining table, and a stress-free lifestyle can bring about happiness, health, and longevity. Much of this begins with digestive health and the digestion of a methylation diet.

4.06 Physical Examination for Optimum Food and Digestive Aid

There is no way to maintain good health without a set of periodic diagnostic testing. We must opt for annual medical examinations, which should include the following tests for digestive functions and health indicators. Our physicians can give details when asked for them.

- presence of food antibodies
- methyl melonic acid to check for defficiency of vitamin B_{12}
- stool examination for unhealthy bacteria
- magnesium and zinc levels (a low level implies poor acid production)
- H. pylori test to check for stomach ulcers
- urine test for mannitol and lactulase
- allergy tests
- presence of parasites
- breath test as a check for lactose intolerance

We can prevent DNA damage, manage our epigenome, and keep our telomere gene intact for longevity by selecting common and cost-effective foods. Reliance on commercial diet programs has been ineffective because none of them is part of either a time-tested normal dietary habit or a cultural and social norm. None of them is truly individualized or personalized. None of them is balanced and complete, like a methylation diet ought to be in order to support a person with nutrients that build our brain and body, defend our immune system, and keep our genome and epigenome well. Most importantly, we need foods that protect mitochondrial DNA.

The methylation diet that I am proposing not only provides us with all necessary nutrients with appropriate calories, albeit on a restricted basis, but also permits functioning of our second brain—the enteric nervous system. A poor functioning of the enteric nervous system due to poor choices of foods, deficiency of key nutrients, and high calories can no doubt lead to chronic diseases.

4.07 Major Reference Institutions

The following major institutions are involved in research on digestive physiology and gastroenterology.

National Institute of Health; Mayo Clinic; John Hopkins Hospital; Massachusetts General Hospital; Cedars-Sinai Medical Center; Digestive Disease Institute, Cleveland Clinic.

CHAPTER 5

Probiotic Foods and Our Second Genome

"One hundred trillion bacteria in our colon and their genes, our
second genome, matter to our immunity and sustainable health."
—Triveni P. Shukla

Evolution has mandated a unique biological construct for the human body in that 10 trillion of its cells and 100 trillion bacteria in the colon have to live together in order for perfect human biology. The biological truth is that the bacterial gene pool constitutes the second genome for humans.[102] This uniqueness requires that we eat both our common foods with dietary fiber and the right kind of bacteria. The bacteria help in digestion of plant fibers; provide the vitamin B complex, vitamin K, and other nutrients; complement 50 percent of our the immune system by talking to mucosal cells and maintaining and balancing an anti-inflammatory response in the intestinal ecosystem; and help control autoimmune diseases, diabetes, and hypertension. The bacteria on our skin and scalp, and in saliva and the colon, not only create a more powerful genetic mix; they create a new organ for symbiotic immuno-control. The foreign bacteria become friendly to immune cells; they train them and evade the wrath of the immune system's T-cells by inducing regulatory T-cells. GPR43, a molecule expressed by immune cells, binds short-chain fatty acids and acts as an anti-inflammatory receptor.[103] Good strains of probiotic bacteria regulate both homeostasis and immunity. Fifty to 75 percent of our immune system comes from approximately 1 kg of these bacteria that make up our *personal microbiome*. Whereas our parental

---

102 http://www.secondgenome.com/microbiome-and-human-health/.
103 http://www.nature.com/nature/journal/v461/n7268/edsumm/e091029-11.html.

genes do not change, the bacterial genes can and do change with time. We can select prebiotic dietary fiber and proper probiotic foods in order to improve our immune system, health, and well-being.

The gut bacteria belong to bacteroids in the case of people with a high-meat diet, prevotella and ruminococcus in the case of a vegetarian diet.[104] Depending on the foods we eat, these entero types differ in the vitamins they produce and the means they use to process energy. The gut bacteria can be both good and bad. Bad bacteria can produce heart disease via L-carnitine and can cause fatty liver disease, inflammatory bowel syndrome, cancer, and even attention deficit disorder.[105] We have better homeostasis and an enhanced immune system when good bacteria dominate and talk to our immune cells. They help maintain a good ratio of anti- and pro-inflammatory cytokines. Probiotic foods, in view of how human beings evolved, are just as necessary as other major and minor food nutrients.[106] Bacteria in our colon play a huge role in helping digest proteins and carbohydrates, managing micronutrient homeostasis, and boosting immunity.[107]

The last five feet of the alimentary canal is the colon. It is a living factory by and in itself. It produces water-soluble vitamin B complex, biotin, and vitamin K, enhances the immune system, completes digestion of food, has influence on hormones, controls constipation, prevents bowel diseases, reduces cholesterol and sex hormones, suppresses cancer development, and reduces arthritic inflammation. Bifido bacteria in particular help maintain a good ratio of anti- and pro-inflammatory cytokines, the key immunomodulating agents. Known as not truly synergistic and commensal bacteria, they are region-specific in the intestine. The key feature is the cell-wall composition of these bacteria. First they become friendly to our immune cells and then train them. Colon bacteria thus evade the wrath of T-type immune cells and survive in symbiosis. The mechanism of this action is not known, but a balance of bugs in our colon is critical to our health.

Bacteria in our gut are our friends. There are 100 trillion of them, compared to only 10 trillion cells in our body. Some 3.3 million genes of bacteria matter to us for routine cancer-free health and a robust immune system. Fifty to 75 percent of our immune system thus resides in our digestive

104 http://www.ncbi.nlm.nih.gov/pubmed/22182464.

105 http://gaps.me/?page_id=20/

106 http://www.oprah.com/health/Raw-Food-Probiotic-and-Macrobiotic-Diets.

107 http://renalnutrition.org/files/uploads/RNFSum2013FeatVol32No3.pdf.

tract. This is a great evolutionary design for immuno-control of an open system. The probiotic bacteria manufacture key vitamins for us, break down toxins, and produce protective and healthful metabolites. Their collective metabolic activities in support of our living can be equated to an organ. Our life thus depends on these bacteria and their gene pool. As a matter of fact, with the right microbiome or gut bacteria, one may not be malnourished compared to another individual with an improper complement of gut flora, claims Dr. Gordon of Washington University in St. Louis [108].

The microbiome, gut bacteria specific to an individual, begins to form immediately after birth, making our personal microbiome with added genes. As we progress in life, consumption of probiotics (yogurt, dahi, sauerkraut, kefir, kimchi, tempeh, natto, and other fermented foods) improves our health greatly by anti-inflammatory metabolites of our friendly bacteria. Probiotics maintain our digestive, urogenital, and immune health systems.

5.01 Type of Bacteria

Some 5,000 species of bacteria live in our colon, around 1 trillion per gram of fecal matter. To live well and healthy, we have to know the bugs in our colon and treat them well. Although sterile at birth, the colon populates itself with bifidobacteria by the fourth day via mother's milk. At later stages, 50 percent of the dry weight of the stool becomes bacterial mass. Lactobacilus acidophilus, L.casei, L fermentum, L. salivarius, L. brevis, L. lichmanni, L. plantarum, L. cellobiosis, Bacilus adolescentis, B. longmum, and B. lactis are common probiotic bacteria. There are others like L rhamnosus and L reuteri. New research is revealing that folks who eat plant foods have more Firmicutes in their gut as opposed to bacteriods found in the case of people who do not consume plant foods. It is almost certain that these bacteria shall soon be listed in detail on probiotic food labels.[109]

Bifidobacterium infantis against inflammatory bowel syndrome and Lactobacillus gasseri against fat belly are great choices.[110,111]. Lactobacillus rhamnosus is present in regular plain yogurt.[112]

---

108 https://www.aamc.org/initiatives/awards/307224/research-gordon.html

109 http://jn.nutrition.org/content/137/3/850S.full.

110 http://www.powerofprobiotics.com/Bifidobacterium.html.

111 http://suppversity.blogspot.com/2013/04/do-you-have-guts-to-lose-8-belly-fat-in.html.

112 http://www.livestrong.com/article/321660-the-benefits-of-lactobacillus-rhamnosus/.

5.02 The Good Bacteria

Bacteria can be healthful in many ways. For instance, they can convert pomegranate and strawberry compounds into anticarcinogenic urolithin and convert daidzein isoflavone from soy foods into equol, which is anti-inflammatory.[113,114] Other routine functions include fermentation, treatment of the immune system, fighting pathogens, developing and maintaining the gut, producing vitamins K and biotin, and producing hormones for fat storage. Right stomach bacteria have been found to stop development of type 1 diabetes in lab mice.[115]

5.03 The Bad Bacteria

Bad bacteria promote body fat that in turn increases inflammation and risk of diseases. We know that bacteria have a lot to do with nonalcoholic fatty liver disease, inflammatory bowel disease, some types of cancer, and neurological disorders like attention deficit disorder. *Bacteria that produce L-carnitine can cause a heart attack.*

Bacteria influence our body's ability to break down food and drugs, and this may explain why food affects people differently. For instance, some people can't derive benefit from one of soy isoflavone daidzein conversion to uquol, an isoflavandio, because they lack the gut microbes necessary to process it.[116,117]

5.04 Prebiotic foods

Prebiotics, such as onion, leeks, banana, apple, asparagus, chicory, garlic, Jerusalem artichoke, wheat, rye, and barley, can be consumed daily as part of soup and salad. Still other prebiotics can come from fruit pectin, gums, and

113 http://www.ncbi.nlm.nih.gov/pubmed/23710216.
114 http://www.ncbi.nlm.nih.gov/pubmed/22113864.
115 http://www.genengnews.com/gen-news-highlights/engineered-gut-bacteria-reverse-type-1-diabetes-in-experimental-mice/81246608/.
116 http://www.the-scientist.com/?articles.view/articleNo/28891/title/Gut-bacteria-are-what-we-eat/.
117 http://www.ncbi.nlm.nih.gov/pubmed/17579894.

inulin. In fact, raw fruits and vegetables serve as a source of both prebiotics and probiotics. After all, sauerkraut evolved from fermentation of sugars in cabbage leaves by natural bacteria in them. [118]

Eggplant is a great natural laxative due to its glycoalkaloid content.[119] All dietary fibers, including cellulosics, pectin, and beta-glucan, are good for regularity because of their mild laxative effect. Polyphenols, anthocyanins, and chlorogenic acid provide high antioxidant activity.[120]

5.05 Probiotic Foods

Sauerkraut, yogurt, kefir, brined olive, India's dahi and idli, Korean kimchi, buttermilk, Asian miso and tempeh, cheese, and Tanzanian beverage togwa are age-old probiotics that have built and boosted immune systems in various cultures for centuries. These homemade paleo-fermented vegetables give a good dose of both prebiotic dietary fiber and probiotic lactic bacteria. Lactic bacteria are present on most any vegetable leaf.[121] During fermentation, growth of other epiphytic bacteria is slowed down by the brine helping the lactic bacteria to dominate fermentation of vegetable sugars to lactic acid. Sauerkraut can have up to 1 billion bacteria per gram. For a daily dose of 6 billion cells of live probiotic bacteria per day, we need only a tablespoonful of sauerkraut per day.

Probiotic bacteria regulate and modulate homeostasis and immunity, and they talk to our immune cells. Specific bacteria can be engineered for enhancing immunity for fighting chronic diseases. Specific strains have varying power in this respect, and many new therapeutic probiotics have yet to come to retail market.[122] Actually, probiotics may have given rise to bacteria-like mitochondria in our cells.[123]

---

118 https://www.foodpreservationmethods.com/sauerkraut-kimchi-pickles-relishes/sauerkraut/fermentation

119 http://www.cancerdefeated.com/newsletters/The-Eggplant-Cure-for-Skin-Cancer.html

120 http://www.naturalways.com/medValFd.htm.

121 https://bu.digication.com/chase_gorland_boston_university/The_Utility_of_Lactic_Acid_Fermentation.

122 http://www.uphs.upenn.edu/news/News_Releases/2012/03/allergies/.

123 http://www.greenmedinfo.com/blog/probiotics-and-mitochondria-bacteria-are-not-other.

5.06 Probiotics and Obesity

New frontiers of research on microbiome are now unraveling correlations of bacterial metabolites with health and bug-based drug development. For example, obese people may have fewer fermicutes (energy restoration bacteria) and more bacteriods. The Western diet promotes fermicutes. Obese people, it is now well-established, have a different makeup of symbiotic bacteria in their intestines.[124,125] The second genome attributable to these bacteria can be analyzed for diagnostic purposes.[126] As a matter of fact, fecal transplants, whereby a community of bacteria from one person is transplanted to another for treating clostridium difficile, colitis, Crohn's disease, or diabetes, have been done with FDA approval. Fecal bacteriotherapy or fecal transplant is found very effective in treating clostridium difficile infection.[127] Dr. Robert C. Peale was right in saying that the best and most efficient pharmacy is within one's own system, and in this case, I add within his or her guts.[128] Sir Peel's insightful quote is "The best and most efficient pharmacy is within your own system". The famous conservative from Bury, Lancashire, England (5 February 1788- 2 July 1850) who became Prime Minister of England, had much liberal views of healtrh and well being. The gut bacteria, we are learning now, a lot to do with the health of our body and mind.

5.07 Major Reference Institutions

The following institutions are leaders in cutting edge research on gut bacteria and microbiome.

Brown University, Mayo Clinic, Yale University School of Medicine, National Institute of Health, Washington University in St. Louis, University of Colorado in Denver, Harvard Medical School, the Broad Institute, Stanford University, J. Craig Venter Institute, International Human Microbiome Consortium, Enterics Research Investigational Network, and Virginia Commonwealth University.

---

124 http://www.sciencedaily.com/releases/2013/08/130828131932.htm.
125 http://www.ncbi.nlm.nih.gov/pmc/articles/PMC3448089/.
126 http://www.sciencedaily.com/releases/2013/02/130228093831.htm
127
128 http://justforyouth.utah.gov/health/.

PART THREE
FOOD AND DISEASE

Fifty percent of diseases we suffer from can
be prevented by the methylation diet, which
best keeps mitochondrial DNA stable.

CHAPTER 6

Food Functions

"Metabolism, a dynamic exchange of matter and
energy, is an act of electron and proton hopping over
molecules of life that keep us alive and well."
—Triveni P. Shukla

Metabolism in the human body means exchange of matter into energy. We are born with genes, and gene-expression mechanisms operate well only with the right nutrients. The body and mind are built as we grow at the expense and with the help of food we eat. Food includes carbohydrates, fat, proteins, minerals, vitamins, and all essential nutrients. Our daily food is for the energy necessary to make new vital molecules for body functions, growth, and routine physical performance. On an average, we need 2,000 calories per day, which is equivalent to about 97.01 watts. Different organs in our body have different energy demands. We use more energy during exercise than in running our vital functions. Next to water, protein is the most important major nutrient because proteins confer human body metabolic sustainability and control by enzymes, neurotransmitters, and receptors. Of course, proteins give our body its structure.

We are born of our mother's metabolism, and we live by daily processes of exchange of matter and energy in our body. Reactions for gaining energy from what we eat and for spending energy in making new molecules are coupled for efficiency. Our body creates order out of disorder rather efficiently. A healthy body optimizes this gain and loss of energy in terms of what we call the *basal metabolic rate*. Our daily diet must balance metabolism for good physical and mental health and optimum body weight. It should positively affect the health of the digestive system, bones, eyes, and muscles and support all vital functions. Thus, our daily food should promote disease-free longevity. Such can be the miracle of the methylation diet, loaded with essential fatty acids and amino acids, vitamins, minerals, antioxidants, anticarcinogens, and phytochemicals of therapeutic value. This can obviate the need for supplements. New research shows that there is no need of high calcium supplements beyond 600 mg per day and that highly touted supplement L-carnitine is downright dangerous. L-carnitine is now known to cause heart attack[129,130,131] . Foot note 122 is reinserted here for emphasis.

We live because our cells live and work with foods we eat and liquids we drink. What we eat and drink is used for producing energy and other molecules of life. The body's activities—including digestion, breathing and lung function, heart beating, muscle contraction, walking, running, and most important of all, conscious thinking—all require energy. The metabolism underlies all reactions and functions of all organs of the human body. The calorie requirement varies from person to person, and it depends on age, sex, and health. As a rule, a 160-pound person with moderate activity doesn't need more than 2,000 calories per day. A person with 200-pound weight may need 2,500 calories per day. The range can be from 1,600 to 2,500 calories. The Food and Drug Administration requires listings of 2,000 and 2,500 calories per day on the nutritional information panel for all retail food items. Our major caloric ingredients in foods are proteins, carbohydrates, and fat or oil. Fat is the most caloric, at nine calories per gram. Proteins and carbohydrates carry only four calories per gram. Glucose is the energy molecule of choice, and it can be readily produced from carbohydrates. However, the human body does have the capability of making glucose from protein or fat, although with different energy gain and different consequences to the metabolism.

6.01 Definition of Calories

A nutritional calorie is actually measured in kilocalories, denoted as calories. One calorie equals 4.2 kilojoules or 3.98 BTUs. A person consuming 2,000 kilocalories per day is getting 7,961.52 BTUs or 331.17 BTUs per hour. In

129 http://ods.od.nih.gov/factsheets/Calcium-HealthProfessional/.

130 http://ods.od.nih.gov/factsheets/Calcium-HealthProfessional/.

131 http://www.health.harvard.edu/blog/new-study-links-l-carnitine-in-red-meat-to-heart-disease-201304176083.

engineering vernacular, it is 0.133 horsepower per hour or 97.01 watts. So an average human being is roughly equivalent to a one-hundred-watt bulb.

All organs in our body use energy in order to function.[132] The liver and spleen use twenty-three watts, the brain sixteen watts, the skeletal muscles fifteen watts, the kidney nine watts, the heart six watts, and other organs in the body sixteen watts.[133] Since energy expenditure is directly proportional to oxygen use in milliliter per minute, the corresponding oxygen use per minute is calculated to be sixty-seven, forty-seven, forty-five, twenty-six, seventeen, and forty-eight milliliters. One watt equals 0.86 kilocalories per hour, and the preceding numbers are averaged for ease of comparison. The table below, excerpted and modified by me using data from multiple sources,[134] lists power used by different body organs.

Proportionality of Power to Oxygen Consumption			
Organ	Watts Consumed	ml/Min Oxygen	% of BMR
Liver & spleen	23	67	27
Brain	16	47	19
Muscle	15	45	18
Kidney	9	26	10
Heart	6	17	7
Others	16	48	19
BMR or Resting Metabolic Rate is defined as the rate at which heat is produced by the body at rest, twelve to fourteen hours after eating, measured in kilocalories per square meter of body surface per hour.			

Sleeping uses eighty-three watts. Sitting at rest and standing relaxed use up to 125 watts. But as physical activity becomes strenuous, energy use increases. Walking uses 280 watts, and cycling at 15 Km per hour uses 400 watts—almost similar to shivering and swimming. *Climbing stairs and running are great exercises because of their high energy consumption, in the range of 685–740 watts.* Basketball playing uses a high 800 watts. In order to burn extra calories so that we may end up eating once in a while, physical activity,

132 http://ajcn.nutrition.org/content/92/6/1369.full.pdf.
133 http://cnx.org/content/m42153/latest/?collection=col11406/latest.
134 http://cnx.org/content/m42153/latest/?collection=col11406/latest.

planned exercise, and other forms of expedited mobility should be part of our lifestyle.

6.02 Metabolism

Our cells die when our metabolism stops. Metabolism begins when we are conceived and ends when we die. Metabolism includes catabolism or breaking molecules down and anabolism or making life molecules of DNA, proteins, hormones, and neurotransmitters. There is more anabolism (making of new molecules) after eating and more catabolism during periods of physical activity and during eating. Catabolism is spontaneous in that it releases net energy. Anabolism is nonspontaneous, and it uses energy. Catabolism generates ATP energy, and anabolism uses ATP energy. Our body is designed to couple anabolism and catabolism for efficiency. It is an open system, and it exchanges matter and energy. [135]The metabolic systems are dissipative and not at equilibrium; they maintain order by creating disorder.

Digestion, the initial breaking down of the food matrix, is a catabolic process and requires energy expenditure. Transport of nutrients and regulation of metabolism also use energy. Actually, the faster the metabolism, the quicker the use of calories. We now know that people have variable rates of metabolism and efficiency of fuel-to-energy conversion. Also, we know that metabolic rate depends on age, gender, lean body mass, and our genes. Metabolism is slow if body weight can be easily reduced and maintained by calorie reduction or restriction. ATP is the bridge between anabolism and catabolism, and it is continuously recycled.

Metabolic pathways and systems are conserved. Salmonella bacteria, elephants, and human beings have similar pathways of energy production, which appeared early in evolution and have been conserved. Once in a while, mutations can change metabolic pathways, though. The participants in metabolic reactions are amino acids and proteins, fats and oils, carbohydrates, and nucleotides, which make DNA and RNA. DNA and RNA can be made from amino acids. Fats are used to make cholesterol. We use carbohydrates that are originally made by plants with sunlight for the production of energy. Our metabolism needs vitamins, nutrient molecules that the body can't make by itself. Our metabolism also needs minerals and ions for muscle contraction (calcium, sodium, and potassium) and action potential gradient

---

135 http://www.myvmc.com/anatomy/metabolism-and-energetics/

for energy production. Ions move around and across cells with proteins via ion channels. Although metal minerals are required in very small amounts, their transport mechanism is very specific.

Electrons are removed from organic foods in metabolic pathways and finally transferred to oxygen. Hydrogen or proton is pumped out of mitochondria in the cell, where electron transfer takes place and enzyme ATP transferase drives the pumping action by force of proton gradient or differential. All metabolism is meticulously regulated for homeostasis. This includes regulation of enzymes and their activity, regulation of a group of enzymes at the same time, hormones, and growth factors. Insulin is a great example of a hormone needed for regulation and control of our body's glucose metabolism.

6.03 Basal Metabolic Rate (BMR)

Basal metabolic rate refers to the rate of energy production from foods we consume. Factors that affect BMR are genes and heredity, body composition with respect to muscle and fat tissue, enzymes, and hormones. A number of enzyme-based health conditions that affect BMR are (1) low enzyme glucose-6-phosphate dehydrogenase, (2) lack of galactose-producing enzyme or lactose intolerance, (3) low or high thyroxin, (4) no enzyme for breaking down phenylalanine, and (5) diabetes due to either no insulin production (type 1) or insulin resistance (type 2) or abnormal insulin response by our cells. A good diet can boost metabolism and help reduce weight. *Low-BMR people gain weight, and high-BMR people maintain or lose weight.* Exercise increases the basal metabolic rate (BMR) and forces the body to produce more energy. BMR has a lot to do with routine functioning of our mitochondria.

6.04 The Diet

Diet, as habitual nourishment, means a regimen of eating foods and drinking beverages regularly. It is the manner of living for regulating general health and weight. Our daily diet includes proteins, carbohydrates including dietary fiber, fats and oils, minerals, vitamins and antioxidants, anticarcinogens, and phytonutrients that are constituents of common daily foods. The diet should be selected and designed for balance. It should be consumed at fixed intervals with increments at breakfast, lunch, a snack, and dinner. Incremental eating is good for health.

The daily diet must be good for bone, eye, muscle, and bowel health; it must prevent colds and flu, weight gain, digestive problems, and loss of energy and performance, and it should create good mood, good sleep, and good behavior. It should supply all major and minor nutrients and prevent any deficiencies.

6.04.01 Carbohydrates (300 grams nonfiber and 25–30 grams dietary fiber per day): Carbohydrates, sugar in particular, are ready sources of energy. Starchy foods have to be broken down to glucose in the intestine, which is then absorbed in the bloodstream. Refined sugars are glucose in corn syrup, sucrose, fructose, and lactose. The human body converts all these sugars to glucose for producing energy. Glucose is the master energy molecule. When necessary, the human body can make glucose from fat and protein also. Starch, hemicellulose, cellulose, and resistant starch are all carbohydrates. Molecules of cellulose, hemicellulose, pectin, and lignin are indigestible or less digestible and, therefore, are not caloric, but they are very beneficial to our digestive health as food for colon bacteria. They are called prebiotic dietary fiber.

6.04.02 Proteins and Bioactive Peptides (40–60 grams per day): After water, protein is the next extremely critical daily nutrient. Just like water, we can't live too long without protein. A protein molecule may have up to 200 to 5,000 amino acids. Proteins supply nine essential amino acids that our body can't make and eleven others. These amino acids are critical for muscle building, enzyme synthesis, DNA and RNA synthesis, repair of body cells, bone building, cell signaling and messaging, immune response, neurotransmission, cell adhesion, and many metabolic, structural, and mechanical functions in our body. Protein intake in excess of 2 g per kilogram body weight can cause overload for the kidneys, though. A 70 Kg (154 lb) person, if consuming 140 g protein per day, is for sure inviting kidney problems. Two Burger King Whoppers at 25 grams of protein each books one up for a daily requirement of 50 grams.

Conversion of proteins to amino acid is completed at the end of the duodenum, and amino acids are absorbed in the jejunum. Essential amino acids for our routine diet are *histidine, isoleucine, leucine, lysine, methionine, phenylalanine, threonine, tryptophan, and valine.* The others—including *alanine, arginine, aspartic acid, cysteine, glutamic acid, glutamine, glycine, proline, serine, tyrosine, and asparagines*—are present in common foods that we eat (meat, poultry, cheese, eggs), but with different amino acid profiles. It is the profile that we need to balance by selection of a variety of protein foods. Balanced protein from beans and legumes, including soy, whey protein, and egg protein, should become part of our daily diet because they provide dietary fiber also. Prebiotic fiber and high-protein diets together can be very satiating.

Unlike egg and milk proteins, plant-derived proteins are not complete with all essential amino acids, and therefore, we need to consume a variety of plant proteins. Retail foods often contain soy protein, egg protein, and milk proteins such as casein and whey protein. Pea protein, as a good source of arginine and lysine, is getting popular in Canada and Europe. Those of us who commit to regular exercise can benefit from whey protein, which is the best source of branched amino acids leucine, isoleucine, and valine. Whey proteins act as an ACE inhibitor and control both systolic and diastolic blood pressures. Proteins containing arginine, tyrosine, and tryptophan must be part of our daily diet because (1) tyrosine is a precursor for neurotransmitters like epinephrine, norepinephrine, and dopamine; and (2) tryptophan is a precursor for serotonin and melatonin. Both move across the blood-brain barrier and can reach brain cells. Arginine is necessary for production of heart-healthy vasodilator nitric oxide.

Many peptides, partly broken-down proteins, from our common protein foods (sour cream, yogurt, Indian dahi, and many other fermented products like buttermilk and soy sauce) are now known to have health-enhancing properties. Specific examples are casein phosphopeptides, soy peptides, lupine peptides, milk proteins lactalbumin and lactoglobulin, lactoferin, and corn and wheat gluten peptides. Peptides in our daily diet help manage weight, diabetes, and cardiovascular problems.[136,137] They help us feel full and satiated. Examples of peptides in human food include oryzatensin, soy and pea peptides, and wheat protein peptides. Peptides are also known to function as antioxidants and anticarcinogens.[138,139] They help increase of bone mass, immune system modulation, blood pressure reduction, nervous system regulation, and satiation. More energy is required to convert amino acid to glucose, and therefore, net energy gain from proteins is lower than

136 http://www.sciencedaily.com/releases/2011/07/110720142346.htm.

137 http://www.bachem.com/news-media/news/detail/article/new-monograph-peptides-and-diabetes/.

138 http://www.hindawi.com/journals/jaa/2013/939804/.

139 http://www.ncbi.nlm.nih.gov/pubmed/17430183.

from carbohydrates and fat. Proteins help preserve fat-free body mass and lower the risk of obesity.

6.04.03 Fat and Oil Called Lipids (40 grams of fat per day): Fat and oil are what physicians and chemists call triglycerides, an essential nutrient, especially the fats or oils containing omega-3 fatty acids. Triglycerides, made up of glycerol and three fatty acids, are for energy storage. In formulated foods that we buy at the retail store, they improve texture, impart lubricity, act as solvents for flavors, nutraceuticals, and vitamins A, D, E, and K. Fats or lipids are part of our cell membrane and make up 60 percent of our brain. They are part of many messenger molecules. Fats regulate insulin and other hormones; they are responsible for memory and fertility.[140,141] Lecithin, a phospholipid, supplies choline for neurotransmission.[142]

Fat digests slowly and regulates appetite. It is the main source of energy and is used daily in upkeep of cellular membrane systems, brain health, the immune system, and the cardiovascular system. Digestion of fat begins with the help of bile acid from the gall bladder in the duodenum, and its absorption is complete in the jejunum. Bile acid functions as an emulsifier for making four- to eight-nanometer micelle or chilomicrons, which get transported via the lymphatic system. This is very different from the direct absorption of glucose and amino acids.

Extremely low- or no-fat foods can cause blindness, osteoporosis, depression, dry skin and hair, a poor immune system, impaired cell growth, and reproductive problems. We need fat to stay alive. In short, one will die without the benefits of fat. Omega-3 unsaturated fatty acids are very important because they are part of our brain cells. The problem today is the 10:1 ratio of omega-6 and omega-3 consumption. We need to select foods such that this ratio is 1:1, or at the most 2:1. We have known for a long time that omega-6 and omega-3 fatty acids are essential nutrients, but only in the right ratio. Fats and oils contribute to sensory and nutritional attributes of foods we consume. Snack foods fried in partially hydrogenated soybean oil are the biggest culprits because of high semi-saturated fat containing trans fat which is twice as unhealthy as simple saturated fat.

140 http://www.ncbi.nlm.nih.gov/pubmed/23820633.
141 http://nourishedkitchen.com/foods-for-fertility/.
142 http://www.whfoods.com/genpage.php?tname=nutrient&dbid=50.

6.04.04 Prebiotic Dietary Fiber (25–30 grams per day): Our digestive system depends on dietary fibers for peristalsis, growth of probiotic bacteria, production of vitamin K and short-chain fatty acids, controlled movement of fluid in our digestive tract, general digestive health, and enhancement of our immune system. We need only 400 grams of total carbohydrates per day, 30 grams of which should be dietary fiber. The rest should be low-glycemic, complex carbohydrates and not refined sugar. A good selection of low glycemic carbohydrates with high dietary fiber reduces net carbohydrate and thus net energy production from corresponding total glucose

Dietary fiber comes from fruits and vegetables, beans and lentils, flaxseed, black currants, and even herbal condiments like fenugreek. Fruit and vegetable fibers contribute characteristic flavors, mouth feel, and texture to prepared foods. They confer processability and shelf-life extension to many retail cereal, snack, and baked foods. A few common vegetable foods like asparagus, chicory, garlic, Jerusalem artichoke, leek, and onion are great sources of the dietary fiber inulin. Other sources of soluble fibers are apple, orange, lima bean, blackberry, figs, avocado, and flaxseed.

Fibers are polysaccharide, oligosaccharide, lignin, tannin, and associated plant-cell-wall materials. Other food ingredients associated with dietary fiber (polyphenolic antioxidants, flavonoids, carotenoids, and anthocyanins) make dietary fiber an even more powerful nutrient.

The average American is fiber-deprived, with daily consumption of less than fifteen grams. This needs to increase to a daily consumption in a range of twenty-five to thirty grams per day for managing weight, controlling type 2 diabetes, reducing cholesterol level, preventing colon cancer, and reducing risk of cardiovascular diseases. Dietary fibers also function as laxatives, improve bowel function, regulate insulin response, add capillary resistance, and act as antioxidants that tie up free radicals. Furthermore, they offer protection of vitamin C, inhibition of cataracts and diabetes, and prevention of glucose spiking. Most important of all is the immune-boosting property of antioxidants associated with dietary fibers.

Galacto-oligosaccharides (GOS) and fructooligosaccharides stimulate growth of probiotic bacteria, enhance calcium absorption, offer relief from constipation, support natural immunity, and induce satiety. They are low-calorie carbohydrates, with only two calories per gram. The food labels may list other oligosaccharides, including trans-galactosaccharides (TOS), isomalto-oligosaccharide (IMO), xylooligosaccharides (XOS), soy-oligosaccharides (SOS), gluco-oligosaccharides (GOS), and lactosucrose. FOS, TOS, and lactulose are used as effective and rather popular prebiotics in Europe.

6.04.05 Probiotics for Digestive Health and Immunity: Probiotics are symbiotic microorganisms (bacteria) capable of fermentation in our colon. As a matter of fact, they constitute our second genome. Common foods containing probiotic bacteria are yogurt, cottage cheese, natural cheeses, sauerkraut, tempeh, and S. Korean kimchi. India has used dahi (yogurt) and idali for centuries. Kefir and kimchi have a very old history as daily foods. Probiotics promote digestive health, immunity, and bowel integrity. Fruits and vegetables add their resident probiotic bacteria to our colon regularly, which increase to billions of probiotics per gram when we use their fermented versions based either on vegetables (sauerkraut) or milk (yogurt). Leafy green salads have lactic bacteria that survive the high acidity of the stomach and establish as probiotic in the colon.[143] Such bacteria have a direct relationship with mitochondrial function in our cells.[144]

6.04.06 DNA and RNA: The best sources of nucleic acids are fish, nuts, vegetables, mushrooms, yeast, and beef broth and soups. Nucleic acids are made in our body from phosphate and ribose sugar. They are not absorbed as such. Such foods help increase good cholesterol (HDL) and reduce inflammation. Use of 1.5 ounces of tree nuts as a daily afternoon snack is a much better alternative to ribose supplements.[145,146]

6.04.07 Minerals as Vitamins for Optimum Metabolism: Vitamins and minerals make up 7 percent of our body weight. They are essential and critical to myriads of reactions in our daily metabolism. As a matter of fact, this book summarizes details on foods essential to mood and DNA methylation. Table 6.01 gives a list of nonexotic common foods rich in minerals, proteins, fiber, antioxidants, phytonutrients, and vitamins. They can be incorporated in our daily methylation diet.

---

143 http://www.tdx.cat/bitstream/handle/10803/7932/trtm.pdf;jsessionid=A5CAD86DACA48 EE11786CED6928380EA.tdx2?s.

144 http://www.epidemicanswers.org/tag/mitochondrial-dysfunction-2/equence=1.

145 http://www.mayoclinic.com/health/nuts/HB00085/NSECTIONGROUP=2.

146 http://jn.nutrition.org/content/138/9/1736S.full.

Table 6.01 Common Foods for Trace Minerals and Antioxidant Vitamins

Food	Minerals	Other Nutrients	Vitamins
Almond	Copper and magnesium, manganese, vitamin E	Protein and dietary fiber	Almond delivers vitamin E also
Banana	A single banana provides 460 mg potassium, good for our daily need.	High in dietary fiber, copper, manganese, magnesium	Vitamin B_6, C, and antioxidants lutein, zeaxanthin, and carotenoids
Beans	Copper, magnesium, manganese, molybdenum	High tryptophan protein, dietary fiber, antioxidants	Folic acid
Brazil nuts	Selenium, magnesium, manganese	Protein and dietary fiber	Folic acid and vitamin E
Cashew nut	Copper, magnesium, zinc, selenium, and tryptophan	Monounsaturated fatty acids	B_5, B_6, riboflavin, thiamine
Cereals	Copper, manganese, magnesium, selenium, vitamin B_1	Dietary fiber	Vitamin B_1
Chia seed	ALA omega-3 fatty acid	Protein and dietary fiber for appetite control	Helps burn fat, improves metabolism
Coconut oil	Medium-chain lauric acid	Healthy fat	
Cottage cheese	Selenium	Protein	Vitamins and minerals
Garbanzo beans	Molybdenum, copper, zinc, phosphorus, manganese	Protein and dietary fiber	Choline, vitamins A and C

Food	Minerals	Other Nutrients	Vitamins
Horseradish	Calcium, magnesium, vitamin C phosphorus	Allyl isothiocyanate	Vitamin C
Mushroom	Copper, potassium, iron, selenium, zinc	Protein, dietary fiber	Vitamins A, B, C
Mustard	Selenium	Protein and oil, good emulsifier	Vitamins B_2, B_3, B_5, C, K
Potato	Potassium, and copper	Starch	B vitamins
Pumpkin seed	Zinc	Good source of protein	Vitamins A, C, E, K, and folic acid
Sesame seed	Molybdenum, calcium, copper, iron, zinc, manganese	Protein and fiber	Vitamins B_6, C, choline
Sprouted grains			B vitamins, especially B_{12}
Sunflower seed	Selenium, copper	Protein, dietary fiber	B vitamins, especially B_{12}
Sweet potato	Copper, iron, magnesium, manganese	Protein, good fat, dietary fiber	Vitamin A and carotenoids
Oysters	Zinc	Protein	Vitamin A

Mineral nutrition is essential for good health and longevity, and under conditions of deficiency, pain and diseases are physiological consequences. Minerals are elements that originate in the soil and cannot be created by living beings such as plants and animals. Plants get them from soil, and animals get them from plants. Yet plants, animals, and humans all need minerals in order to survive. Calcium and phosphorous being the most abundant, minerals stay ionized and make up our electrolytes, whose active transport across our cells is fundamental to life processes. Actually, half of our cells' energy is spent in transferring ions and molecules around and across. Calcium is involved in bone health and phosphorus in energy metabolism. Actually phosphorus is part of the structure of DNA, RNA, and various molecules that transfer electrons and protons in our body. The sources and functions of minerals are described in tables 6.02 and 6.03 respectively. Micronutrient minerals are cobalt, chromium, copper, iodine, magnesium, manganese, selenium, and zinc (see table 6.02).

Table 6.02 Common Foods for Specific Trace Minerals

Boron (6 mg/day): Pears, plums, grapes, chickpeas, peanut butter, hazelnuts, and kidney beans

Chromium (120 mcg/day): Broccoli, garlic, and onion

Cobalt (7 mcg/day): Leafy green vegetables, crab, lobster, meat, dairy products

Copper (1.5 to 3 mg/day): Kiwi, dried beans, nuts, crab, and lobster

Iodine (150 micrograms/day): Iodized salt; fruits, vegetables, and nuts grown in high-iodine soils

Iron (18mg/day): Dried figs, fortified cereals, kidney beans, lima beans, molasses, spinach, peaches, pinto beans

Manganese (2–5 mg/day): Roasted soybean, flaxseed, sesame seed, wheat germ, cocoa powder, hazelnut, cinnamon, cloves, coriander, and turmeric powder

Molybdenum (75 mcg/day): Black beans, lentils, and walnuts. A cup of beans may offer 120 mcg of molybdenum.

Selenium (70 mcg/day): Brazil nuts, other nuts, and whole grains. Cereals can provide an incremental dose.

Zinc (15 mg/day): Oysters, meat, poultry, garbanzo, baked beans, whole grains, and nuts. If necessary, take multivitamin and mineral supplements. Diuretic drugs interfere with zinc.

Table 6.03 Health Maintenance by Minerals in Common Foods

Mineral (daily need)	Health Maintenance Functions	Common Food Sources
Boron (6 mg/day)	Bone, eye, and joint health; critical for use of vitamin D	Citrus fruits, pears, plums, grapes, chickpeas, peanut butter, hazelnuts, and kidney beans

Calcium (600 mg/day)	Bone, teeth, nerve, and heart health; prevents ovarian cancer; improves muscle and nerve function; regulates nutrients across cell walls; involved in muscle contraction, blood clotting, message conduction through nerves, hormone function	Fruits, broccoli, almonds, white beans, cheese, yogurt, ice cream, and seeds have reasonable amounts.
Chromium (120 mcg/day):	Enhances effects of insulin in glucose metabolism	Broccoli, garlic, onion
Cobalt (7 mcg/day):	Core of B_{12} structure, necessary for making red blood cells, DNA, and erythroproteins; involved in protein, carbohydrate, and fat metabolism	Leafy green vegetables, crab, lobster, meat, dairy products
Copper (1.5–3 mg/day	Iron metabolism, formation of red blood cells, oxygen supply, enzyme superoxide dismutase for free radical protection	Kiwi, dried beans, nuts, crab, lobster
Iodine (150 mcg/day	Regulates energy production, body weight, and proper growth; promotes healthy hair, nails, teeth, and skin; prevents hypothyroidism and goiter	Iodized salt, fruits, vegetables, nuts
Iron (18mg/day	Promotes cognition and learning, helps maintain immune system, transports oxygen through hemoglobin	Dried figs, cereals, kidney, lima, and pinto beans, molasses, spinach, peaches
Magnesium (420 mg/day	Magnesium is involved in 300 enzyme reactions in our body; immunity, bone and heart rhythm, DNA building; involved in signals to heart, muscles, and nerves; involved in energy release and muscle contraction; needed for bone, protein building, activation of B vitamins, relaxing nerves, clotting blood, insulin secretion and function	Whole grains, brussels sprouts, spinach, almond, nuts. Deficiency leads to type 2 diabetes.

OUR GENES, OUR FOODS, OUR CHOICES

Manganese (2–5 mg/day	Cofactor in in enzyme control of blood sugar, metabolism, enzyme arginase in particular, and thyroid function	Roasted soybean, flaxseed, sesame seed, wheat germ, cocoa powder, hazelnuts, cinnamon, cloves, coriander, turmeric powder
Molybdenum (75 mcg/day):	It is part of the sulfite oxidase enzyme, and it prevents gastrointestinal cancer. A deficiency is rather rare.	*Black beans*, lentils, walnuts
Phosphorus (1,000 mg/day):	Formation of bones, teeth, and nerves. Second to calcium in abundance in our body and is involved in energy metabolism. It is part of the DNA structure and the structure of energy molecule ATP.	Milk, meats, eggs, cheese, soft drinks; nuts, seeds of pumpkin, sesame, and sunflower, rice bran, wheat germ
Potassium (400mg/day)	Critical for heart, kidney, and nerve function; essential for growth and maintenance; involved in water balance between cells and body fluids; lowers blood pressure	Sweet potato, parsnips, figs, celery, avocado, artichoke, asparagus, carrot, raisins, cantaloupe, kidney beans, lima beans, pinto beans, tomatoes
Selenium (70 mcg/day	Part of many enzymes and can protect from cancer; a potent antioxidant with vitamin E; part of antioxidant glutathione and peroxidase	Brazil nuts, other nuts, whole grains, cereals
Sodium (500–1,000 mg/day):	Key to glucose and amino acid absorption across intestine; prevents cell damage; regulates fluid balance, blood pressure, and volume; involved in electrical signals of nerves and muscles—all outside the cells	Whole grains, legumes, fruits, vegetables

Sulfur (not official but 800 mg per day is generally accepted)	Protects joints, is part of enzymes and other proteins like collagen, helps oxygen balance	Edamame, garlic, onion, mustard
Zinc (15 mg/day):	Prevents colon cancer; improves brain development, motor skills, cellular communication, enzyme activities, and immune system; maintains blood glucose and metabolism	Sesame and pumpkin seeds, lentils, legumes, meat, poultry, scallops

6.04.08 Vitamins

We have known for years that deficiency of vitamin A causes blindness, of vitamin C causes scurvy, of D causes rickets, and of B vitamins causes poor energy and protein production. We consume vitamin and mineral molecules either bound to other big food molecules or in our daily water. Mechanical grinding in the mouth and stomach renders them free in the stomach juice. Vitamins A and vitamin B members are absorbed in the duodenum; a number of B members and C are absorbed in the jejunum; and the ileum absorbs the majority of C, D, E, K, B_1, B_2, B_6, and B_{12} vitamins. Vitamins A, D, E, and K are absorbed with fat. Vitamin B_{12} is absorbed with a carrier intrinsic factor protein produced in the stomach with the specific job of carrying vitamin B_{12} through the digestive tract. We need B_1 (niacin) and B_2 (riboflavin) for metabolizing carbohydrates, proteins, and fat for energy. These are water-soluble vitamins available in fortified wheat flour-based cereals and baked goods, and green vegetables are a major source. B vitamins and vitamin C are water-soluble. The best source for B_7 or biotin is egg yolk. Sprouted grains and legumes are good sources of B vitamins.[147] Vitamin B_{12} is not found in plants, and dietary sources for it are fish, milk, milk products like cheese and yogurt, poultry, and eggs. Vegetarians should consume yogurt-, tempeh-, and kimchi-like fermented foods because probiotics in them produce B_6, B_{12}, and K2 in our intestine.[148] Cereals fortified with B vitamins are a must for most of us. Functions of various vitamins are described in tables 6.04 and 6.05. Vitamin B complex including B_6, B_{12}, and

---

147 https://www.usaemergencysupply.com/information_center/using_whole_grain_foods/nutritional_study_in_sprouts.htm.
148 http://www.ncbi.nlm.nih.gov/pubmed/21933312.

intrinsic factor, vitamin K, and folic acid are very critical in health maintenance because of their involvement in energy production and electron transport. Biotin is necessary for making glucose from noncarbohydrate foods like proteins and fat. Vitamins B_6, B_{12}, and folic acid recycle homocysteine into harmless amino acids and prevent narrowing of the arteries (atherosclerosis).[149,150]

Whereas B_{12}, the brain and blood vitamin, can last for several years,[151] most other B vitamins need to be replenished every four to five days. Fat-soluble vitamins are storable and can be time-released on demand. Forty percent of Americans show vitamin B_{12} deficiency. Vitamin supplements are ineffective because of very poor (only 10 percent) absorption. Poor absorption notwithstanding, supplements are a very good health insurance for maintaining heart and bone health and preventing cancer. We need to get our vitamins from natural wholesome sources, though. Vitamins are the key nutrients for gene expression.

Table 6.04 Water-Soluble Vitamins
(Note: Year when discovered is given in parentheses.)

Vitamin	Health Maintenance Functions	Source
B_1, thiamine **(1.5 mg/day)** (1910)	Deficiency causes beriberi. Involved in energy production from carbohydrates; helps in functioning of heart muscles and nervous system	Asparagus, avocado, wheat germ, whole wheat flour, fruits, vegetables
B_2, riboflavin **(1.7 mg/day)** (1920)	Deficiency causes cracked lips, tongue inflammation, and edema.	Wheat germ, whole wheat flour, fruits and vegetables, baker's yeast
B_3, niacin **(20 mg/day)** (1936)	Deficiency causes pellagra. Involved in body growth, digestive system's function, health of skin and nerves, reproduction, red blood-cell production, energy production from food	Baker's yeast, whole wheat flour, fruits, vegetables

---

149 http://www.hsph.harvard.edu/nutritionsource/vitamin-b/.

150 http://www.ncbi.nlm.nih.gov/pubmed/9791839.

151 http://ods.od.nih.gov/factsheets/VitaminB12-QuickFacts/.

Vitamin	Health Maintenance Functions	Source
B_5, pantothenic acid (10 mg/day) (1931)	Deficiency causes acne. Involved in food metabolism, formation of hormones, and cholesterol control	Baker's yeast, wheat germ, whole wheat flour, fruits, vegetables
B_6, pyridoxin (1.5 mg/day) (1934)	Maintains red blood cells; deficiency causes anemia, homocysteine problems, and problems with heart tissue; involved in antibody formation; enhances nervous and immune system; involved in blood-cell metabolism and protein formation; helps convert tryptophan to serotonin	Baker's yeast, whole wheat flour, fruits, vegetables
B_7, biotin (100–300 mcg/day) (1931)	Deficiency causes neurological disorders; it is involved in sugar control, carbon dioxide transfer, energy production, fat and protein metabolism, energy production, cell growth, and nail and hair health. Probiotics produce biotin for our daily requirement.	Baker's yeast, fruits and vegetables, wheat germ, whole-wheat flour
B_9 or folic acid (400 mcg/day)	Deficiency causes anemia and neural tube defect in the case of pregnant women; involved in DNA formation and repair, red blood cell formation, and formation of components of the nervous system. Converts homocysteine to other amino acids.	Asparagus, avocado, fruits and vegetables, brussels sprouts
B_{12} or cyanocobalamine (5 mcg/day), discovered in 1948.	Deficiency causes loss of cognition and memory, peripheral neuropathy, and elevated homocysteine; involved in metabolism, prevention of pernicious anemia, red blood cell formation, DNA synthesis, and neurological functions.	Beef, trout, baker's yeast, sunflower seed

Vitamin	Health Maintenance Functions	Source
Vitamin C **(200 mg/day)** (1930)	A vitamin with antioxidant power and therapy against cancer and cardiovascular diseases; it is antiviral and controls infection. It is involved in immune health, ligament health, and tissue repair.	Citrus fruits and vegetables
Coenzyme Q (1960); up to 600 mg per Day.	Ubiquinone or Q_{10} and energy (ATP) production	Wheat germ, whole-wheat flour

Common vegetables like asparagus (vitamins A, C, K, and folic acid), avocado (vitamins B_1, B_2, B_6, and folic acid), beets, broccoli, brussels sprouts (folic acid, vitamins A, C, K), collard green (vitamins A, C, K), cauliflower (vitamins A, C, K), celery (vitamins C and K), green beans (vitamins A, C, K), lettuce, mango (vitamins A, C, E), sunflower seed (vitamin E), and winter squash (vitamins A, C, K) are great natural nutrients for unhindered gene expression.

Table 6.05 Fat-Soluble Vitamins

Vitamin	Health Maintenance Functions
Vitamin A **(1,000 mg or 500 IU/day)** (1913)	Critical to vision and white blood cells; involved in cell division/reproduction and differentiation; enhances immune system; helps hormone formation; enhances vision, bone, and skin health; works against measles. Vitamin A lowers cancer cell growth.
Carotenoids **(6,000 mcg/day),** **discovered early** **in 1847.**	This group includes beta-carotene, lycopene, lutein, alpha-carotene, and zeaxanthine. Carotenoids are precursors for Vitamin A and very critical to eye health.
Vitamin D **(7.5 mcg or 1000 IU/day):** (1920)	Helps absorb calcium and phosphorus. Signals kidney to preserve calcium and phosphorus. Improves heart health and helps prevent diabetes, autoimmune diseases, and cancer. It is involved in neuromuscular functions and enhancement of the immune system. Vitamin D is a hormonelike anti-inflammatory nutrient.

Vitamin	Health Maintenance Functions
Vitamin E (10 mg or 1,000 IU/day) (1922)	Antioxidant, prevents cell damage, protects LDL from oxidation, prevents heart diseases and cancer. It is involved in the immune system, cell signaling, and gene expression.
Vitamin K (80 mcg/day): (1943)	Critical to blood clotting by managing thirteen proteins; also involved in bone health and calcium maintenance.

Vitamin D is necessary for cardiovascular and bone health. I include a detailed description of the therapeutic value of vitamin D in chapter 16. Milk, beef, egg yolk, liver, and fatty fish are good sources of rare and essential vitamin D. All other vitamins have to come from our daily food. Nutritional guidelines are based on no sunlight availability for production of active vitamin D in our skin, which is further activated in the liver and kidney. Vitamin K, the blood-clotting vitamin, is also involved in bone health. Vitamin D is a hormonelike anti-inflammatory nutrient (see table 6.05).

6.05 Omega-3 Fatty Acid Foods with Anti-Inflammatory Prostaglandins

Prostaglandins are twenty carbon molecules produced in the vicinity of body cells from fatty acids in our daily diet. They bind with receptors on the cell for their regulatory activities, like curing of injuries, smoothing muscle-cell contraction, and modulating blood flow, blood clotting, and inflammation. A diet that delivers a good daily dose of gamma linoleic acid or omega-3 fatty acid is necessary for the production of inflammation-fighting prostaglandins.

Foods that must be avoided are trans fat, polyunsaturated fatty acids, and too much animal protein. Managing inflammation by a good diet is the first step in avoiding heart disease, diabetes, aches and pains, and stress.

6.06 Foods with Antioxidants and Phytochemicals

Major antioxidants such as lycopene, leutin, vitamins C and E, flavonoids (anthocyanins), quircetin, and catechin are liberated from the food matrix and are absorbed in the bloodstream. Flavonoid quircetin is a potent antioxidant with antihistamine-like effect. Fruits and vegetables—particularly citrus fruits, apples, onions, parsley, sage, tea, and red wine—are the primary dietary sources of quercetin. Olive oil, grapes, dark cherries, and dark berries—such as blueberries, blackberries, and bilberries—are also high in flavonoids, including quercetin.

Herbs and spices listed in table 6.06 have been found by way of long-term direct experience to be digestive aids. They are part of modern-day gastronomic delights. Those of demonstrated therapeutic value are asafoetida, cloves, ginger, oregano, mango powder, nutmeg, and turmeric.

Table 6.06 Herbs and Spices for Digestive Health

Common Foods	Active Ingredients	Ailments and Diseases Helped
Asafoetida	Ferulic acid, asaresinotanol	Antibacterial, antiflatulent, anti-constipative
Buttermilk, plain yogurt	Lactic bacteria, probiotic	Improves digestive health
Cloves	Eugenol, omega-3 fatty acids, manganese	Promotes digestive health, antibacterial, antidiabetic,
Coriander	Phthalides, coumarins, terpenol	Boosts digestive health, prevents cancer, reduces cholesterol, antibacterial
Cumin	Luteolin, piperine, pipene, terpine, iron	Prevents prostate cancer and improves digestive health
Ginger (great with chili pepper and garlic)	Gingerol	Antioxidant, anti-inflammatory, digestive aid, fights arthritic pain, helps blood flow
Mango powder	Citric, oxalic, and malic acids	Aids digestion
Millet	Fiber	Absorbs cholesterol and increases bile-acid secretion.
Nutmeg	Atropine-like bioactive	Digestive aid, antifungal

Common Foods	Active Ingredients	Ailments and Diseases Helped
Oregano	Vitamin K, carnasol, quircetin, thymol, carvacol	Antioxidant, antifungal, digestive aid
Salt, black		Anti-carminative, antiflatulent, digestive aid, laxative
Sanguisorba (China, Japan)		Cures diarrhea, duodenal ulcer, has antioxidant and anticarcinogenic effects.
Turmeric	Curcumin	Anti-inflammatory
Whole-grain foods	Polyphenol antioxidants, folic acid, soluble and insoluble fiber, minerals	Promote digestion
Yogurt	Probiotic lactic bacteria, B_{12}	Digestive aid

Table 6.07 lists additional commonly used plant foods that are now known to deliver antioxidant, anticarcinogenic, and cholesterol-reducing foods used for centuries against specific ailments and diseases.

Table 6.07 Phytochemicals from Common Foods

Common Foods	Active Ingredients	Ailments and Diseases Helped
Artichoke	Silimarin, dietary fiber	Antioxidant protects skin cancer.
Barley (pearled)	Beta-glucan fiber, selenium	Reduces cholesterol; an antioxidant
Beet roots	Betaine, folic acid, manganese, potassium, dietary fiber	Antioxidant, quenches free radicals, prevents DNA damage, prevents lung and stomach-tumor growth
Brown rice	Polyphenol	Antioxidant, low-glycemic food

Common Foods	Active Ingredients	Ailments and Diseases Helped
Buckwheat	Polyphenol	Antioxidant, quenches free radicals, prevents DNA damage, prevents macular degeneration, prevents cancer, improves cognition
Carrots (provitamin A-rich)	Beta-carotene, vitamins A, K, C	Antioxidant, reduces risk of cancer and heart disease
Chocolate	Polyphenols	Antioxidative effects
Cranberry, grape-seed extract	Anthocyanins	Cures urinary infections, antibacterial, anticarcinogenic
Dill seed	Quircetin, kaempferol	Powerful antioxidant
Flaxseed	Lignan and omega-3 fatty acid	Antioxidant and anti-inflammatory
Grapes, raw and dried	Resveratrol	Heart-healthy
Green peppers	Beta-carotene, lycopene, vitamin C	Antioxidant, has 4X vitamin C content compared to citrus fruits.
Green tea	Epigallocatechin-3-gallate, vitamins A and C	Heart health and cancer protectant
Parsley	Apigenin, apiin, cresoviol, luteolin	Antioxidants
Rosemary	20 antioxidants	Anti-inflammatory, antibacterial, improves cognition and memory
Sage	Rosemaric acid, cornosol, thymol, carvacol	Inhibits the enzyme that destroys acetylcholine. It is neuroprotective, a cure for Alzheimer's disease.
Thyme	Geraniol, borneol, thymol, carvacol, vitamin K	Antimicrobial, antioxidant, antispasmodic, carminative

Common Foods	Active Ingredients	Ailments and Diseases Helped
Turmeric	Cucumin, folic acid, vitamins C and K, B_6	Antioxidant, antibacterial, anti-inflammatory, prevents free radical damage, good for osteoarthritic pain relief, prevents enlargement of the heart
Yumberry and honeyberry		Antioxidants

6.07 Anti-Constipative Foods

Constipation is a big problem for 80 percent of Americans[152]. As a matter of habit, we go for magnesium hydroxide-based milk of magnesia, psyllium-based soluble fiber in Metamucil, and polystyrene glycol-based MiraLAX is literally an American staple. The easier solution is proper diet, containing anti-constipative foods with high magnesium (almonds, cashew nuts, cooked spinach, lentils, and legumes) and high dietary fiber (brown rice, barley, oat, vegetables, flaxseed, and beans). A daily dose of 30 grams of prebiotic dietary fiber from legumes, lentils, and spinach is a much better solution than milk of magnesia and MiraLAX. Eggplant contains glycoalkalois, which has a mild laxative effect.

6.08 Satiety Foods

Protein foods increase satiety by delaying stomach emptying, and as such, cause fullness. Digestion of proteins uses almost 30 percent of daily metabolic energy and thus reduces available net calories. High-fiber and high-protein diets offer the triple advantage of satiety, low net energy intake, and supply of critical phytonutrients.

Most vegetables and fruits add fiber for satiety. Oatmeal accounts for the highest satiety index.[153] Other high-satiety foods are bulgur, rye bread, and vegetables. Factors responsible for inducing satiety are high-volume low-density foods, low energy density, and large particle size. All such diets must have optimum micronutrient composition and must offer palatability.

---

152 http://www.fascrs.org/patients/conditions/constipation/
153 http://wholegrainscouncil.org/whole-grains-101/health-benefits-of-oats.

In physiological terms, peptide YY and peptide I induce satiety, and the hormone ghrelin increases hunger, stimulates leptin, and decreases satiety.

6.09 Volumetric Eating Plan

Low-energy foods are more filling. Volumetric eating induces satiety and helps reduce weight. Both at one hundred calories, two cups of grapes are better than one quarter cup of raisins. where water in the grapes accounts for the fullness. A large portion of pasta with vegetables is more filling than pasta with Alfredo sauce. Other examples of high-volume, low-calorie foods are pizza with broccoli, peppers, and asparagus and a sandwich with lettuce, tomato, cucumber, and zucchini. Soups, salads, vegetables, and fruits are high-volume, low-calorie foods. Also, fiber in oatmeal, bulgur, whole-wheat bread, and high bran cereals is what makes them low-calorie foods. Fruits and vegetables have the same effect due to their fiber and high moisture content. Six servings daily of fruits and vegetables accounts for approximately a large glass of satiating water.

6.10 Therapeutic Values of Everyday Foods

Whole grains, fresh fruits and vegetables, stir-fried foods, lean meats, plant proteins (beans, legumes, soy), and plant fats from flax and sesame seeds and nuts like walnuts and almonds are great inclusions to our daily diet. They deliver fiber, good fat with omega-3 fatty acids, and key minerals, vitamins, and phytonutrients. We can manage our gene expression and even reverse aging simply by consuming the methylation diet and consuming methylation diet on time.

6.11 Value of Age-Old Alternative Medicine

Specific dietary practices in different countries include (1) elderberries for fighting cancer by antioxidant anthocyanins and vitamins A, C, and B_6 in Austria; (2) breast cancer-fighting anticarcinogens (cordycepin) from caterpillar fungi in China; (3) bitter gourd with iron and vitamins A, B_1, B_2, and C used to fight diabetes by the Okinawa diet in Japan; (4) durian fruit for boosting the immune system in Malaysia; and (5) Indian mulberry for fighting inflammation. Over-the-counter pain relievers and Metamucil based on the psyllium husk for regularity are examples of very prevalent folk medicine. Antioxidants, anticarcinogens, phytosterols, phytonutrients, high-satiety protein foods, amino acids like arginine, tyrosine, and tryptophan, and omega-3 essential fatty acids have been part of folk medicine for a long time.[154,155] Commonly used herbs and spices—great sources of antioxidants, anticarcinogenic, and anti-inflammatory compounds[156]—have been part of time-tested ethnic diets in various countries for ages, as included in the summary table 6.08.

Table 6.08 Therapeutic Benefits of Everyday Foods

Disease	Food Cures
Against allergies	Onion, rich in antioxidants
Against cancer	Artichoke with folic acid and zinc has been used against skin cancer for a long time. Fruits and vegetables such as avocado, asparagus, apple, barley, broccoli, brussels sprouts, cabbage, cauliflower, collard greens (leutin, beta-carotene, and vitamin C), carrot, celery, kale, and kiwifruit, flaxseed, Gouda cheese, licorice, mushrooms, nuts, oranges (vitamin C and beta-carotene), parsley, seaweed, kale, soy foods, strawberries, tomatoes, and turmeric can offer our daily remedy of anticarcinogens.
Anticoagulant	Cinnamon, clove, melon, tea
Anti-constipative	Apple, beets
Antidiabetic	Cinnamon promotes insulin activity; fenugreek seed, white potato with protease inhibitor
Anti-inflammatory	Apple, clove, flaxseed oil, garlic, ginger, onion, pineapple
Antiviral foods	Kimchi, turmeric, garlic, chocolate, green tea, omega-3 fatty acids, vitamin E
Against urinary infection	Blueberries and cranberries
Bone health	Celery (delivers silicates)
Digestive health	Ginger

---

154 http://www.omicsonline.org/evaluation-of-antioxidant-activity-of-various-herbal-folk-medicines-2155-9600.1000222.pdf.

155 http://ymj.or.kr/Synapse/Data/PDFData/0069YMJ/ymj-45-776.pdf.

156 http://www.ncbi.nlm.nih.gov/books/NBK92763/.

Disease	Food Cures
Energy balance	Pumpkin seeds
Brain health	Lecithin, beans and legumes, white beans in particular
Heart health	Omega-3 fatty acids to prevent inflammation, improve heart health. and for better cognition, fish and fish oil, garlic, mushroom, olive oil; tomatoes, walnuts, other nuts, flaxseed, beans, lentils, eggplant, grapes with resveratrol and quircetin, grapefruit juice, oats, pumpkin with antioxidants, rice bran, sweet potato (beta-carotene), turmeric
Nervous system regulation	Wheat gluten, lecithin
Immune response	Rice bran
Osteoporosis	Turmeric
Painkiller	Chili pepper (capsaicin), clove, date with natural aspirin, raspberry, turmeric
Satiety	Protein foods for satiety and weight control; chickpeas against bacterial infection; soy foods
Stimulation	Coffee
Fiber & phytonutrients	Barley, beets, dates, fruits, vegetables, whole wheat
Probiotics	Yogurt, sauerkraut, kimchi, tempeh

Describing a diet for good metabolism by the color of foods must have been the selection criteria for optimal paleo diets centuries ago. Red, purple, and blue vegetables mean anthocyanins, vitamin C, lycopene, fiber, and folic acid. Beets, watermelon, strawberries, and kidney beans mean lycopene and leutin supply for eye and heart health. Cherries, figs, and tomatoes are high in potassium, which can lower blood pressure. Orange and yellow vegetables like carrots are a great source of beta-carotene, vitamin C, potassium, and folic acid.

Bell peppers, sweet potato, cantaloupe, peaches, papaya, and mangoes are great sources of vitamins A and C and folic acid; pumpkin, sweet potato, and butternut squash are high in potassium, which lowers blood pressure; and the enzyme bromelain from pineapple is a great remedy for indigestion. Green vegetables in general, such as broccoli, brussels sprouts, kale, spinach, lettuce, and collard greens, are high in vitamin A; greens, including peppers and broccoli, are good for vitamin and mineral nutrition. Evolution, it appears, proceeded with color as a selection guide for foods.

Common foods as cures for specific ailments are listed in table 6.09. The foods listed deliver adaptogens, anti-allergens, antidepressants, antioxidant vitamins, enzymes, trace minerals, and stimulants. The cures are based on science, the science of vitamin and mineral nutrition in particular. Added remedies come from omega-3 fatty acids, dietary fiber, and mineral- and vitamin-based antioxidants (see table 6.10).

Nutrition is the provision to cells and organisms of materials necessary to support life. What we call alternative medicine today has made our good health possible for years. It is now proven to be effective as antioxidants that repair DNA, anticarcinogens that cure cancer, immune-boosting anti-inflammatory agents, and foods that help regulate estrogen, insulin, and other hormone metabolism. Good nutrition is simply intended to confer genomic and epigenetic stability. *Vitamin D$_3$ is key to DNA transcription and gene expression.*[157] We now know that key to digestive health are prebiotic dietary fibers and probiotic foods like yogurt and sauerkraut and that our routine diet can furnish molecules of neurotransmission and memory. Food variety is critical to the methylation diet, which is good for routine metabolism, gene expression, DNA maintenance, and many other therapeutic purposes as listed below.

- Foods such as chili pepper and cinnamon that fight pain and chronic diseases
- Foods that fight cancer: avocado, garlic, horseradish, mint, mustard, onion, pepper, soy milk, tomato, tofu, texturized vegetable protein, rosemary, sage, tea, turmeric, watercress
- Foods that fight DNA damage: artichoke, beet, bok choy, brussels sprouts, cauliflower, collard, grapes, green beans, kale, mustard, peas, radish, spinach, strawberry, and turnip
- Foods for proper blood pressure, such as ginger, omega-3 fatty acids from flaxseed, salmon, and acid foods that help iron absorption, minerals, and vitamins
- Foods for vitamin D and calcium for bone health
- Foods that lower cholesterol, such as good fats in olive oil and walnuts

---

157 http://www.ncbi.nlm.nih.gov/pubmed/11435613.

- Foods that maintain brain function and mental health and foods that fight inflammation.

The value of a chosen diet against various ailments (see tables 6.09 and 6.10) can be confirmed by soon-to-be-available cost-effective tests. It is a good idea to ask your physician *for the C-reactive protein test*. Finding of a high reading means low HDL (good cholesterol) and propensity to inflammation. Soon our physicians will be able to monitor performance of adipose tissue in terms of leptin, adiponectin, plasminogen activator, resistin, sex steroids, and glucocorticoids. We will be able to validate the value of our methylation diet simply by keeping track of what we eat, our fitness and weight profile, and periodic test results from our clinics. We will become more concerned about our health; we will become our own therapist.

Table 6.09 Common Foods against Specific Ailments

Foods That Fight Common Ailments		
Common Foods	**Active Ingredients**	**Ailments and Diseases Helped**
Antidepressant Foods		
Broccoli and eggs	Calcium, vitamin D	Prevent depression
Guarrana	Caffeine-like stimulant	
Red clover	Isoflavone	
St. John's wort		Reduces depression
Indian ginseng ashvagandha	Adaptogen	Promotes sleep
Tart cherry	Melatonin	Insomnia
Foods that Fight Allergies		
Dark chocolate	Polyphenol, theobromine	Cures cough, antioxidative
Marjoram	Flavonoids and polyphenols	Cures sinus congestion, indigestion, stomach pain, headache, dizziness, colds, coughs, and nervous disorders

Foods That Fight Common Ailments		
Common Foods	**Active Ingredients**	**Ailments and Diseases Helped**
Nettle with onion	Antiallergenic	Effective against allergies with quircetin
Red and yellow onion; Baked onions	Quircetin, vitamin C, chromium, flavonoids; antioxidative and anticarcinogenic	Cures eye, nose, lung, and intestine allergies, reduces risk of Alzheimer's and Parkinson's diseases
Eye Health and Vision		
Bilberry	Pycnogenol	Improved eye blood flow, improved oxygen level, reduced pressure
Sprouts	Enzymes	Soft tissue recovery, heart and joint health
Spinach	Iron, folate, other phytochemicals	Prevent macular degeneration
Green beans	Leutin and zeaxanthin	Improve eye health
Kale	Beta-carotene, vitamins A, C, E, and K, folate, leutin, manganese	Powerful antioxidant, protects against macular degeneration
Swiss chord	Fiber, leutin, carotenoids, vitamins A, B_6, C, K	Lutein protects against macular degeneration, cataracts
Gall Bladder and Liver Health		
Milk thistle	Sylamarin	Prevents gall bladder disease

Table 6.10 Healthful Nutrients from Common Foods

Nutrients	**Common Foods**
Omega-3 fatty acids	Flaxseed, kiwifruit, purslane, and salmon are good sources of alpha-linolenic acid, EPA, and DHA. A newborn infant's brain is 50% DHA.

Nutrients	Common Foods
Beta-carotene	Carrot, spinach, sweet potato, kale
Vitamin A	Sweet potato
Vitamin B complex	Green vegetables
Folic acid	Beans, legumes, lentils, green vegetables
Vitamin B$_6$	Trout, salmon
Vitamin B$_{12}$	Troutand salmon, s; Can come from supplements also.
Vitamin C (400 mg)	Broccoli, red bell pepper, brussels sprouts, papaya
Vitamin D (500 IU)	One should take supplements of at least 5000 IUsake because it is a major defficient vitamin among American consumers.
Vitamin E (400 IU)	Almond, sunflower seed, hazelnut, peanut butter
Calcium (1,000 mg)	Milk, cheese, yogurt, potato, spinach, collard green, leavening agents
Minerals	Copper, chromium, magnesium, selenium, and zinc from seeds, nuts, fruits, and vegetables.
Minerals and vitamins	Soy, artichoke, dried tomatoes, dried fruits, limnoids, mixed seeds of pumpkin and sunflower, mixed nuts, wasabi peas, peanuts. Artichoke is rich in antioxidants and suluble fiber also.

6.12 Rules for a Good Daily Diet

Our daily diet should deliver vitamins from asparagus, avocado, beets, broccoli, brussels sprouts, collard greens, carrots, cauliflower, celery, green beans, lettuce, mango, sunflower seeds, tomatoes, and winter squash.[158]; Baker's yeast is a great source of B$_{12}$, pineapple with bromelain is a great digestive aid for those on high-protein diets. Garlic and onion for inulin-like dietary fiber should be part of our diet, and they are increasingly becoming so. High-magnesium foods like almonds, cashews, and spinach, along with dietary fiber, can prevent constipation.

Licorice (DGL or deglycyrrhizinated licorice) and good fat (pistachios and walnuts) enhance digestive health. Asian mushrooms, yeast, tree nuts,

---

158 http://www.ncbi.nlm.nih.gov/pubmed/11435613.

and beef broth can be used for high-DNA intake for ribose sugar. Actually Asian and Indian foods can be classified as DNA methylation foods.[159,160]

Alpha-lipoic acid from spinach and chromium from broccoli and onion should be used for blood-glucose management in the case of people suffering from type 2 diabetes. Daily iodine can come from milk, fish, and seaweed for metabolism, muscle function, and energy.

Calcium supplementation has been over-specified by supplement marketers. Calcium and magnesium in a 2:1 ratio (600 mg and 300 mg) serve best in the daily diet. Similarly the ratio between omega-6 and omega-3 should be 2:1. Coenzyme Q_{10} or antioxidant ubiquinone intake can be increased with eggs, nuts, beans, fish, and plant oils. This is for quenching free radicals and improving energy production and proper metabolism. Coenzyme Q_{10} is vital to the functioning of our cells' powerhouse, the mitochondria, and its electron-transport chain. It is a fat-soluble vitamin-like molecule present in heart, liver, kidney, and pancreas, organs that are routinely involved with maximum energy production by cellular respiration and gene expression on demand. It appears that E. C. Segar rightfully conceived in the 1930s the power of spinach for Popeye, the sailor.[161] I should equate it to mitochondrial power for reasons of its vitamin, mineral and antioxidant values..

Tabulated information has been classified as ailment and common food, key nutrient and common food, and phytochemicals and common food for the explicit purpose of emphasis, even at the risk of duplication and redundancy. The author believes that readers will find the classification useful.

6.13 Major Reference Institutions

The following institutions are great information sources on minerals, vitamins, and food nutrients in general including antioxidants.

- US Food and Drug Administration for Dietary Guidelines for Americans 2010, 2012, and the new one to come in 2015. http://www.health.gov/dietaryguidelines/dga2010/DietaryGuidelines2010.pdf

---

159 http://www.scgcorp.com/pdf/scg_written_11.pdf.

160 http://www.ncbi.nlm.nih.gov/pubmed/22957669.

161 http://ods.od.nih.gov/factsheets/VitaminB12-QuickFacts/.

- Cancer Research Center of Hawaii, University of Hawaii, Honolulu, Hawaii, USA
- Baylor College of Medicine, USDA Children's Nutrition Research Center, Houston, Texas, USA
- American Cancer Research Institute

CHAPTER 7

Food and Disease Connection

"The doctor of the future will give no medicine but will
interest his patients in the care of the human frame, in
diet and in the cause and prevention of disease."
—Thomas Edison

D isease is a malfunction of any one of the body systems or the system's organ. Since food maintains the body's structure and function, it is tied to all malfunctions and a majority of all chronic diseases. Although relevant science has been accumulating for the last thirty years, the links between diet and disease became well-accepted during the early nineties by way of the Food and Drug Administration's issuing of thirteen major diet-based health claims on retail food nutrition labels. Regular exercise, physical activity, a plant-based diet, and a stress-free lifestyle were revealed by extensive scientific and clinical research in many cohort studies to prevent heart disease and many types of cancers, but only when we choose to live on low-refined sugar, low-saturated fat, no-trans fat, low-sodium, low-cholesterol, and low-total calorie diets in regular and routine complement with exercise and a physically and mentally active lifestyle.

Five hundred thousand Americans suffer from chronic fatigue due to immune dysfunction. Chronic diseases such as diabetes, cancer, high blood pressure, heart disease, and lung diseases are now more common than ever before. Diabetes, it is now proven, is the mother of many chronic diseases [162]. It leads to heart disease, high blood pressure, blindness, kidney disease, nervous system problems, and dental disease. Seven percent of all Americans

---

162 http://www.health.ny.gov/diseases/conditions/diabetes/

die of chronic diseases. Diet alone, it is well-documented now, can not only prevent 50 percent of such chronic diseases, but also reverse them.

We now know that the wrong foods and contaminants (industrial chemicals and additives) present in them by way of additives and chemicals can cause DNA damage and change gene expression by epigenetic effects. Foods that we choose to eat must maintain and enhance genomic and epigenetic stability, and thus good health for a longer life span; they must maintain the stability of our mitochondrial DNA. Such foods are methylation foods with appropriate B vitamins, minerals, proteins, and omega-3 fatty acids.

The science underlying the diet and disease connection has been in the making over the last three to four decades. Five of ten causes of death have been linked to diet (US Surgeon General, 1988, http://profiles.nlm.nih.gov/NN/B/C/Q/G/), and seven out of ten deaths today are due to chronic diseases [163]. That our diet can be an effective prevention against chronic disease is great news because a majority of problems of long-term disability and health-care cost, it is now documented, can be solved by our daily diet and nutrition [164].

Current statistics (in millions of people) on the incidence of various diseases include 46.0 for arthritis, 27.0 for osteoarthritis, 5.0 for fibromyalgia, 23.6 for diabetes, 26.6 for heart disease, 12.5 for cancer, 10 for osteoporosis, 5 for rheumatoid arthritis, and 3.0 for glaucoma. This totals to a high 158.7 million, or 49.6 percent of the US population.[165,166] The figure goes even higher if Alzheimer's, Parkinson's, autism, ADHD, depression, and chronic pain incidences are included. It is now widely believed that diet-based remedies should be an integral part of health maintenance and health-care policy.[167,168] We have to modify foods that contain unwanted ingredients like high fat, high trans fat, high sugar, high salt, and high cholesterol. Trans fat should be completely eliminated, and omega-6 fatty acids consumption reduced.

Good diets must include fruits, vegetables, dietary fiber, plant fats, oils and fats containing monounsaturated fatty acids and vitamin E, and essential fatty acids omega-6 and omega-3 in a 2:1 ratio. Consumption of

---

163 http://www.cdc.gov/chronicdisease/overview/index.htm?s_cid=ostltsdyk_govd_203

164 http://www.who.int/dietphysicalactivity/publications/trs916/summary/en/

165 http://www.cdc.gov/chronicdisease/stats/.

166 http://www.cdc.gov/chronicdisease/.

167 http://www.ncbi.nlm.nih.gov/pubmed/17824853.

168 http://www.euro.who.int/__data/assets/pdf_file/0005/74417/E82161.pdf.

fruits, vegetables, whole grains, plant fats, nuts, and seeds adequately ensures availability of vitamins A, C, E, and B, along with minerals, antioxidants, and phytonutrients. Antioxidants quench free radicals and prevent DNA damage, and phytonutrients prevent cancer. Such a selection of diet will be not only in compliance and agreement with all four US laws listed in table 7.01, but will also constitute what I hope to establish as the methylation diet. Table 7.02 is more specific and categorizes chronic diseases as diet-dependent and diet-protectable. It is too bad that we came to realize all this rather late, and a good deal of science has been misleading. The recent, and not-too-old chronology, I emphasize, of these laws is a good indicator of our poverty in understanding the critical role of nutrition for public health.

Table 7.01 Chronology of Major US Food Laws

Nutrition Labeling and Education Act (1990)
Dietary Supplement Health and Education Act (1994)
Food Modernization Act (1999)
Bioterrorism Act of June 12, 2002
Food Safety Modernization Act (2011)

Table 7.02 Diet-Dependent and Diet-Protectable Chronic Diseases

Disease	Diet-Dependent	Disease	Diet-Protectable
Diabetes	Very diet-dependent	Rheumatoid arthritis	Diet-protectable
Cardiovascular diseases	Very diet-dependent	Glaucoma	Diet-protectable
Cancer	Diet-dependent	Compromised immune system	Diet-protectable
Osteoporosis	Diet-dependent	Chronic pain	Diet-protectable

7.01 Major Chronic Disease and Diet Connections

Dental caries and diabetes came about by high sugar consumption in the early seventies. Cardiovascular diseases became rampant with high total and saturated fat consumption by the early eighties. Trans fat became a recognized problem because of excessive use of artificially hydrogenated fats in our fried foods and sweet snacks. Food scientists, it now appears, forgot that nature is largely CIS (not TRANS). The high blood cholesterol problem stemmed mainly from excessive use of meat, egg, and dairy-based products.

A low-fat, low-calorie, low-cholesterol, low-saturated and trans fat diet with proper doses of vitamin A, vitamin C, vitamin D, vitamin E, folic acid, B complex, selenium, zinc, dietary fiber, and phytochemicals such as carotenoids, coumarins, dithiolthiones, flavonoids, glucosinolates, indoles, isothiocyanatesm phenolics, and terpenes, we now know, can boost our health and longevity. These are true components of the methylation diet based on common grocery store foods. Although the major focus lately has been on cancer and heart diseases, other diseases that are part of twelve health claims listed in table 7.03 include dental caries, neural tube birth defect, and osteoporosis. The remedial dietary ingredients are antioxidants, dietary fiber, trans fat-free unsaturated fat, and vitamins including water-soluble B vitamins and antioxidant vitamins C, E, and K. We can secure the health values of these claims right in our kitchen by selecting proper foods. The cure is right in our kitchen.

Table 7.03 Major Health Claims Approved by US Food and Drug Administration

Calcium and vitamin D for osteoporosis, 21 CFR 101.72 (2008)
Dietary fat and cancer, 21 CFR 101.73 (1993)
Sodium and hypertension, 21 CFR 101.74 (1993)
Dietary fat and cholesterol/coronary heart disease, 21 CFR 101.75(1993)
Fiber, Fruits, and Vegetables and coronary heart disease, 21 CFR 101.76 (1993)
Fruits and vegetables and coronary heart disease, 21 CFR 101.77 (19993)
Fruits and vegetables and cancer, 21 CFR 101.78 (1993)
Folic acid and neural tube defect, 21 CFR 101.79 (1996)
Carbohydrates and dental caries, 21 CFR 101.80 (2008)
Soluble fiber and coronary heart disease, 21 CFR 101.81 (1997)
Soy protein and coronary heart disease, 21 CFR 101.82 (1999)
Stanol/Sterol and coronary heart disease, 21 CFR 101.83 (2010)
Oatrim (2003), psyllium (19996), and oats (1997)

7.02 Fetal Connections of Health and Wellness

Diet, environment, and lifestyle influence our well-being and our response to medicine from one generation to the next. Actually our health and well-being begins the day we are conceived because the mother's diet, environment, stress level, emotions, and exposure to carcinogens affects the health of the fetus, which shares everything with the mother.[169,170] *Nine month of gestation is a very critical period, as it determines the wiring of the fetal brain, functioning of vital organs such as heart, liver, and pancreas, susceptibility to disease, appetite, metabolism, intelligence, and even temperament.* Chemicals and toxins in the air that a mother breathes affect the health and propensity to chronic disease of the offspring. *The stress and mental state that a mother has to live with seems to be responsible for intelligence, temperament, and behavior of the offspring.* Scientific work has exploded in areas of fetal origin of arthritis, osteoporosis, cognitive decline, mental illness, cancer, hypertension, diabetes, cardiovascular disease, allergies, asthma, and obesity. Pregnancy is a frontier for scientists, and it needs to be researched and managed wisely for a healthy next generation [171], [172]. Mitochondrial DNA comes from the mother, and too many chronic diseases today can be traced to mitochondrial DNA.[173] Major fetal connections are

1. Heart disease: Individuals weighing less at birth are prone to having heart diseases in later years because the undernourished fetus diverts nutrients to the brain first.
2. Obesity: Obese mothers pass on their condition to their offspring, and the effect seems to be nongenetic. The seeds of obesity begin, it appears, right from the mother's uterus.
3. Diabetes: High blood sugar of a diabetic mother disrupts the metabolism of the fetus and predisposes the offspring to diabetes.
4. Schizophrenia: Mothers under stress or under undernourished conditions may have have their babies develop schizophrenia.
5. Depression: Depressed women give birth to premature babies with lower weight. The mother's mood can be responsible for sensitivity of the fetus to stress.
6. Mental illness: The intrauterine environment seems to be a major pathway to mental illness. Intrauterine factors are part of the epigenetic modifications that affect gene expression without altering DNA.

---

169 http://learn.genetics.utah.edu/content/epigenetics/nutrition/.

170 http://content.time.com/time/magazine/article/0,9171,2021065,00.html.

171 http://www.ncbi.nlm.nih.gov/pubmed/21684471

172 http://blogs.plos.org/neuroanthropology/2010/09/30/fetal-origins-in-the-womb-in-the-news/

173 http://www.ucsdbglab.org/mmdc/brochure.htm.

7.03 Common Dietary Constituents of Therapeutic Value

Plant and animal foods have medicinal value. We have evolved by eating many of these plant-food materials. Our digestive canal is designed for high-fiber plant foods. The origin of this medical value stems from the origins of antibiotics, peptides, and alkaloid drugs from plants and marine animals; cancer-fighting drugs from filter feeders that stick to rocks and corals; drugs from microalgae and cyanobacteria; migraine treatment by preparations from frog skin; the cancer drug Taxol from the Pacific yew tree; and old-fashioned aspirin. The sun is the original source of energy for us because we depend for our nutrition and medicine on plant products and plant metabolites. Only plants harvest the sun's energy directly. This relationship, the bedrock of evolution, is only poorly communicated to the consumer. The American food industry needs to educate consumers and follow US laws meant exactly for this purpose with due diligence. We need better food products that straddle more than one category and more than one aisle in the retail food store. Table 7.04 offers excerpts from the FDA's qualified health-claim data on therapeutic foods and nutrients against cancer, heart disease, diabetes, and atopic dermatitis that clearly suggest that such foods can and do maintain and promote good health.

Table 7.04 Qualified FDA Claims

Cancer	Heart Disease	Miscellaneous
Antioxidants and vitamins	Calcium	100% whey protein hydrolysate for atopic dermatitis
Selenium	Nuts	
Green tea	Walnuts (vitamin E)	Chromium for diabetes
Calcium and colon cancer	B vitamins, B_{12} and folic acid	Fluoridated water for dental caries
Tomato sauce and cancer	Omega-3 fatty acids	
Whole grains	Monounsaturated fat/oil	
	Corn oil	
	Calcium for hypertension	
	Whole grains	
	Potassium for hypertension	

Consumption of foods even with qualified claims must begin early in childhood, and we need to promote kid-friendly foods, enforcing in the children's minds good dietary habits from the very beginning. We should design a lifelong lifestyle for children inclusive of exercise, physical activity, yoga, meditation, and social interaction integral to day-to-day living.

Such a food-consumption pattern can help manage health by maintaining healthy weight and a stress-free life. The pattern should include consuming monounsaturated and polyunsaturated oils (flaxseed, walnut, avocado, olive), antioxidants, and vitamins like niacin (vitamin B_3), and doing everything possible under a physician's advice to raise HDL to a range of 40 to 60 mg/deciliter because it is critical for heart and brain health. Our daily food should be antiangiogenic, anticarcinogenic, and anti-inflammatory; it must contain common adaptogens that help control anxiety and stress. We should consume foods that are highly hydrated, and as such, offer stress control, optimum energy, a balanced weight, and an enhanced mental health. Again, these are foods for proper gene expression. Such foods are part of the methylation diet.

7.04 Major Reference Institutions

The following institutions of repute have issued many reports on diet-disease connection.

World Health Organization; Council of Agricultural Science and Technology; Codex Alimentarius; US Food and Drug Administration's Health Claim Approvals; Institute of Medicine in United States; Health and nutrition claims by European Union; and foods with nutritional function and foods for specified health uses claims in Japan followed FDA in USA. The FOSHU (Foofd for Specified Health Use) system on dietary health claim is oldest of all such inventions [174]

---

174 https://cspinet.org/reports/functional_foods/japan_regltry.html

CHAPTER 8

Foods That Build Immunity

"The immune system, connected directly to our brain and
every other cell in our body, routinely defends our health."
—Triveni P. Shukla

One of the routine functions of food is to maintain the health and rebirth
of our immune cells in the bone marrow, all 1 million of them.[175] Food
is tied closely to their rebirth in the bone marrow. Born every week, the new
cells should be capable of recognizing all 10 trillion cells of our body and
any and all foreign and external protozoa, bacterial, fungal, or viral invader.

Our worst enemies are microscopic organisms like viruses, bacteria,
fungi, and parasites. They can get to our cells via the air we breathe, the
water we drink, and the foods we eat. However, we have evolved with a
formidable cellular defense system called the immune system—a marvel of
cellular and molecular recognition, intelligence, and defense works, and a
kind of sensory control. The lymphatic system, the pathway of immunity, is
one way drainage via heart's blood circulation system without a contractile
heart like pump. It is our defense to diseases caused by bacteria, protozoa,
fungi, and viruses above and beyond the physicochemical defense systems
of skin, stomach acid, and enzymes in our tear, saliva, and intestine. Foods
we eat, our state of mind, and the environment around us have a definite
effect on our immune system's ability to detoxify, destruct, and defend. A
number of bad things happen, including hormonal dysfunction, immune
suppression, and chronic degenerative diseases when it fails. Advanced
age, overuse of corticosteroids, and HIV subvert and subdue the immune

---

175 http://faculty.ccbcmd.edu/courses/bio141/lecguide/unit4/innate/bloodcells.html.

system. Balanced diet, a daily dose of exercise for moving the lymph-muscular system, daily sleep, and freedom from pollution are necessary for an enhanced and functional immune system, which is just as important as breathing, digestion, and metabolism. As a matter of fact, nutrition and immunity are closely linked, and our digestive tract in and of itself is a large immune organ.

Foreign invaders like bacteria and viruses provoke the immune system. Tissues and organs get inflamed and produce prostaglandins and histamines, which in turn tell neutrophile white blood cells to engulf the pathogens. Lymphocytes that are produced by stem cells are of two types: B cells that produce antibodies on demand that neutralize the invading organism, and T cells that attach directly by lymphokines and kill the pathogen. Each B cell produces a specific antibody. Lymphocytes circulate in the blood, and are present in the bone marrow, spleen, thymus (mature T cell), lymph nodes (B cells), and lymph per se. We succumb to autoimmune diseases and allergies when the immune system attacks our own body. B cells and T cells remember all prior invaders, but in the case of autoimmune diseases, they see our own organs as foreign. Most of the time, our immune system knows all our 10 trillion cells and their functions because B cells have memory.

This chapter is about how to maintain and enhance natural immunity by our daily foods, methylation food in particular, that help gene expression and build the immune system for better health. The mitochondria in our cells are involved in many diseases.[176]

The human body is protected by three types of immunity: (1) natural immunity codified in our DNA, (2) active or 500-million-years-old adaptive immunity that develops throughout our lifetime, and (3) passive or borrowed immunity from mother's milk. We get in trouble when our immune response is less than total and complete.

The immune system has different kinds of cells. White blood cells are present in the thymus, spleen, and bone marrow. Then there are neucleophiles, which chew up bacteria. Also, there are B-cell and T-cell lymphocytes that remember all prior invaders. Whereas B cells of bone marrow constitute a critical intelligence system, T cells of the thymus gland are soldiers that kill by destroying foreign organisms. B cells produce

---

176 http://www.nature.com/scitable/topicpage/mitochondria-and-the-immune-response-14266967.

antibodies that lock on to antigens (invader organisms) and neutralize toxins. Immune-system problems can be many and varied.

1. We may suffer from immune deficiency. Sometimes medicines we take can cause this.
2. We may develop autoimmune diseases because our own immune system attacks our cells. Examples are lupus, rheumatoid arthritis, type 1 diabetes, celiac, allergies, and asthma. Autoimmunity, a matter of complex inheritance, is controlled by multiple genes and environmental factors.[177]
3. We may succumb to allergies when our immune system becomes over-responsive.
4. We may suffer from cancer of the immune system (leukemia and lymphoma).
5. We may have free radical-damaged DNA as we age under a stressful environment.
6. We may not have the full complement of our second genome via probiotics.

Regulators of embryonic origin, interferons and interleukins, modulate our immune system, which is much more than just fighting against pathogens and foreign antigens. It keeps the proper balance of inflammation by controlling master molecules C_6 and C_{10} cytokines, by repair, and by growth; balance and the response depends on our daily nutrient intake.[178]

We should consider our immune system to be our body's various units of military. The entire system of organs (thymus, bone marrow, and lymphoid tissue, a network of cells, and tissues and organs) defends us from pathogens in a very organized manner. The immune system has to recognize millions of enemies quickly, differentiate them from our own cells, kill them by antibodies or natural killer T cells, and communicate about it all the time.[179] T cells can edit their own DNA, are responsible even for cognition, and communicate to the brain by thousands of signals. All begins with a protein molecule and encompasses the sets and subsets of cells. Our immune system must recognize an infection, produce invader antibodies and killer cells, and

177 http://www.ncbi.nlm.nih.gov/pmc/articles/PMC1480569/.
178 http://jn.nutrition.org/content/126/6/1515.full.pdf.
179 http://jonlieffmd.com/blog/immune-t-cells-are-critical-for-cognitive-function.

direct them to the target site. It has to recognize billions of foreign proteins. It needs to kill only the bad cells and not the good cells of our own. Most importantly our immune system must know each and every one of our 10 trillion cells and their unique functions and responsibilities.

Skin, mucus membranes, and enzymes in our tears, enzymes in nasal passages, eyes, and respiratory and digestive tracts are the first line of defense, providing a physical barrier against bacteria and viruses. The immune system with its novel sensory and recognition capability is our second line of defense. The system detects pathogens by molecular mechanisms embodied in 10^{15} different receptors on immune cells. One lymphocyte can bind many pathogens, and it regenerates constantly to kill invaders. It learns about the invaders very quickly. Actually B cells have memory.[180] The immune system includes the thymus, spleen, lymph system, bone marrow, white blood cells, antibodies, complement system, and hormones. Internally, specialized white blood cells fight antigens that make it past the skin. To emphasize, there are three types of immune cells.

1. T-lymphocytes continuously patrol the body in search of antigens; they fight infections. NK (Natural killer) cells are born as killer cells. NK phagocytes roam all through the bloodstream, scavenging what they recognize to be foreign and then killing them.
2. B-lymphocytes manufacture antibodies, special blood proteins that neutralize or destroy foreign bacteria and viruses.
3. Neutrophile and macrophages scavenge antigens from the blood for delivery to the lymphatic system, which disposes of them.

In order to work smoothly, though, these cells depend on the good health of our body. Essential nutrients from a balanced diet are critical for the production and maintenance of key germ-fighting cells in the immune system. A balanced diet has an effect on vascular function, and in turn the blood flow, dependent on the immune function. Circulating blood from the heart to every other cell is the route along which infection-fighting cells travel throughout the body to where they're needed. We need a diet, say the methylation diet, that is balanced and RDA (Recommended Daily Allowance)-wise complete, rich, and sufficient.

---

180 http://www.ncbi.nlm.nih.gov/pubmed/23660557.

8.01 Symptoms of a Poor Immune System

Sinus congestion, colds and allergies, bad breath, excessive flatulence, constipation, red itchy eyes, skin rashes, indigestion, day-to-day stress, and fatigue are signs of a weak or compromised immune system. Fifty to 70 percent of our immune system is part of the digestive system. *What we call probiotics is, in fact, immunobiotics.* We need probiotic bacteria, the source of our second genome, in our colon to maintain a balanced and strong immune system. Allergies and autoimmune diseases are caused by overly active immune systems, notwithstanding the fact that the liver, kidney, lungs, and gastrointestinal tract work 24/7 for detoxification. Autoimmune diseases are often due to out-of-control response. Immune-compromised people succumb easily to heart disease, HIV (human immunodeficiency viruses), and diabetes.

We must have a complete understanding of how robust or weak our immune system is. Our doctors should give us information on blood count (red blood cells, white blood cells, platelets, and red blood cell-to-blood volume ratio), IgG antibodies against common allergens, T Cells (CD_4 and CD_8), suppressor T cell count, chemistry of blood, liver and kidney tests, urine tests for checking diabetes, infection, autoantibodies, and blood culture. When all is good, our methylation diet and lifestyle can boost the immune system.

8.02 Genetics of the Immune System

A good immune system helps improve longevity, transplant capability, obesity control, and even phobias. Multiple sclerosis, type 1 diabetes, and rheumatoid arthritis—common autoimmune diseases—are chronic because genes behind the immune system cell receptors don't work properly.[181] Cancer, in particular, is a disease where cells begin to multiply out of control. This happens due to DNA damage. In the not-too-distant future, gene therapy may become an option for boosting the immune system. A good example is gene P53, a tumor-suppressor gene. This can be inactivated by simple aging, environmental triggers, metals, chemicals, radiation, smoking, and physical inactivity. Our daily food is involved in turning off tumor genes P16, MGMT, DAPK, RASSF1A, GATA4, PAX5a, and PAX5β.

---

181 http://www.plosgenetics.org/article/info%3Adoi%2F10.1371%2Fjournal.pgen.1000322.

Gene TP53, the guardian of our chromosome, on the short arm of chromosome 17, encodes tumor protein p53, which is responsible for tumor suppression and regulation of the cell cycle. *Red meat and high consumption of refined sugar suppress TP53, causing breast and colon cancer.*[182] The mitochondria are directly involved in antiviral immunity.[183]

8.03 Lifestyle for Boosting Immunity

The first line of defense is simple sanitation in day-to-day living. We need to boost white blood cell production by permanent habits of sanitation, bathing, hand washing, exercise, daily water intake of two liters per day, daily intake of fruits and vegetables as prebiotics and yogurt and sauerkraut as probiotics. Exercise and eight hours of sleep must be part of the chosen lifestyle.

8.04 Immune-Boosting Foods

Common immune-boosting foods include carrot and lentil soups, roasted vegetables, blueberry smoothies, curry-glazed tofu, orange-glazed sweet potato, and mushrooms and spicy tomato sauce. Vitamin A comes from yellow and dark vegetables. Vitamin C comes from watermelon, citrus fruits, and berries. Vitamin E comes from almonds and whole grains or wheat germ, and minerals selenium, zinc, and iron come from a good choice of nuts, grains, and legumes. The nutrients in these foods constitute a methylation diet necessary for a good immune system. Other common foods of choice are broccoli for glutathione, NK cell-boosting coenzyme Q, and vitamins A and C[184]; cheese for vitamin D and anticarcinogens; kale, mushrooms, spinach, cabbage, cauliflower, carrots, cinnamon, and red bell pepper for B vitamins; grapefruit for vitamin C; wheat germ for zinc and choline; and pumpkin seed for high-tryptophan protein.

Colored fruits and vegetables provide antioxidants, flavonoids, isoflavones, lycopene, dietary fiber, and other phytochemicals. Zinc (legumes, meat), selenium (Brazil nuts, oysters, fish), omega-3 fatty acids (flaxseed and salmon), probiotics (yogurt, sauerkraut), prebiotic dietary

---

182 http://www.bioinformatics.org/p53/introduction.html.
183 http://www.ncbi.nlm.nih.gov/pmc/articles/PMC3193288/.
184 http://www.sinovedic.in/role-of-nutrition-.

fibers (fruits, vegetables, barley, oats), vitamin A (sweet potato), amino acid cysteine (for acetyl cysteine functionality) from chicken soup spiced with garlic and turmeric, and antioxidants (dark chocolate and green tea) should be included in our weekly diet for antioxidative, anticarcinogenic, and immune enhancing effects. A switch to spicy foods for flavor and taste is advisable because spices and condiments added to our daily food provide anticarcinogens, antimicrobials, antioxidants, and antistress adaptogenic foods. Food is involved in cytokine production that regulates interferons. Great examples are yogurt and sauerkraut, containing at least a billion lactic acid bacteria per gram. *Such foods can help produce interferon alpha, beta, gamma, and beta-interlukin, which promote tumor death.*[185]

Use of vitamin D and multivitamin tablets as supplements is advisable for senior citizens. We need to avoid deficiencies of zinc and selenium. Table 8.01 emphasizes daily consumption of dietary fiber, probiotics, omega-3 fatty acids, B vitamins including B_{12}, carotenoids, vitamins A, C, E, and K, minerals, iron, and antioxidants leutin, zeaxanthin, polyphenols, and phytochemicals like lipoic acid, isocyanates, and flavonoids. Almost every food listed in this table is nonexotic and easily available. Necessary for gene expression, they have been part of our food chain for centuries. Notable examples are chickpea hummus in Israel and the Middle East, herbs and spices in the Orient and Latin America, and fruits and vegetables in tropical countries.

Food for multiple sclerosis (MS) patients includes vitamin D_3; a well-balanced, low-fat diet, low in saturated fat, full of fiber and colorful fruits and vegetables is likely the best place to start. B_1, B_{12}, folic acid, coenzyme Q_{10}, iodine, creatine, and omega-3 are musts for an MS diet. Add yogurt and sauerkraut to the daily diet. Use a variety of a lot of green vegetables recommends Mayo Clinic of Rochester, Minnesota.[186] Immune system-wise, type 1 diabetes and multiple sclerosis are almost the same.[187] Vitamin D_3, omega-3 fatty acids, and low-fat high-fiber foods, essentially a methylation diet, are good to combat multiple sclerosis.[188,189,190]

---

185 http://ajcn.nutrition.org/content/73/6/1142S.long.

186 http://www.overcomingmultiplesclerosis.org/html/blob.php?attach=true&documentCode=6708&elementId=20084.

187 http://www.sciencedaily.com/releases/2001/03/010322074643.htm.

188 http://www.overcomingmultiplesclerosis.org/html/blob.php?attach=true&documentCode=6708&elementId=20084.

189 http://circ.ahajournals.org/content/9/3/335.

190 http://www.thelancet.com/journals/lancet/article/PII0140-6736(90)91527-H/fulltext.

Also, we need to realize that eating is a process. Chewing, the upstream process, activates, extracts, and releases phytochemicals. Use of quick-roasted foods is simple practice in pasteurization and sterilization and great for flavor development. Use of a hot breakfast is strongly advisable. Lunch can be roasted sesame and ginger over our daily salad of stress-relieving adaptogen astragalus. Fruits and vegetables can be designed into tasty cakes, dips, and icings. Tables in appendixes II, III, and IV have detailed compositions of common foods. Foods with maximum nutrients can be used regularly for designing cakes, smoothies, dips, icings, salads, and soups on a day-to-day basis.

Table 8.01 List of Immune-Building Methylation Foods

Common Foods	Active Ingredient(s)	Ailments and Diseases Helped
Banana	High potassium	Against breast cancer
Bell pepper	High vitamin C and carotenoids	Build immune system, prevent cardiovascular diseases
Brazil nut	Vitamins and minerals	Cell signaling; vitamins C & E and mineral selenium help turn on gene Nrf2, which regulates hundreds of other genes
Chickpeas and red pepper hummus	Iron, vitamin C	Build red blood cells
Dark chocolate	Polyphenols, antioxidants	Prevent DNA damage
Dietary fiber, 30–35 grams/day	Cellulose, hemicellulose, beta-glucan, psyllium, lignin, and inulin	Boost immune system, enhance humoral immunity, increase immunoglobulin concentration, ferment to supply short-chain fatty acid
Green tea, black tea	Antioxidants gallate and epigallocatechin.	DNA repair and gene stability
Flaxseed, walnuts, salmon	Omega-3 fatty acids	Prevent inflammation

Common Foods	Active Ingredient(s)	Ailments and Diseases Helped
Herbs and spices: garlic, cayenne pepper, basil, cayenne pepper, cilantro, mint, oregano, rosemary, turmeric	Phytochemicals	Infection-fighting foods,
Jarlsberg and Gouda cheese	Vitamin B$_{12}$	Active principle is vitamin K against lung cancer
Parsley	Antioxidants, flavonoids, vitamins A, C, and K	Build immune system and enhance white blood cells
Pumpkin seed	Vitamin E	Work with anabolic receptors, protect bladder health
Sweet potato orange vegetables including carrot, squash, and sweet potato	Carotenoids	Eye and bone health health
Squash	Zeaxanthin and lutein	Prevents Non-Hodgkins lymphoma
Fruits & Vegetables: broccoli, yellow squash, red bell pepper, leafy greens, tomato, potato peels, and citrus	Antioxidants, fiber, sulphoraphane, isothiocyanates; are both prebiotic and probiotic	Cancer prevention by arresting cell growth, immune system enhancement, cellular communication, preserve gene expression and digestive health; phytochemicals that bind protein tubulin for cellular mobility
Spinach	Alpha-Lipoic acid	Boosts energy (ATP) production
Cabbage, kale, and cauliflower	Isocyanates	Affect genes that produce telomerase, effective in preventing cancer cell growth

Common Foods	Active Ingredient(s)	Ailments and Diseases Helped
Broccoli, brussels sprout, citrus fruits, garlic, red bell pepper, spinach, and straberries	Bioflavonids	Bioflavonoids block the receptor sites and prevent entry of invaders.

Common foods like prunes, blueberries, strawberries, kale, spinach, brussels sprouts, and red pepper have a high level of antioxidants.

Tables 16.02 and 16.03 in chapter 16 on antioxidant therapy can be consulted for an extensive list of antioxidants in fruits and vegetables and their oxygen radical absorbance capacity values. Vitamins C and E are under rigorous study as antioxidants for cancer prevention by strengthening the immune system.[191] And then there is the endogenous antioxidant glutathione, our body's natural antioxidant. Fruits and vegetables as gastronomic delights of smoothies and soups are the best source of daily antioxidants. Antioxidants, anticarcinogens, anti-inflammatory phytochemicals, and stress-fighting adaptogens from fruits and vegetables are our best defense against a poor immune system.

8.05 Major Reference Institutions

The following institutions have ongoing research in the area of immune boosting effects of common foods.

US Food and Drug Administration; American Diabetes Association; Harvard Immune Disease Institute; Christopher H. Browne Center for Immunology and Immune Diseases; Blizzard Institute, Barts and the London School of Medicine and Dentistry; Autoimmune Disease Research Center, Johns Hopkins.

---

191 http://www.cancer.gov/cancertopics/factsheet/prevention/antioxidants.

CHAPTER 9

Foods That Fight Osteoporosis and Osteoarthritis

"DNA designs our body as a temple; the temple is made
from methylation foods that we must consume daily."
—Triveni P. Shukla

Methylation foods that prevent inflammation, build a good immune system, and provide for daily calcium, magnesium, and vitamin D can help us avoid osteoporosis and osteoarthritis.

Human beings evolved with the best of facilities to move around. Bone supports the body, and bones and muscles in complement provide us with a lever system for moving around. The skull protects our brain. The spinal cord, the pathway for electronic messages, is protected by the backbone. Ribs shelter our heart, lung, liver, and spleen. The pelvis protects the bladder and intestine. Bone building, therefore, is a lifelong process, and human beings have 206 of them to build and repair. A joint is where bones meet. Knees and elbows represent a hinge, the head moves sideways on a pivot, and hips and shoulders represent a ball-and-socket mechanism. All such bones have compositional, structural, and functional specificities. All of them must remain in good health for pain-free living.

Muscles are elastic tissue. Along with tendons, cartilage, and ligaments, they make up our musculoskeletal system, empowering us with physical activity and movement. Cartilage provides flexibility, and ligaments act as fasteners. *Movement of all of our 650 muscles is coordinated by the brain's nervous system and the cerebellum.*

Osteoporosis and rheumatoid arthritis are autoimmune diseases. Poor health of bones and joints stems from chronic inflammation caused by a hormone called prostaglandin 1, and it is perpetuated by the enzyme carboxygenase 2 (COX-2). Anti-inflammatory substances inhibit COX-2 without inhibiting COX-1, the good prostaglandin. Over-the-counter remedies against joint pains are aspirin, Tylenol, Vicodin, and ibuprofen. However, the best long-term remedy is a good diet that supplies vitamins D, K, B_6, and B_9 or folic acid and minerals calcium, magnesium, boron, and zinc. Such methylation foods can inhibit the COX-2 enzyme and promote COX-1 for better bone health,[192] including nutrients like flavonoids, calcitrol (vitamin D), and vitamin E.

The health of bone marrow, which comes from its own stem cells, is critical to bone health also. This is where blood cells are made. Regular exercise can help reduce pain from osteoporosis, and so can a stress-free lifestyle. Omega-3 fatty acid nutrition is a key factor in managing prostaglandins, which underlie all inflammation. Vitamin B_6 prevents homocysteine formation for optimal collagen formation. Genes behind vitamin D, collagen production, and production of estrogen receptors are involved in bone health and, therefore, in osteoporosis. Mitochondrial gene deletions may be a cause of osteoporosis.[193,194]

Two hundred million people around the world are affected by osteoporosis. In the United States, 10 million people are diagnosed as affected, 27 million suffer from osteoarthritis, and 34 million have low bone density.[195,196] Two of every three diabetic people have osteoarthritis. Five million Americans seriously suffer from fibromyalgia. This group of chronic diseases costs $46 billion a year today, and it is bound to soon rise to $1 trillion a year.[197]

9.01 Osteoporosis

Postmenopausal women are most vulnerable to osteoporosis, which represents weak, fragile, and porous bones of the hip, spine, and wrist. The cause is low bone density. In the case of advanced osteoporosis, overactive thyroid and kidney diseases are also common.

---

192 http://www.chiro.org/nutrition/FULL/Natural_COX-2_Inhibitors.shtml.

193 http://www.ncbi.nlm.nih.gov/pubmed/10501795.

194 http://www.ncbi.nlm.nih.gov/pubmed/21762117.

195 http://www.aaos.org/about/papers/position/1113.asp.

196 http://www.healthline.com/health/osteoporosis-2011?toptoctest=expand.

197 http://www.sciencedirect.com/science/article/pii/S1361311102001309.

9.02 Genetics of Osteoporosis

Research shows that osteoporosis may be an autoimmune disease. Autoimmune effects arrest formation of the *protein osteoprotegenin,* which builds bone mass. Rheumatoid arthritis is also an autoimmune disease, and it is linked to development of osteoporosis. Both regulatory and structural genes that control bone mass are involved in the development of osteoporosis. It is a polygenic disease possibly involving the estrogen receptor gene, vitamin D receptor gene, and genes for collagen production. So we are born with genes for osteoporosis. There is a lot of research in this area, but at the present, our genome-wide understanding of genes involved is minimal.[198] Epigenetic factors also play critical roles. Vitamin D deficiency and poor physique of the mother can lead to osteoporosis in offspring. DNA methylation and inappropriate gene expressions, we are now finding out, may be the major causes. In other words, the methylation diet matters in enhancing bone health.

9.03 Risk Factors

Risk factors include the following:

- excessive alcohol consumption
- low calcium intake (which should be no more than 600 mg per day). Excessive calcium supplementation is a bad idea.
- lack of exercise
- excessive use of antacids and steroids

9.04 Lifestyle

Our lifestyle must include daily exercise, a lot of physical activity, proper weight for a given height, and low stress. Meditation, yoga, or tai chi are known to reduce pains of osteoporosis and even fibromyalgia.[199]

---

198 http://www.hebrewseniorlife.org/research-genetics-of-osteoporosis.
199 http://www.arthritistoday.org/arthritistreatment/natural-and-alternative-treatments/meditation-and-relaxation/meditation-eases-symptoms.php.

9.05 Diet for Bone Health

The key to reducing osteoporosis and improving bone health is calcium and vitamin D nutrition, along with proper intake of the minerals boron, magnesium, and zinc.[200] However, overdosing with calcium above 600 mg per day is not advisable because it tends to promote osteoporosis when consumed in excess. Calcium should come from dried herbs as opposed to supplements. *Good examples of such herbs are savory (2,132 mg/100 g), celery seed (125 mg per 100 g), and thyme (57 mg per 100 g).* Sesame seed, almond, flaxseed, corn tortilla, yogurt, green leafy vegetables, broccoli, kale, collard greens, Brazil nuts, whey powder, whole-wheat bread, chili powder, and fortified cereals are other sources of calcium. *Broccoli is great for fighting osteoporosis because of vitamins C and K and minerals potassium and calcium.* We should consume asparagus, green peas, broccoli, lettuce, cabbage, and spinach for vitamin K, vitamin B_6, and folate for reducing problems of osteoporosis. Our daily requirement of vitamin D of 600 IU or 15 mcg can be met easily by proper choice of foods.

Cod liver oil, an old well-known remedy, is a great source of vitamin D, followed by eggs, milk, tofu, mushrooms, oysters, caviar, and fish. Vitamin D promotes absorption of calcium and phosphorus in the intestine. Vitamin B_6 inhibits homocysteine formation and thus promotes collagen formation. Minerals other than calcium necessary for bone health are magnesium, boron, and zinc. Almonds, apples, and peanuts supply boron, a mineral required in very small amounts but necessary for bone health. Lentil protein, gamma linoleic acid, and omega-3 from salmon, flaxseed, and walnuts are good traditional therapies for preventing inflammation. Other foods that help are citrus and lime for vitamin C, Brazil nuts for selenium, onions, leeks, and green tea. In essence, these are the nutrients of a methylation diet. Vitamin supplements are advisable if dairy products are not part of the daily diet. A few traditional remedies against osteoarthritic pain are:

1. An ayurvedic mixture of turmeric, ginger, Boswellia serrata resin gum, and Indian ginseng asvagandha has been found just as good

---

200 http://www.mayoclinic.com/health/osteoporosis/DS00128/DSECTION=prevention.

as glucosamine and chondroitin sulfate in reducing osteoarthritic pain.[201]

2. Five grams/day of candied ginger can reduce osteoarthritic pains.

3. A beverage made from orange juice, ginger, and pureed sweet potato is also helpful.

4. A ginger lemonade made with grated ginger in honey and lemon juice diluted with water to taste is another pain-reducing beverage.[202]

5. Ginger can be used in many ways in our daily diet. Examples are a ginger-based salad dressing made out of ginger, soya sauce, olive oil, and garlic and sauteed vegetables with grated ginger as salad. Furthermore, ginger extract can go in many household recipes.

6. Ginger and turmeric can be designed into beverages, bars, chews, and fruit loops. Such recipes should include black pepper for bioperine and turmeric for cucurmin. Nuts and seeds can serve as a source of zinc and boron.

Table 9.01 details the effects of key ingredients of many common foods that can be used in day-to-day recipes for bone health. Examples are vegetable soup made out of leafy vegetables and avocado spiced with black pepper, turmeric, and ginger. This then supplies vitamins K, C, folic acid, calcium, and other minerals. Black currants and prunes can be used as digestive aids that also increase bone density. Coenzyme Q_{10} from chicken, meat, and fish and the mineral magnesium can improve muscle and bone health. Roasted seeds and nuts can be used as routine snacks for a daily supply of minerals, antioxidants, and tryptophan.

Table 9.01 Foods for Bone Health

Food	Ingredient	Ailments and Diseases Helped
Avocado	Saponifiables	Promote cartilage production
Bark extract	Phytodolor, twice as effective as aspirin[203]	Nonsteroidal anti-inflammatory agent; suppresses enzyme involved in swelling

---

201 http://www.integrativepractitioner.com/article.aspx?id=19384.

202 http://umm.edu/health/medical/altmed/herb/ginger.

203 http://rheumatology.oxfordjournals.org/content/40/7/779.full.

Food	Ingredient	Ailments and Diseases Helped
Black currant	Gamma linoleic acid, vitamin C, and antioxidants	Improves bone health
Black pepper	Bioperine	Reduces pain
Bosswella serrata	Oil and terpenoids	Cures osteoarthritis
Chicken, beef, trout	Coenzyme Q_{10}	Promotes muscle health
Compfrey ointment	Allantoin	Reduces inflammation, promotes growth of healthy tissue
Dark chocolate	Flavonoids	Reduces pain
Hop extract from beer	Perluxan	Fast pain reliever
Olive oil	Oleocanthal	Anti-inflammatory agent
Pine bark extract	Pycnogenol	Prevents osteoporosis
Prune/dried plum	Boron, vitamin K, antioxidants	Bone density improvement
Red grape skin	Resveratrol	COX-2 inhibitor, anti-inflammatory
Pumpkin and sunflower seeds	Tryptophan	Reduces pain
Turmeric	Curcumin	Reduces pain

Note: Other than pine and poplar bark, the rest of the list includes easy-to-use common foods.

Bioperine from black pepper and curcumin from turmeric in spices and seasoning, black currants and prunes in salads, minerals boron, selenium, magnesium, and zinc from roasted nuts and seeds for snacks, and calcium and vitamin D from green vegetables can be used to design daily recipes for better bone health. Certain mushrooms are a good source of vitamin D. All such foods contain methylation nutrients.

Medications such as methyl sulfonyl methane (MSM), alpha-phenylalanine that sustains endomorphins and enkephalin in the brain, and enzyme nattokinase have been found useful in treating osteoporosis. I have personal experience of reducing knee pain by glucosamine and chondroitin sulfate by a two-month-long use.

9.06 Major Reference Institutions

The following research institutions are involved in leading works on effect of daily diet on osteoporosis and osteoarthritis.

- Melbourne University study with 2,000 women over age seventy years.
- Mayo Clinic treats osteoporosis by diet and exercise.
- University of Maryland Health Center promotes food and exercise.
- Creighton University in Omaha promotes foods in the treatment of arthritis.

CHAPTER 10

Foods That Fight Chronic Pain

"Every nerve that can thrill with pleasure, can also agonize with pain."
—Horace Mann

Pain is born of stress, and stress is a big inflammatory risk factor in the onset of chronic pain. Foods with calcium and magnesium, and adaptogenic foods like astragalus that reduce stress are pain-fighting foods. Exercise, yoga, and meditation for stress-free living greatly enhance the intrinsic value of our pain-fighting daily foods, foods that can help avoid inflammation. Always a protective, pain is ever present as our bodyguard. It is a symptom and not a disease. If anything, it is a disease of the central nervous system affecting sensory, emotional, and motivational pathways of perception and cognition. We now know that external magnetic circuits can reset neural circuits involved in pain and that electronic impulses do mask the body's brain signals.[204]

Ordinary foods like broccoli, blueberries, boiled eggs, cherries, coffee, cranberries, cucumbers, garlic, edamame, ginger, green tea, hot peppers, kale, salmon, sweet potatoes, tomatoes, turmeric, and yogurt, components of a methylation diet, reduce inflammation and, therefore, reduce pain. Magnesium, the muscle mineral, is the second most important muscle-relaxant mineral.[205,206] Both Whole Foods and Science Board are organizations devoted to dietary value of foods we consume daily. It decreases nerve pain

---

204 http://backpain.theworldhealth.org/back-pain-electrical-impulses.html.

205 http:http://www.whfoods.com/genpage.php?tname=nutrient&dbid=75//www.scienceboard.net/community/perspectiv.

206 http://www.scienceboard.net/community/perspectives.185.htmles.185.html.

by settling down the pain-carrying neurotransmitter N-methyl-D-aspartate. Good sources of magnesium are Swiss chord, pumpkin seed, sunflower seed, sesame seed, almonds, cashew nuts, and Brazil nuts.

We know no more today than two hundred year ago about this debilitating inflammatory problem that afflicts 70 million Americans. Brains in pain act differently. Circuits gone haywire cause the brain to register permanent pain, a matter of an electronic impulse. The methylation diet is very helpful in reducing and controlling pain by protecting mitochondrial DNA.[207]

Eighty percent of Americans are magnesium-deficient.[208] Another problem is excessive use of calcium, which contracts muscles and puts them in a state of spasm. A good ratio of calcium to magnesium is 2:1, with respective recommended RDA of 600 and 300 mg per day. The two minerals work together for muscle tone.[209] High calcium supplementation up to 1,000 mg per day during the last twenty years may have been counterproductive.

Pain, unfortunately, is a horrible necessity of life. It protects us by alerting us to things that might cause injury, although some long-term pains have nothing to do with any obvious injury.

Stress, which preoccupies our lives today, is a big factor in experiencing pain. Seventy million Americans suffer from daily pain, more than two thirds of which is lower back pain. Arthritis—rheumatoid or osteoarthritis—affects close to 40 million Americans. Pain leads to anxiety, depression, and body-mind disconnect. The answer to pain thus has to come from examining all these three conditions. Barring clear-cut nerve damage, we know nothing about the why, how much, and when of pain. *The best remedies are stress control, exercise, stretching, and massage.* Anti-inflammatory drugs like Bengay, ibuprofen, Tylenol, and codeine are only temporary fixes. Much more can come from daily nutrition, taming the mind by meditation and yoga, and even by prayer in faith. A proper ratio of 600 mg calcium and 300 mg per day of magnesium seems to be a good dietary remedy.[210] Cortisone injections and surgery should be considered as options of the last resort.

207 http://www.ncbi.nlm.nih.gov/pubmed/24151337.
208 http://www.jigsawhealth.com/resources/magnesium-mineral-deficiency.
209 http://www.ncbi.nlm.nih.gov/pubmed/2153470.
210 http://www.osteopenia3.com/Magnesium-supplement.html.

10.01 Symptoms of Pain

Muscle pain and cramping, loss of appetite, nausea, weakness and fatigue, twitching of eyelids, headaches, migraines, and osteoporosis are good indicators of *magnesium deficiency*. Magnesium resides inside all body cells, and it is involved in many vital functions like energy production by mitochondria, bone and teeth formation, relaxation of blood vessels, heart muscle action, bowel function, regulation of blood sugar, and detoxification. It is strongly advisable to control daily calcium intake. Glutathione, the most powerful antioxidant found in broccoli, in our body works with magnesium. A balanced daily calcium and magnesium intake is key to pain-free muscular health. Other developments to watch for in relation to pain are:

1. Antibody for a protein called nerve growth factor (NGF), which is vital for new nerve growth during development. NGF, it turns out, has a critical role in regulating sensitization to pain in chronic conditions. Although pain sensitivity depends upon our state of mind, NGF is critical to transmission of the pain signal.[211], [212] . Footnote 210 is by American Society for Surgery of the Hand and as an organization it is devoted fundamental disclosures on pain transmission.
2. The gene known as *SCN9A* codes for a protein that allows the channels along which nerve signals are transmitted to remain active for a longer time and thus transmit more pain signals. Variations in *SCN9A* may also explain why some patients prefer different classes of painkillers.

10.02 Genetics of Pain

Differences in individual tolerance and response to drugs clearly suggest that pain has a genetic origin and that pain has a polygenic inheritance. Understanding of the genetic basis of pain today is in its infancy at best. Epigenetic contributions, environmental effects, and changes in gene

---

211 http://www.assh.org/Professionals/ProdsSvcs/journalclub/Pages/ThePotentialRoleof NerveGrowthFactor(NGF)inPainfulNeuromasandtheMechanismofPainReliefbyTheir RelocationtoMuscle.aspx.
212 http://www.ncbi.nlm.nih.gov/pubmed/23157347

expression may be involved. Mitochondrial gene deletions and duplications may be responsible for abdominal and other chronic pains.[213]

10.03 Pain-Fighting Foods

In theory, all anti-inflammatory foods can fight pain. Oranges, apricots, nectarines, tangerines, papaya, peaches, plums, and watermelon have anti-inflammatory betacryptoxanthine. Omega-3 fatty acids from flaxseed and walnuts are also anti-inflammatory, and so is vitamin D. Grapes and mulberry are COX-2 inhibitor-type painkillers. Resveratrol in grapes and wine works like aspirin. Hazelnuts, peanuts, sesame seeds, sunflower seeds, and tofu help manage pain, most likely because of their high magnesium and tryptophan content. The latter is a precursor for serotonin. Similar painkilling effects are experienced by consuming Brazil nuts, pecans, pumpkin seeds, and spinach. Pain-fighting flavonoids are also present in apples, asparagus, avocado, berries, cabbage, kiwi, grapefruit, grapes, green tea, and soy-based foods. Flavonoids from fruits and vegetables are known to slow pain-causing degeneration.[214] Soybean products are the common source of isoflavones. The aforementioned foods deliver methylation and gene-expression nutrients. We can easily avoid over-the-counter medications such as glucosamine, chondroitin sulfate, and MSM (methyl sulfonyl methane) as painkillers by use of pain-fighting foods present in methylation diets.

10.04 Vitamin D for Back Pain Therapy

Vitamin D deficiency is now linked to back pain, and proper vitamin D intake and good old low doses of aspirin can cure this pain problem.[215]

Table 10.01 Pain-Fighting Common Dietary Phytochemicals— A List Almost Identical to Foods That Fight Osteoporosis

Table 10.01 clearly shows that trace minerals, antioxidants, coenzyme Q_{10}, muscle relaxant mineral magnesium, and bioflavonoids, cucurmin from

---

213 http://www.umdf.org/site/pp.aspx?c=8qKOJ0MvF7LUG&b=7934627.

214 http://www.sciencedaily.com/articles/i/isoflavone.htm.

215 http://www.mayoclinic.com/health/vitamin-d/NS_patient-vitamind/DSECTION= evidence.

turmeric, omega-3 fatty acids fight pains. All such phytochemicals can be delivered to our body via daily methylation diet.

Foods	Key Nutrients
Asparagus, avocado, berries, broccoli, cabbage, cauliflower, grapefruit, kiwi, oranges, peaches, tart cherries, tomato, eggplant	Antioxidants, vitamin C, lycopene, vitamin B complex, coenzyme Q$_{10}$,
Almonds, cashews, other high-magnesium foods	Muscle relaxant, good boron source
apricots, bell peppers, nectarines, oranges, papaya, peaches, watermelon	Betacryptoxanthine
Apples, grapes, green tea, onion, soy	Flavonoids
Berries	Antioxidants
Bilberries, blueberries, grapes, mulberries, peanuts	Resveratrol
Black currant and primrose oil	Gamma linoleic omega-6 fatty acid
Boswellia serrata (Indian gugal), ethanol extract	Boswellic acid
Celery and celery seeds	Vitamin C, phalides, coumarins
Chili and cayenne, up to 40,000 Scoville units	Capsaicin
Coffee	Caffeine, theobromine
Edamame	Isoflavones
Flaxseed	Omega-3 fatty acids
Fruits, vegetables, nuts, legumes, and pulses that have high boron; avocado, asparagus, almond, hazelnuts, apricots, peanut butter	S-adenosylmethionine, calcium, magnesium
Ginger	6-gingerol
Grapes	Resveratrol, COX-2 inhibitor
Olive oil	Monounsaturated fat
Licorice	Antioxidants
Probiotics	Glycirrhizin
Salmon, flaxseed	Omega-3 fatty acids

Foods	Key Nutrients
Turmeric (more powerful than phenylbutazone)	Cucurmin
Walnuts	Omega-3 fatty acid, magnesium
Body's natural painkillers	Serotonin and norepinephrine
Outer skin of crustaceans (a supplement)	Glucosamine
Horse chestnut, chestnut	Escin
Pine bark extract	Pycnogenol

10.05 Foods for Routine Consumption

We should prepare our daily dishes from a combination of cherries, coffee, cranberry, edamame, ginger, grapes, green vegetables, hot pepper, mint, olive oil, salmon, soy foods, strawberries, turmeric, whole grains, wine, and yogurt. The Mediterranean diet, a close cousin of the methylation diet, is in line with this type of combination. The daily diet should include fiber, protein, omega-3 fatty acids, antioxidant vitamins and minerals, and plenty of fluid by way of beverages and water. Avoid excessive use of eggplant for reason of its alkaloid content that can be a purgative.

10.06 Alternative Therapies

Consume foods high in magnesium and vitamin D. Maintain mitochondrial functions by methylation diets rich in antioxidants from vegetables and Brazil nuts and avoid use of too many prescription drugs unless deemed necessary by your physician.

- Regular exercise for boosting natural endorphins.
- Deep breathing or meditation for chronic pain relief.
- Yoga, meditation, and tai chi for reducing stress.
- Acupuncture and touch therapy message.
- Weight reduction and energy building.
- Avoiding alcohol and sleep deprivation.

10.06.01 Arthritis can be controlled by anti-inflammatory agents from ginger, beta-cryptoxanthin from pumpkin, and carotenoids from red bell pepper.[216]

---

216 http://www.sbcoachescollege.com/articles/Powerpoints/Arthritic_Nutrition.pdf.

10.06.02 Migraine headaches can be controlled by quinoa for magnesium and riboflavin, anti-inflammatory flaxseed, spinach for riboflavin and magnesium, pineapple, and almonds.[217] Malfunctioning mitochondria are involved in many neurological diseases,[218] and these food items help mitochondrial health.

10.07 Major Reference Institutions

Osteoporosis and osteoarthritis are major chronic diseases. The following institutions are devoted to research on their treatment by nutrient therapies.

- American Chronic Pain Association, California.
- American Geriatric Society, New York.
- Rehabilitation Institute of Chicago.
- Center for Pain Management at Stanford School of Medicine.
- Pittsburgh Center for Pain Research, University of Pittsburgh.
- University of Michigan Chronic Pain and Fatigue Research Center.

---

217 http://www.migrainetrust.org/assets/x/50129.

218 http://www.painresearchforum.org/news/9570-make-way-mitochondria.

CHAPTER 11

Foods for Skin and Hair Health

"As we grow old, the beauty steals inward."
—Ralph Waldo Emerson

Skin health is a good barometer of good health and well-being. There can be no beauty without vitamins, minerals, antioxidants, amino acids from proteins, and the right blend of omega-3 and omega-6 essential fatty acids from good plant fats. To preserve skin and hair for beauty is to inhibit aging, and good foods for the health of skin and hair are practical antiaging foods.[219]

Water, protein, good plant fat, minerals iron, selenium, zinc, and copper, vitamins A and beta-carotene, biotin, and vitamins C, D, E, and B_{12} are necessary for skin health, along with a daily water intake of at least two to two and a half liters. Nutrient-rich common methylation foods include salmon for omega-3 fatty acids; oyster and poultry for zinc; sweet potato for vitamin A; eggs for selenium and zinc; spinach for carotenoids, folate, B_6, and vitamin C; lentils for protein, iron, zinc, and biotin; Greek and plain yogurt for vitamin D; tomatoes for lycopene; and blueberries for vitamin C and antioxidants.

Proper intake of vitamin D is a must for skin health.[220] Vitamin C, biotin, zinc, and copper are critical to elastin formation in skin.[221,222] Antioxidant foods should be included in the weekly diet for vitamin C, phytoceramides 3 and 6, bromelain from pineapple, and lycopene from tomato paste. Indian

---

219 http://www.huffingtonpost.com/2013/12/11/anti-aging-foods_n_4419906.html.
220 http://lpi.oregonstate.edu/infocenter/skin/vitaminD/.
221 http://www.elastagen.com/media/The_Science_of_Elastin.pdf.
222 http://onlinelibrary.wiley.com/doi/10.1016/0307-4412(90)90121-4/pdf.

ginseng aswagandha is good for skin health, and so are moisturizing creams containing lactic acid and alpha-hydroxy acid exfoliants for removing dead skin. Hyaluronic acid, good for the health of connective tissues of the skin, from poultry, turkey, and fish oil (good sources of retinol) helps maintain skin, cartilage, and joint health.[223,224] Mitochondria, it is now known, are much involved in skin health.[225,226] A high-antioxidant methylation diet is good for mitochondrial DNA stability.[227]

The human skin, the five-layer cover of the human body, is varied in color compared to other mammals. Color and skin development is under diet-dependent complex genetic control.[228]

Skin health is a barometer of health and beauty, and there can be no beauty without good food. This chapter then is on a beauty diet, and I would like the readers to note that molecules that matter for upkeep of a beautiful skin are the same as what is required for good health, vitality, and day-to-day well-being. One can save money by creating a daily methylation diet for beauty and longevity by proper use of vitamins, minerals, antioxidants, amino acids from proteins, and omega-3 and omega-6 essential fatty acids from plant and marine fats. Antioxidant-, mineral-, and vitamin-rich foods can for sure help us avoid wrinkles and skin cancer; they can help us live longer and postpone death.

Skin is the largest organ of our body. Its tennis-court-size surface area is the first line of defense against bad environmental factors. It serves as a sense of touch and protects all other internal organs. The biggest problems in skin health are wrinkles and dryness. There is an $8.4 billion-per-year business in the United States alone dedicated to making our skin beautiful. Men want a wrinkle-free face, freedom from gray hair, and as little abdominal fat as possible. Women want to look twenty years younger than their chronological age. Nobody wants to age.

Japan is ahead of everybody else, with annual sales of $923.70 million for personal beauty care food and supplement products. A worldwide $3.7 billion market is out there for beauty foods and supplements. Omega-3

---

223 http://inhumanexperiment.blogspot.com/2009/02/hyaluronic-acid-for-skin-hair.html.
224 http://www.naturalhealthlibrarian.com/ebook.asp?page=Hyaluronic%20Acid.
225 http://www.dermascope.com/chemistry/cell-science-the-mighty-mitochondria.
226 http://www.seahorsebio.com/resources/tech-writing/techbrief-fibroblasts.pdf.
227 http://ajcn.nutrition.org/content/93/4/897S.full.
228 http://lpi.oregonstate.edu/infocenter/skin.html.

fatty acids alone, which are key to health and longevity, have become a $13 billion-a-year business.

11.01 Genetics of Skin Diseases

Mutations in a single gene p63 cause a lot of problems in skin health, including cancer, cracking, and discoloration.[229,230] *This gene regulates the Satb1 gene responsible for genome-wide organization.* As a master regulator of epidermal development, it controls gene expression and chromatin remodeling. It is also involved in development of T cells of the immune system.[231,232]

11.02 Remedial Approaches for Skin Health Recommended by Mayo Clinic

Mayo clinic of Rochester, Minnesota is a leader in diet based remedies for skin and hair health. Organizations such as WebMD, cosmopolitan, and Healthy Foods of New zealand also promote key components of methylation diet (Biotin, iron, Omega-3 fatty acids, protein, selenium,vitamin A, Vitamin E, and zinc) for routine skin and hair care. Here is a list.

1. Protect yourself from the sun during the hours of 10:00 a.m. to 4:00 p.m.
2. Avoid smoking.
3. Limit bath time and use less-strong soap. Keep skin pores open.
4. Eat healthy and maintain a hormonal balance, estrogen-progesterone in particular.
5. Manage stress with almonds, other tree nuts such as Brazil nuts, hazelnuts, and macadamia nuts, green tea, and high-glutamine foods for gamma amino butyric acid—the main inhibitory neurotransmitter.[233]

---

229 http://newscenter.lbl.gov/news-releases/2011/09/20/key-genes-skin/.

230 http://www.newscientist.com/article/dn7856-master-gene-for-skin-may-hold-key-to-youth.html.

231 http://newscenter.lbl.gov/news-releases/2011/09/20/key-genes-skin/.

232 http://www.google.com/url?sa=t&rct=j&q=&esrc=s&frm=1&source=web&cd=6&cad=rja&ved=0CF8QFjAF&url=http%3A%2F%2Fintimm.oxfordjournals.org%2Fcontent%2F12%2F3%2F281.abstract&ei=vnKnUqCeDoeTrQGDkoHgCQ&usg=AFQjCNH327ZBvToiSSYG2InsyMeptcl6JQ&sig2=1A0axm8dGTlr3wG6XB56hg.

233 http://skincare-cosmeceuticals.com/formulations/index.php?topic=52.0.

6. Consume salmon and sardines for dimethylaminoethanol (DMAE) and watercress and arugula vegetables as salad. Watercress prevents DNA damage. DMAE works with the cell membrane.[234] It is an antiaging molecule. It stops the making of arachidonic acid and thus prevents wrinkles. DMAE is promoted as a memory-boosting food compound for enhanced cognition.

7. Consume vitamin C, zinc, and copper, critical to elastin formation in the skin.

8. Have a good night's sleep.

11.03 Nutrition for Personal Care and Skin Health

Managing emotions and daily nutrition is the key for skin health. Our foods must deliver daily requirements of antioxidants from fruits and vegetables, omega-3 fatty acids and other anti-inflammatory foods such as astaxanthine from salmon, and other key nutraceuticals listed in tables 11.01 and 11.02.

Table 11.01 Therapeutic Foods for Skin and Hair Care

Therapeutic Foods	Active Ingredients	Ailments and Diseases Helped
Abyssine	Bacterial exopolysaccharide	Antiaging ingredient in beauty and blemish (BB) cream
Green coconut	Cococin with Magnesium	Prevents wrinkles
Kokum (balm formulations) from India	Luteolin, flavones, apigenin, hydroxycitric acid	Therapeutic to foot and skin, softens skin, promotes cell oxygenation
Lipowheat ceramide	Fatlike molecule	Smoothens skin surface
Pomegranate	Antioxidant	Speeds skin-cell synthesis
Beauty balm cream	Medium chain triglycerides	Antiaging cream

---

234 http://www.google.com/url?sa=t&rct=j&q=&esrc=s&frm=1&source=web&cd=41&ved=0CCsQFjAAOCg&url=http%3A%2F%2Fdevitastyle.com%2Fhome%2F%3Fwpdmact%3Dprocess%26did%3DMS5ob3RsaW5r&ei=d3enUoLNFYaJrQGh9oHYAQ&usg=AFQjCNHgrjWbh-djRhEckuQ2wZb1STctvQ&sig2=i8eZO2g3SOBnuqToMAzgVQ.

Table 11.02 Common Foods for Skin and Hair Care
(Avoid saturated fats, trans fats, and refined carbohydrates.)

Almond oil	Good for hair care
Alpha-Lipoic acid	Good for skin quality
Astaxanthine	An antioxidant
Avocado	Delivers 27% RDA of vitamin B_3
Coenzyme Q_{10}	Consume broccoli, fluid milk, and other beverage products. Coenzyme Q_{10} is involved in energy production.
Fruits and vegetables	Fruits and vegetables are also a great source of antioxidants. Also, lean protein, whole grain, beans, and legumes are good sources of both proteins and antioxidants.
Glucosamine supplement	Against wrinkles and for wound healing
Hesperidins	Hesperidins as antioxidants come from citrus fruits.
Hyaluronic acid	A negatively charged amino-carbohydrate good for promoting longevity. Design good soups with chicken bone and cartilage for glucosamine and hyaluronic acid.
Omega-6/ Omega-3 balance	Flaxseed and walnuts can balance omega-3 and omega-6. Too much omega-6 is bad news.
Resveratrol from wine	Against skin cancer. Have one serving of red wine every day.
Quircetin	Flavonoid antioxidant from onion, red wine, and tea
Peptides	Isoleucine-proline-proline and valine-proline-proline peptides are incorporated in Calpis sour milk and Evolus sour milk. These are milk-derived peptides. Polypeptide-albumin powder Tiens is sold in the Chinese market.
Phytosterols and tocotrienols from rice bran	Rice bran is a great source of balanced nutrition, with antioxidants and trace nutrients.
Polyphenols	anthocyanidins, pro-cyanidins like pycnogenol, catechins, flavonoids, tannins

Vitamin A	Against Acne. Consume precursors of vitamin A—carotenoids, lycopene, and leutin. Mango delivers 80% Recommended Daily Allowance of vitamin A. Frozen berries are also a good source of vitamin A.
Vitamin B complex	Against scaly skin. Mushrooms for B_2 and other B vitamins. Consume wheat germ for wrinkle-free skin. Biotin or niacin is essential for hair health.
Vitamin C	Antioxidant against wrinkles. Consume cherries.
Vitamin E	A great remedy against psoriasis and autoimmune diseases of patchy skin. Vitamin E boosts immune-cell activity. It is also an antioxidant. Almonds are a good source.
Vitamin K	Leafy green vegetables are very important for skin health—broccoli, kale, asparagus, and brussels sprouts.
Minerals	Selenium as an antioxidant is good for promoting skin health. Other good minerals are zinc and copper. DMAE (dimethylaminoethanol for neurotransmitter acetyl choline) is also good for skin health.
Copper	Baked potato is a good source.
Selenium	Brazil nuts, cottage cheese
Zinc	Oysters are a great source.
Zeaxanthin	Antioxidant

11.04 Foods for Routine Consumption

Mineral nutrition is very important in skin and hair health. Our daily dishes should be designed around 1) garlic and onion for chromium, barley, and mustard seed; 2) Brazil nuts and cottage cheese for selenium; 3) sunflower seeds for selenium, copper, and vitamin E; 4) mushrooms for selenium and copper; 5) tomatoes for chromium and molybdenum; and 6) pumpkin seed for zinc to balance your daily diet. Consuming green vegetables and fruits, a variety of beans and legumes, whole grains, cauliflower, and mustard seed preparations is necessary for B vitamins and minerals for skin and hair health. Cinnamon and dark chocolate should be part of routine teatime snacks. There is a trend in the food industry for designing beauty drinks, beauty snacks, beauty bars, and beauty confections based on tea leaves, tomatoes, grape seed, berries, citrus, algal and fungal preparations, and cocoa.

Apricots, blueberries, carrots, dark chocolate, edamame, kidney beans, lentils, kiwi, almonds and walnuts, oatmeal, oysters, peppers, pomegranate, salmon, snapper, soy foods, spinach, sunflower seeds, tomatoes, and turkey provide nutrients listed in table 11.01. Avoiding alcohol, refined carbohydrates, and salt; controlling DNA damage by free radicals with high antioxidant foods; and avoiding internal inflammation with omega-3 fatty acids and vitamin D should be the criteria for daily food selection and preparation. Build a strong immune system and reduce stress by meditation for emotional control if you want to have beautiful skin.

11.05 Commercial Products for Skin Health

Twenty percent of the antioxidant business in Europe is directed to hair care, and another 20 percent is for skin care. Antiaging foods and topical skin applications constitute the rest of the utilization of antioxidants. Inneov, a joint venture between Nestle and L'Oreal, sells anti-hair loss, anti-dry skin, anticellulite and antiaging products in Europe. Nestle and L'Oreal are marketing cosmoceutical nutrition and dermatology-based pills in the European market.

Also, phytosterols like Benecol, Flora Proactive (UK), and Danecol in margarine are becoming slowly recognized as nutritionally relevant. Digestive health and an enhanced immune system via our second genome in the gut goes with skin and hair care.[235] Examples in this category are Danone's Essential Yogurt and an old established probiotic, dahi, common in India. It is now known that peptides in these products have health values beyond simple nutrition.[236]

Beware of bad foods: One should avoid skim milk, which promotes acne. Antiperspirants should also be avoided because of their aluminum content, which blocks skin pores and prevents sweating. Skin and hair diseases are signs of mitochondrial diseases and aging[237]; these are preventable by the methylation diet.

---

235 http://www.sciencedaily.com/releases/2013/02/130228093831.htm.

236 http://ajcn.nutrition.org/content/71/4/861.full.

237 http://pediatrics.aappublications.org/content/103/2/428.abstract.

11.06 Major Reference Institutions

The following research institutions conduct research and innovative developments on diet based therapies for skin and hair care.

- Cleveland Clinic's Multicultural Hair and Skin Care Center
- Gold Skin Care Center in Nashville, Tennessee
- Hair and Skin Care Research and Treatment Center in Dallas, Texas
- Mayo Clinic Hair and Skin Care Center in Rochester, Minnesota
- L'Oreal's Global Hair Research Center in France

CHAPTER 12

Foods for Diabetes Therapy

"Glucose, the very source of energy for our cells, if not in
balance, can hurt them, causing chronic diseases."
—Triveni P. Shukla

Glucose is necessary for each and every one of our 10 trillion cells. It is much more critical for the life of our brain cells[238], says the Franklin Institute in Philadelphia, one of the oldest science education center in the United States.. Electric signaling alone uses 10 percent of all body energy, and our neurons use two times more energy than the rest of the cells.[239] Good foods can moderate and modulate glucose entry in our blood, and then its management by the hormones insulin and glucagon. A good rule to remember is that whereas consumption of refined carbohydrates is like an injection, complex carbohydrates function like time-release nutrients for glucose. This is critical to managing glucose in the bloodstream, although the brain can live on lactate and keones also.[240]

Diabetes is a disease of metabolic mismanagement of glucose by the multitasking master hormone insulin that controls glucose metabolism and glucose-related homeostasis. Many mitochondrial DNA mutations are linked to diabetes.[241] Glucose, when not in balance, can corrupt every cell in our body, including the nerve cells in our brain.[242] Although caused by both genetic and environmental factors, the inheritance pattern of diabetes is

238 http://www.fi.edu/learn/brain/carbs.html.
239 http://www.ncbi.nlm.nih.gov/pmc/articles/PMC2859342/.
240 http://www.frontiersin.org/Journal/10.3389/fnene.2011.00004/full.
241 http://diabetes.diabetesjournals.org/content/53/suppl_1/S103.full.
242 http://learn.genetics.utah.edu/content/begin/cells/badcom/.

complex because even among two twins, one may have type 1 or type 2 diabetes, and the other may not have either.

Type 1 diabetes, 10 percent of total incidence of diabetes, is an autoimmune disease caused by destruction of insulin-producing beta cells in the pancreas by our own immune system. The immune system may play a role even in type 2 diabetes cases due to insulin resistance, which accounts for 90 percent of diabetes incidence. Insulin resistance is caused by blind and blunt receptors on our cell surfaces preventing glucose entry into our cells and diverting it for storage. Inflammation is a major cause of type 2 diabetes. Harmful chemicals in the body alert B cells of the immune system to cause inflammation. An excess of B cells in the fat cells of obese people are linked to type 2 insulin resistance.[243] The condition is aggravated by unregulated overconsumption of refined and processed carbohydrates.

Other causes of type 2 diabetes beyond genetic and environmental factors are social, lifestyle, and personal choices of foods and drinks. A highly sedentary lifestyle and day-to-day chronic stress add to the problem enormously. Hypothyroidism leads to weight gain, low metabolism, and finally diabetes. Insulin-resistant people suffer from low muscle mass and a propensity to cardiovascular diseases. Diabetes is the cause of many other chronic diseases of the eyes, kidneys, heart, and worst of all, brain.[244] Suppose a person on a two-thousand-calorie-per-day diet gets 1,520 calories from carbohydrates. This translates to 380 grams of glucose per day, or 0.26 gram glucose per minute. Management of this apparently miniscule amount of glucose by insulin is critical to diabetes-free health. Also, equally critical is maintaining a balanced microflora in our gut, for which a simple answer is daily intake of fermented foods, such as yogurt, sauerkraut, and kefir.

Diabetes statistics are alarming, revealing 7.8 percent of Americans suffer from type 2 diabetes, 66 percent are overweight, and 33 percent are obese. Nine out of ten Americans are prone to becoming overweight, and 300,000 of them die of obesity. Worst of all, close to 33 percent of children are overweight or obese; they are bound to have a shorter life expectancy[245] Much of it is because of the consumption of high-glycemic refined and processed carbohydrate foods[246] because our body is not designed to manage literal injections or spikes of refined sugar.

---

243 http://www.ncbi.nlm.nih.gov/pmc/articles/PMC3270885/.
244 http://diabetes.niddk.nih.gov/dm/pubs/stroke/.
245 http://www.diabeticmctoday.com/HtmlPages/DMC1105/DMC1105_Ross.pdf.
246 http://www.diabetes.org/diabetes-basics/diabetes-statistics/.

The propensity for type 1 diabetes is high for people with high levels of an enzyme known as glutamic acid decarboxylase. Our immune system sees this molecule as a foreign antigen and begins to kill insulin-producing beta cells of the pancreas.[247] The human body has evolved to maintain a blood glucose level of 76 to 140 mg/dl by maintaining levels of two hormones—insulin and glucagon. When beta cells of the pancreas are destroyed, there is less insulin produced. Insulin is not produced at all in the case of type 1 diabetes, and not enough to compensate for the impaired ability of cells to use glucose due to resistant and uncooperative cell receptors in the case of type 2 diabetes. Diabetics often lose 50 to 60 percent of pancreatic beta cells.[248]

Rampant type 2 diabetes in the United States is no doubt an epidemic, with one in ten Americans vulnerable to it. It has increased by 600 percent since 1950, and one third of Americans are insulin-resistant today. Sixty-four percent of Americans are more prone to it because of being overweight or obese.[249] Type 2 diabetes is also responsible for depression and poor eye, nerve, kidney, and heart health. It costs us $116 billion a year.

On an average, a 70-kg adult with 1.5 gallon or 5.5 liters of blood has no more than four grams, or a teaspoonful, of glucose,[250] Glucose enters the bloodstream from the digestive system at a rate of 0.5 gram per minute, and its level is under a very sophisticated control system by the hormones insulin and glucagon.[251] Eating every four hours, there is a potential of 125 grams of glucose entering the blood. But glucose disappears also because it is constantly being burned into energy in our cells. The stress hormone cortisol acts as a glucose sensor in beta cells of the pancreas[252]; it can bind to DNA, inducing genes to help produce more glucose-producing enzyme amylase.[253,254]

Blood glucose control is critically important to life. Let us keep in mind that neurons can't store glucose, and our brain runs largely on a steady supply

247 http://www.ncbi.nlm.nih.gov/pubmed/12530520.
248 http://www.ncbi.nlm.nih.gov/pubmed/16306347.
249 http://medicalcenter.osu.edu/patientcare/healthcare_services/diabetes_endocrine/about_diabetes/statistics_about_diabetes/Pages/index.aspx.
250 http://ajpendo.physiology.org/content/296/1/E11.
251 http://www.diabetes.org/living-with-diabetes/treatment-and-care/blood-glucose-control/.
252 http://phys.org/space-news/.
253 http://www.ncbi.nlm.nih.gov/pubmed/6751800.
254 http://phys.org/news197256843.html.

of glucose for bioelectric communication. We quickly become unconscious below a 30 mg/dl level of glucose.

Diet and exercise can manage diabetes problems to a great extent. Our body's response to insulin varies depending on diet, exercise, stress, and illness. Much of the modern-day problem is due to a sedentary lifestyle, lack of aerobic exercise, unhealthy refined processed foods, excessive eating, deficiency of Vitamin D, and contamination of foods with herbicide and pesticide chemicals. The 2 diabetes II due largely to failure of beta cells of the pancreas in producing insulin in view of a rising blood glucose level. The environment plays a huge role in the incidence of diabetes, a metabolic condition of high blood glucose and frequent urination.

12.01 Genetics of Diabetes

The cause of metabolic glucose mismanagement in the case of type 2 diabetes is that our endocrine messaging system has gone wrong. Either there is no insulin production due to more or less complete destruction of beta cells as in case of type 1 diabetes, or there is not enough production of insulin that can compensate for the impaired use capacity of body cells due to nonresponsive receptors on the cell surface in the case of type 2 diabetes. Beta cells of the pancreas are unique in sensing the glucose level in blood, secreting insulin on demand, and storing some for emergency needs, such as to maintain a steady-state glucose level in the blood.

Maturity-onset diabetes of the young (MODY) is traced to a single gene.[255] There is an 80 to 90 percent chance that two identical twins may both have type 2 diabetes, but the chance for type 1 diabetes is only 40 percent. Whereas most of the type 1 genes have been identified, the situation for type 2 genes is uncertain. A handful of genes coordinate a part of the immune system called HLA, which helps recognize foreign invaders like bacteria, parasites, and viruses. The ones under study are HLA-DR3, HLA-DR4, HLA-DR7, and HLA-DR9. It is known that breast-fed children are less prone to type 1 diabetes.[256] Finland has the greatest incidence of type 1 diabetes, followed by Sweden, Norway, and Wisconsin in the United States. Type 2 diabetes is more related to family history. Among the Pima Indians in rural Mexico, obesity runs in families, with a high incidence of diabetes.

---

255 http://www.ncbi.nlm.nih.gov/pubmed/21521318.

256 http://www.nrdc.org/breastmilk/benefits.asp.

Mutations of mitochondrial genes are suspected to be involved.[257] Genes implicated in type 2 diabetes are PPARy on chromosome 3, ABCC8 on chromosome 11, KCN11 on chromosome 11, and CALPN 10 on chromosome 2. A variant of Gene $SLC_{16}A_{11}$ in Mexican populations is killing 73,000 type 2 diabetes patients a year and the variant is spreading fast [258] as reported in the Economist, Jan 9, 2014.

Although type 2 diabetes is the more widely studied, it is understood only poorly. The Harvard Stem Cell Research Institute has announced a breakthrough research on the hormone betatropin, a potential remedy in the near future. The hormone induces thirty times higher production from beta cells. It may suffice to have two injections of this hormone twice a week, compared to daily injections of insulin today.[259]

Behavior, exercise, general lifestyle, and diet are known to be much more important environmental factors than genetic predisposition in the case of type 2 diabetes. For most people, nongenetic factors including environment, body weight, and behavioral factors outweigh genetic disposition. Diabetes leads to abnormal metabolism and diseases of the eye, kidney, heart, and neurons or brain cells.

12.02 Obesity

The gut-brain peptide ghrelin is not suppressed in obese and insulin-resistant people.[260] Hunger thus is not suppressed, even after a full meal, and obese people end up consuming more and more food. High ghrelin also correlates with high blood pressure.[261]

12.02.01 A mutated gene: Gene ApOE e4 is linked to dementia, and Alzheimer's disease is implicated in type 2 diabetes. The A1C (hemoglobin A1C or glycosylated hemoglobin) test relates to blood glucose. A level higher than 7 in the case of type 2 diabetics makes them vulnerable to Alzheimer's disease, characterized by amyloid protein buildup and loss of cognition.[262]

---

257 http://www.ncbi.nlm.nih.gov/pubmed/16060290.

258 http://www.economist.com/topics/diabetes-1

259 http://harvardmagazine.com/2008/11/stem-cell-progress-html.

260 http://www.ncbi.nlm.nih.gov/pubmed/15917842.

261 http://www.ncbi.nlm.nih.gov/pmc/articles/PMC2845796/.

262 http://www.agis.com/Document/4536/link-between-diabetes-and-alzheimers-disease.aspx.

12.02.02 Electron leakage: A perfect metabolism keeps electron and proton channels and proton currents in balance. Electrons have been found to leak when energy production and oxygen consumption in mitochondria are not in sync due to inefficient coupling.[263] The problem arises from mutation in the thermogenic protein 3, responsible for uncoupling the two events. Electron leakage underlies production of reactive oxygen species, better known as free radicals, affecting aging, obesity, and malfunction of the thyroid gland.[264]

12.02.03 Poor Eye Health: Retinopathy is common to 100,000 people in the US. Almost 50 percent of diabetics in the United States have the problem of retinopathy in one form or another. Also, diabetics are prone to glaucoma and cataracts. Molecules within the retina convert light and color information into electrical signals to the brain, where they are decoded and converted into images. The retina is nourished by blood capillaries that get damaged in the case of diabetes, causing retinopathy. The cause is not well-understood, although high blood pressure is known to worsen retinopathy.[265]

12.02.04 Too Many Problems from Refined Carbohydrates: Refined carbohydrates by way of processed foods can and do cause insulin resistance. And then conditions of overweight and diabetes lead to other diseases of hyperglycemia, chronic inflammation, heart and coronary artery, impotence, cancer, eyes, kidney, dementia, and Alzheimer's disease. The list below gives the direct and indirect relationship among major chronic diseases.[266]

Direct Association	Indirect Association
Cardiovascular disease	Arthritis
Some cancers	Immune dysfunction
Diabetes and hypertension	Joint pain

---

263 http://www.google.com/url?sa=t&rct=j&q=&esrc=s&frm=1&source=web&cd=5&ved=0CFUQFjAE&url=http%3A%2F%2Fwww.researchgate.net%2Fpublication%2F12660276_Mitochondrial_proton_leak_and_the_uncoupling_proteins%2Ffile%2F504635151e760c546d.pdf&ei=kbSoUqbWELOQyQGgrYGIBQ&usg=AFQjCNGbuE0Vrek66qsgBfEj5YyhbUVWPw&sig2=_jKKSg2f8EUwDbdkf6Pnvg.

264 http://themedicalbiochemistrypage.org/oxidative-phosphorylation.php.

265 http://www.medicinenet.com/script/main/art.asp?articlekey=41998.

266 http://www.aihw.gov.au/chronic-disease-determinants/.

12.02.05 A Good Proof of Diet Effect: The Pima Indians of rural Mexico consume little animal fat and processed foods. They have a low incidence of diabetes. But they become vulnerable to type 2 diabetes when in the United States because of change to high-fat dietary habits.

12.02.06 Chronic Stress: Stress is a big cause of diabetes.[267] Reducing stress by meditation, yoga, and regular physical activity needs to become part of our lifestyle.

12.02.07 Body Fat Itself Is an Added Problem: Excessive fat by way of adipose tissue in and of itself creates metabolic problems. It produces hormones, regulates immune response, and influences energy intake. Fat cells become plumper when we gain fat. They shrink when we lose fat. Fat cells do the following[268]:

1. Fat cells produce *leptin*, a hormone that regulates energy metabolism, appetite, and hunger. The role of leptin in obese people is often flawed; it ceases to act as a satiety control hormone.
2. Fat cells release pro-inflammatory cytokines, which cause chronic inflammation and elevated insulin level, both of which may cause cancer. Abdominal fat is independently linked to metabolic disorders and risks of pancreatic, colorectal, breast, and endometrial cancers.
3. Body mass index above 25 is a cancer risk. This is largely due to big white fat cells. Obese and overweight people are much more prone to cancer. In the United States alone, 100,000 cancer cases per year are linked to body fat.[269]

12.03 Antidiabetic Foods

Type 2 diabetes, as a metabolic disorder, can be managed by diet, weight reduction, exercise, and added physical activity. A Harvard study including

---

267 http://www.diabetes.org/living-with-diabetes/parents-and-kids/everyday-life/managing-stress-and-diabetes.html.

268 http://www.aso.org.uk/wp-content/uploads/downloads/2011/06/Fact_sheet_Childhood_Fat_tissue__cells.pdf.

269 http://www.cancer.org/cancer/news/report-over-100,000-cancers-linked-to-excess-body-fat.

nine thousand nurses established that 91 percent of the diabetic cases were due to unhealthy diet and the disease could be reversed by prudent diet, exercise, and stress management [270]. To be effective, however, the changes in food and lifestyle need to be extensive and long-term.

What we include in our diet is much more important than what we exclude. A lower-than-two-thousand-calories-a-day diet based on whole grains, fruits, vegetables, low-fat dairy products, and minimal fried foods without excessive saturated and no trans fat can do a lot of good. Such a diet is the methylation diet. It should be based on rigorous planning for low-glycemic complex carbohydrates, high soluble and insoluble fiber, high-magnesium vegetables, garlic and onion for chromium, and healthy plant fats from walnuts and flaxseed. First of all, the plan requires avoiding high-sugar soda drinks, refined carbohydrates, alcohol, bacon, lard, cheese, and red meat. It further requires a careful selection of foods (see table 12.01) for antioxidants, anti-inflammatory omega-3 fatty acids, prebiotic dietary fiber and probiotic foods, high-protein satiety foods, carotenoids from carrots and other fruits and vegetables, vitamins A and D, a high-protein breakfast for satiety, quircetin from onion, magnesium and selenium from seeds and nuts, and plant-based monounsaturated fat. Probiotic foods constitute our second genome, and consumption of yogurt, cheese, sauerkraut, and kimchi can aid in managing type 2 diabetes.[271] Added to this list should be at least two and a half liters of daily water.

This plan of diet, exercise, and stress control is successful the moment blood tests get to and stay normal: blood pressure 120/75, total cholesterol of 130, triglycerides of 200 mg/dl. We should eat in order to balance our daily energy expenditure and daily calorie intake by planned consumption of

1. Dietary fiber can help manage physiological satiety by slowing gastric emptying, by suppressing the hunger hormone ghrelin, and

270 http://www.hsph.harvard.edu/news/press-releases/eating-whole-fruits-linked-to-lower-risk-of-type-2-diabetes/

271 http://www.ncbi.nlm.nih.gov/pubmed/23225499.

by enhancing the PYY peptide hormone that regulates response to meals.[272]

2. Whey protein helps release cholecytokinin (CCK) and GLP-1 peptide hormone, which help us curtail hunger.[273]

3. Potato skin's proteinase inhibitor II (P12) induces CCK secretion. (Kemin Industries, Demons, Iowa, has a commercial extract product.)

4. Monounsaturated fats in avocado, peanut butter, olive oil, and salmon and polyunsaturated fats in corn oil, safflower oil, walnuts, sesame seeds. and sunflower oil are good fats. In general, fats from plants (legumes, oilseeds, nuts) are good. Omega-3 and omega-6 promote good body and brain health when in a ratio of 1:3. They contain essential fatty acids that our body can't make.

5. The emulsion of salad dressing, based on pine-nut oils and diglycerides, increases satiety.

Fat cells in our adipose tissue store triglycerides, and there is very little limit to this storage. Low HDL and high triglycerides are common to type 2 diabetics. The recommended level of triglycerides in the blood is 150 mg/dl. Since fat is insoluble in water, it travels in blood as packages called lipoproteins—LDL and HDL, known as the bad and good cholesterol. Lipoproteins carry cholesterol around the body. When we consume high levels of saturated and trans fat, the LDL level goes up. Consumption of polyunsaturated fats lowers them. Monounsaturated fatty acids, like oleic acid in olive oil, *may* boost HDL, the good cholesterol.

Loads of high glucose, saturated and trans fat, and sweets are sure causes of diabetes. We now know that the Mediterranean and vegan diets, low in readily available sugar, correlate to low insulin-resistance.[274] Also, diets based on unrefined carbohydrates of low-glycemic index, including fruits, vegetables, whole grain, legumes, and lentils, reduce blood sugar, LDL, weight, and incidence of diabetes.

---

272 http://www.google.com/url?sa=t&rct=j&q=&esrc=s&frm=1&source=web&cd=2&ved=0CDUQFjAB&url=http%3A%2F%2Fwww.mdpi.com%2F2072-6643%2F5%2F6%2F2093%2Fpdf&ei=dr-oUqDLJqjXyAH_yIDgBA&usg=AFQjCNFjm0qbmyjLCyTt0OcEWojROpliIg&sig2=R3EPUAXwzZFEeKZLSDvsJA.

273 http://link.springer.com/article/10.1007%2Fs12349-013-0121-7#page-1.

274 http://www.ncbi.nlm.nih.gov/pubmed/17880675.

Table 12.01 The Best Antidiabetic (Type 2) Foods Make the Methylation Diet

Common Antidiabetic Foods	Active Ingredient(s)	Ailments and Diseases Helped
Almonds	Low carbohydrates and magnesium	Reduce inflammation
Avocado	Monounsaturated fat	Reduce inflammation
Beans	Low-glycemic carbohydrates	Blood glucose control
Blueberries	Antioxidants and vitamin C, a water soluble antioxidant.	Reduce stress
Buckwheat	Antioxidants	Reduce stress
Cinnamon, dark chocolate, clove, stevia	Cinnamaldehyde, eugenol, cinnamyl alcohol, cinnamyl acetate (Lund University, Sweden[275])	Reduce fat
Chili pepper	Capsaicin, a metabolism booster	Blood-sugar control, improve insulin function, reduce glucose and cholesterol, boost cognition and anticlotting, cancer protective
Cocoa powder	Antioxidants	Reduce stress
Egg whites	Four grams protein per egg	Satiety and low food intake
Grapes and red wine	Resveratrol	Prevents inflammation
Fenugreek	4-hydroxyisoleucine (4-OH-Ile),	Slows glucose absorption; enhances hunger control and satiation
High-protein foods		Antidiabetic
Honey	Low-glycemic sugar	Antidiabetic and satiating

---

275 http://ajcn.nutrition.org/content/85/6/1552.full

Common Antidiabetic Foods	Active Ingredient(s)	Ailments and Diseases Helped
Isomaltulose	Low-glycemic sugar	Weight and obesity control
Lentils	Low-glycemic carbohydrate	Weight and obesity control
Maple syrup	Sweetener and antioxidant	Diabetes Type 2
Natural food colors	Carotenoid antioxidants	Prevent inflammation
Oats	Soluble beta-glucan fiber	Satiety and enhanced immunity
Onion	Quircetin	Antidiabetic and antioxidant
Salmon	Omega-3 fatty acid	Prevents inflammation
Stevia sweetener	Noncaloric sweetener	Antidiabetic, anti-inflammatory, and antioxidant
Strawberries	Antiglycation agents and antioxidants	Antidiabetic
Turmeric	Curcumin	Reduces inflammation
Vinegar	Acetic acid	Antidiabetic, antiatherosclerotic, prevents lung and prostate cancer
Walnuts	Good fat and vitamin E	Anti-inflammatory

12.04 Probiotics (Our Second Genome) and Diabetes

The gut bacteria are now linked to both type 1 and 2 diabetes (the Human Microbiome project of the National Institute of Health [276]). They have many functions, including (1) digesting prebiotic fibers and producing up to 10 percent extra calories, (2) protecting against pathogenic bacteria, and (3) helping build and talk to our immune system. These microorganisms are often different for different people, depending a great deal on individual genetics and the environment. They can affect weight gain and thus type 2

---

276 http://commonfund.nih.gov/hmp/overview

diabetes. Obese people have more of the fermicutes-type bacteria, good at digesting prebiotics. Lean folks have more of the bacteriodetes-type bacteria. The bacterial types change from less efficient to more efficient as we lose weight. Also, lean people have a more diverse population of bacteria, which are beneficial and capable of combating inflammation and boosting the immune system. Remission of type 2 diabetes has been observed in patients who have undergone bariatric surgery that materially affects the gut bacterial population.[277] Gut bacteria affect the autoimmune system also and thus a balanced gut bacterial population may prevent type 2 diabetes. A diverse population of probiotics must be part of our daily diet, including yogurt, kefir, sauerkraut, kimchi, sour cream, India's dahi, and Asia's tempeh-like products.

12.05 Managing Diabetes by Exercise

Those with a family history of diabetes must exercise and maintain a steady blood glucose level. Low or no exercise can lead to metabolic syndrome. Regular exercise of thirty minutes to an hour is known to be very helpful. Increasing frequency, intensity, and length of exercise as one gets used to it is greatly beneficial in reducing visceral fat and helping manage glucose better. For good health and obesity control, a dedicated effort to reduce weight, to do meditation for stress control, and to sleep well for at least seven to eight hours a day can pay big dividends in reversing to a normal and diabetes-free life. Resistance exercises for reducing weight can be instrumental in modifying the expression of genes for obesity.[278] Flex, stretch, and extend most any limb that you can, and you will feel better. Choose both aerobic and resistance exercises and make them enjoyable and fun. Yoga is great for stress management and oxygen therapy. The American Diabetes Association recommends at least 150 minutes of moderate-intensity exercise per week, or thirty minutes per day. We need to follow the rule of (220 - age) X 0.75 = an approximate heart rate of 115 to 120 beats per minute at the peak of exercise.[279] This can be achieved by a practice of deep breathing, stretching exercises, and meditation. The program we choose should be flexible and acceptable to our body and mind.

---

277 http://www.ncbi.nlm.nih.gov/pubmed/21107106.

278 http://www.inherenthealth.com/media/4759/wm_scientific%20summary.pdf.

279 http://www.mayoclinic.com/health/exercise-intensity/SM00113/NSECTIONGROUP=2.

12.06 Major Reference Institutions

The following institutions lead in research on Type 2 diabetes per se and dietary therapeutics of Type 2 diabetes.

- Franklin Institute;
- Harvard Stem Cell Research Institute;
- Mayo Clinic;
- American Diabetes Association;
- Lund University, Sweden;
- Joslyn Diabetes Center;
- Johns Hopkins University Medical School; and
- University of California, Santa Barbara.

CHAPTER 13

Foods for Fighting Cardiovascular Diseases

"The basis of all health is the blood-pumping function
of the heart, which generates its own electricity."
—Triveni P. Shukla

Our heart, a 310-gram electric pump, is the centerpiece of the cardiovascular system. As a biological machine, it is unique in form, structure, and function. It uses 9 ml of O_2 per minute per 100 grams of tissue,[280] which translates to around two watts per second or 115 kilowatts or 154 horsepower in a twenty-four-hour day.[281] In order to pump 100 million barrels of blood throughout our average seventy-five-year-long life, the power of this 310-gram machine calculates to 4.215-million horsepower. No heart function simply means no life, or literal death.

We live by oxygen, water, and nutrients from foods we eat. Our heart functions by oxygen via a large number of mitochondria, the gates of life and death.[282] For efficient oxygen transport, we have one red blood cell for every three cells in our body. Critical to our living is how many red blood cells there are, how fast the blood flows in our arteries, and how punctually it reaches our cells in the body, the brain cells in particular. Cardiovascular disease can come from poor blood flow and obstructions that cause it. Our life processes and brain function depend on blood flow, which is responsible for transport of nutrients and hormones and disposal of metabolic waste.

---

280 http://www.usc.edu/dept/biomed/bme403/Section_3/cardiac_efficiency.html.
281 http://hypertextbook.com/facts/2003/IradaMuslumova.shtml.
282 http://www.annualreviews.org/doi/abs/10.1146/annurev.ph.47.030185.003241?journal
Code=physiol.

The human body uses four pounds of pure oxygen and two pounds of water in order to burn the less than four pounds of food consumed each day.[283] Air gets to the lungs, and 21 percent of the oxygen in it binds the hemoglobin in the lungs. Oxygenated blood gets to the heart, which pumps it to all 7.5 trillion cells in the body. Carbon dioxide is exchanged, and we breathe it out. The oxygen helps burn glucose in order to produce our daily energy equivalent of, say, a one-hundred-watt bulb.[284] This is consumed by our cardiovascular system (heart, arteries, veins, and capillaries), brain, and other organs.

Blood carries nutrients, water, nitric oxide, and oxygen to all cells in the body and takes away carbon dioxide as a waste product. Since *blood is the conduit for all metabolic activities and functions of all other systems,* health of the heart and blood vessels is the key to overall good health. Unfortunately heart disease is the biggest killer in the United States. Even calcium is now regarded as a risk factor.[285,286]

The heart may get tumors, but cancer of the heart is extremely rare. It is designed to remain safe as long as we feed ourselves with enough oxygen, water, and nutrient-dense diet. The diet doesn't have to be exotic and uncommon. Good health and longevity can be achieved by a simple variety of available foods by way of the methylation diet. Mitochondria play a huge role in heart disease.[287,288]

As stated above, heart disease is the biggest killer in the United States. Nearly 2,300 Americans die daily, or one every thirty-eight seconds, from heart diseases.[289] Almost a third of patients find out about their heart disease after they are dead. Three major causes of cardiovascular diseases are oxidized LDL, oxidative stress, and chronic inflammation. We need to prevent it by powerful antioxidants like pyrroloquinoline quinone,[290] a small molecule that supports mitochondrial function. Fava or butter beans, green soybeans or edamame, and sweet potato can supply enough antioxidants for our daily need.[291]

---

283 http://www.freshairx.com/human-oxygen-requirements/.

284 http://www.fao.org/docrep/u2246e/u2246e02.htm.

285 http://www.nhlbi.nih.gov/health/health-topics/topics/cscan/.

286 http://www.sciencedaily.com/releases/2013/05/130502142657.htm.

287 http://www.ncbi.nlm.nih.gov/pubmed/22399426.

288 http://circres.ahajournals.org/content/111/9/1222.full.

289 http://www.cdc.gov/heartdisease/facts.htm.

290 http://www.ncbi.nlm.nih.gov/pmc/articles/PMC1136652/pdf/biochemj00065-0028.pdf.

291 http://hplusmagazine.com/2011/03/21/rejuvenate-your-cells-by-growing-new-mitochondria/.

The heart, located about left-center of the chest, is a self-powered pump and needs oxygen-rich blood for energy production for contraction during each beat. It pumps three ounces of oxygen-rich blood per beat into the aorta and then to the rest of the body. The heart is made of strong cardiac muscle fibers called myocardium. Heart muscles generate their own action potential, mediated by calcium. Muscle contraction causes a wringing type of action for pumping. Heart cells have active energy-producing mitochondria, which can produce ATP energy on demand. The key is blood flow. Low blood supply leads to low oxygen, to low energy, and thus to poor heart function. In its sixty to eighty beats per minute, the heart pumps around 1.6 gallons of blood per minute or 2,304 gallons per day.

The heart supplies oxygen to our brain and all other cells in the body by a pumping force that is no more than the squeeze of a tennis ball. In order to accomplish this, the heart works twice as hard even at rest as leg muscles for circulating blood effectively over a daily travel of 12,000 miles. In our entire lifetime, the heart pumps close to a million barrels of blood. It must stay healthy, and it must have a good supply of oxygen for electrical power generation by healthy and numerous mitochondria capable of producing 30 kg of ATP, the molecule of energy that is recycled, per day.[292] The heart has two separate pumps. The right side of the heart receives blood low in oxygen. The left side of the heart receives blood that has been oxygenated by the lungs. The oxygen-rich blood is pumped out into the aorta and then to all parts of the body. The cycle continues.

13.01 Risk Factors Involved in Heart Disease(s)

Major risk factors responsible for cardiovascular diseases are listed below. Avoiding stress and alcohol and reducing stress by exercise and meditation is now known to be very valuable in reducing heart problems.

- unhealthy diet of high saturated fat and deficiency of vitamins B_5, B_6, B_{12}, and folate
- high coronary calcium
- chronic stress
- smoking
- excessive alcohol use

---

292 http://cardiovascres.oxfordjournals.org/content/88/1/40.full.

- high blood pressure
- type 2 diabetes and high blood glucose
- overweight and obese conditions, with a BMI over 32
- Physical inactivity and lack of exercise
- High homocysteine level of 15 micromoles/l or more

13.02 Diagnosis and Indicators of Cardiovascular Problems

It is now customary to check for the following during annual examination.

- C-reactive protein test
- Total cholesterol: 200 mg/dl is optimum, and more than 240/dl is problematic.
- LDL(low-density lipoprotein): 100 mg/dl is good; 160 to 190 mg/dl is a high range.
- HDL(high-density lipoprotein): 60 mg/dl is a great number; less than 40 mg/dl is bad.
- LDL (low-density lipoprotein)/HDL (high-density lipoprotein) should be a ratio of 3.5.
- LDL particle size: Small and dense creates a problem.
- Apo lipoprotein B, the LDL carrier, needs to be below 80 mg/dl.
- Triglycerides of 40 to 160 mg/dl range is optimum.
- Blood sugar of 95 to 100 mg/dl is optimum.
- Blood pressure of 130/80 is optimal.

A quantitative coronary arteriography, cardiac pet scan, thallium scan, and ultrafast CT (heart scan) for diagnosing plaque in arteries; coronary angiography; and radionuclide ventriculography should be done just to get a firm grip on the diagnosis. Everyone should monitor blood pressure regularly and blood potassium level periodically.

Blood pressure of 140/90 is bad news, and a 156/110 reading denotes serious trouble calling for a physician's help. The irregularities can be due to a combination of gene effects and effects due to our diet and lifestyle. Diet has a lot to do with high blood pressure, and it must be controlled via consumption of the methylation diet.

Coronary arteries carry blood to the heart muscles. Plaque can narrow these arteries and even block them due to atherosclerosis, deposition of plaque of fatty materials. Plaque reduce blood flow and cause chest pain or angina pectoris. Plaque problems are aggravated by too much saturated fat and cholesterol in our daily diet. Trans-fats from highly hydrogenated fats like Crisco are doubly unhealthy.

13.03 Red Blood Cells

The red blood cell, six to eight microns in diameter, in the circulating blood of human beings represents a wonder in evolution. The 95 percent hemoglobin protein in them saturates with oxygen, which can then be carried to every cell in the body. Red blood cells only carry oxygen, but do not use it because they do not have the energy-producing mitochondria. They have no nucleus and no DNA. *Human bone marrow produces 2.4 million red blood cells every second.* One circulation takes twenty seconds, and red blood cells can circulate for 120 days. Red blood cells can't make RNA, and therefore, viruses can't evolve from them. In a way, they are incorruptible.

Adult human beings have 2.5 trillion red blood cells packed with hemoglobin, an iron protein endowed with the power to carry oxygen. It is the iron in hemoglobin that picks up oxygen in the lung. We simply can't exist without oxygen. By evolutionary design, almost a quarter of our body cells are red blood cells, produced with a longer life span for the purpose of oxygen transport and exchange to the rest of the 7.5 trillion cells. Production, nutrition, and health of these cells are essential for a healthy and energetic body.

13.04 Blood Vessels

There are three types of vessels in the human body—arteries, capillaries, and veins. Arteries carry blood from the heart to all cells in the body, small and single-layer capillaries help exchange nutrients and oxygen, and veins return the blood to the heart. The blood pressure in them is expressed in millimeters of mercury, and a good number is 120 for systolic (the upper number) and 80 for diastolic (the lower number; noncontraction cycle). Thus, our heart functions at one sixth of atmospheric pressure.

Malfunctioning of the vessels is involved in many diseases. Good examples of malfunctions are atherosclerosis and plaque formation in arteries, which lead to high blood pressure and heart attack. Free-radical damage to cells within arteries is also a major cause of cardiovascular disease. Blood-flow rate and optimal oxygen supply is critical to brain function.

13.05 Genetics

Apo-A1 protein produced by a variant gene is found in people of common ancestry in Limone and Garda of Northern Italy (University of Milan, Italy[293]). When genetically engineered, Apo-A1 is administered by injection; it lowers plaque buildup (Cleveland Clinic, Cedars Sinai Medical Center[294]). The variant Apo-A1 offers a natural protection for a population that is otherwise very vulnerable to heart problems because of high triglycerides and low HDL.[295]

The homocysteine-mediated problem of high risk of peripheral arterial disease can result from an abnormal gene on chromosome 1, which is involved in the metabolism of the amino acid methionine. Homocysteine is a precursor to methionine and cysteine, and its conversion depends on vitamins B_6 (pyridoxine), B_{12} (cyanocobalamine), and folic acid in our diet. A deficiency of these vitamins leaves unconverted excess homocysteine in our blood that causes damage to the vessels, promotes blood clotting, and accelerates atherosclerosis. Multivitamin tablets are a temporary fix. What we need is a methylation diet that supplies our daily needs of antioxidants and necessary vitamins.[296]

13.06 A Heart-Healthy Diet

Daily food and lifestyle can lower the blood pressure to 120/80 for more than 50 million senior citizens in America. Good proof is the dietary habits of the Seventh-Day Adventists, vegetarians, and vegans, all of whom have a low incidence of hypertension. Bad diet and chronic stress are devastating in raising blood pressure.

We need an anti-inflammatory diet in order to free us of cardiovascular problems. Good diet can reduce C-reactive protein and LDL, which adhere to the lining of the arteries and form plaque that constricts them. High heart rate, low blood flow due to constricted arteries, and formation of collateral arteries around the ones that are blocked are signs of advanced diseases.[297]

The diet and cardiovascular disease connection has been known for a

293 http://circ.ahajournals.org/content/103/15/1949.long
294 http://www.jbc.org/content/288/29/21237
295 http://ghr.nlm.nih.gov/gene/APOA1.
296 http://lpi.oregonstate.edu/f-w99/vascular.html.
297 http://www.ncbi.nlm.nih.gov/pubmed/18469275.

long time (Harvard Nurses' Study[298], China Study[299], Ancel Key Study[300], Farmington Heart Study[301]). The Chinese eat low-fat, high-fiber foods based on whole grains and vegetables. Total cholesterol in the Chinese population is only 127 mg/dl compared to 200 in the United States. Death due to heart diseases in the United States is almost seventeen times that in China because of overconsumption of high-glycemic refined carbohydrates, highly saturated and trans fat, ten times more omega-6 than omega-3 fatty acids, chronic stress, and sedentary lifestyle.

Fruits, vegetables, beans, legumes, lentils, dried beans, and whole grain for fiber, B complex vitamins, vitamin C, minerals, and phytonutrients are common foods for heart health. Omega-3 fat from flaxseed, walnuts, salmon, and trout must be part of our diet, and so should soluble fiber from oat bran, flaxseed, citrus fruits, and vegetables. Vitamins niacin (B_3), pantothenic acid (B_5), pyridoxine (B_6), B_{12}, and coenzyme Q_{10} or ubiquinone are involved in normal metabolism. Good sources of these vitamins are meats (beef or poultry), fish, broccoli, wheat germ, and peanuts. Rice bran and soy contain phytosterols, and Brazil nuts contain the mineral antioxidant selenium. Diets capable of delivering coenzyme Q_{10} and phytosterol improve health of the blood vessels.[302] Pyrroloquinoline quinone, a very powerful antioxidant found in common foods like broad beans. green soybean, sweet potato, kiwi, spinach, tofu, natto, and green pepper, is very beneficial to heart health.[303]

HDL, the good cholesterol that removes fatty deposits from artery walls to the liver, is difficult to increase. Our body does its best to make HDL in order to remove excess cholesterol. This is hardwired in our genes. Daily intake of high saturated fats, trans fats, and cholesterol requires more HDL. In the case of a fruit, vegetable, and whole-grain diet, HDL may not change, but LDL certainly is reduced.[304]

Food supply and food composition, whether retail or food service, changed during the last fifty years progressively from bad to worse. The following nutrients became deficient in our daily diet, causing many cardiovascular

298 http://www.channing.harvard.edu/nhs/
299 http://en.wikipedia.org/wiki/Cardiovascular_disease_in_China
300 http://www.uh.edu/engines/epi2469.htm
301 http://www.framinghamheartstudy.org/about-fhs/history.php
302 http://www.douglaslabs.ca/pdf/pds/99468.pdf.
303 http://www.biochemj.org/bj/307/0331/3070331.pdf.
304 http://www.health.harvard.edu/newsletters/Harvard_Heart_Letter/2009/October/11-foods-that-lower-cholesterol.

diseases. A prudent diet today must have the daily recommended allowance for the following:

1. Omega-3 fatty acids EPA (eicosapentanoic) and DHA (docosahexaenoic). They fight inflammation, maintain cardiovascular health, and control levels of blood triglycerides and bad cholesterol LDL.
2. High dietary fiber (legumes, whole grains, beans, and lentils). Dietary fiber reduces cholesterol and triglycerides.
3. Potassium intake by way of bananas and potato products.
4. Dark chocolate for antioxidants that prevent LDL oxidation.
5. More spices containing garlic and turmeric for reducing cholesterol.
6. Red rice yeast and India's gugal for cholesterol-reducing statins.
7. Phytosterols from rice bran, wheat bran, soy, and wheat germ.
8. Avocado, oats, whole grains, pistachios, flaxseed, yogurt, and grape juice as snacks, smoothies, or quick bakes.
9. Vitamins B group and D from fruits, vegetables, cereals, and milk.
10. Antioxidants beta-carotene, glutathione, lipoic acid, vitamin E, ubiquinol, pyrroquinolin, vitamin C, melatonin precursors, resveratrol, and selenium.
11. Green vegetables for folic acid and other B vitamins to reduce homocysteine.

In addition, the following should be controlled and reduced in the daily diet:

1. Salt.
2. Saturated fat should be kept below 25 percent of the total daily fat intake from meat, butter, and cream.
3. No trans fat from Crisco-type shortening or hydrogenated oils.
4. High-cholesterol foods like meat and dairy.
5. Highly refined carbohydrates from cakes, cookies, and candies.

13.07 Food Variety in Routine Diet

We should consume common foods like apples, bananas, beans, berries, dark chocolate, garlic, green tea, kale, lentils, almonds and walnuts, oranges, oatmeal, pomegranates, salmon and sardines, sweet potato, tomatoes, whole grains, wine, and yogurt for dietary fiber; and omega-3 fatty acids, vitamin and mineral antioxidants, and a balance of B vitamins for routine energy production. Tables 13.01 and 13.02 list nutrients in common foods that help maintain heart health. This in effect is the methylation diet for good gene expression and heart health.

Breakfast, lunch, snacks, and dinner dishes can be prepared for optimal health, well-being, and longevity. If we don't do it ourselves, food companies will do it for us but at a higher cost. GlaxoSmithKline, for instance, sells Lovaza Omega-3-based triglyceride-lowering medication.

Table 13.01 Heart-Healthy Methylation Foods

Common Foods	Active Ingredients	Ailments and Diseases Helped
Almonds		Lower cholesterol
Anti-inflammatory foods—fruits, vegetables, and probiotics	COX-2 inhibitors	Prevent inflammation
Black currant, borage oil	Omega-6, gamma linoleic acid	Build eye, bone, skin, hair health; prevent inflammation
Blueberries	Pterosilbene	Act like resveratrol
Bran of rice and corn	Phytosterol	Lowers cholesterol, health-healthy
Canola oil	Monounsaturated fatty acids	Improves heart health
Cardamom	Pine, linalool	Antioxidant and antibacterial, lowers cholesterol, improves metabolic rate
Cayenne pepper	Capsaicin, vitamin A	Improves cardiovascular and digestive health, reduces pain, improves metabolic rate
Cocoa powder	Antioxidants, arginine, fiber	Prevents cardiovascular problems
Collard greens	Calcium and fiber	Prevents cardiovascular problems
Dark chocolate	Antioxidants	Promotes heart health
Edamame (soybeans)	Antioxidants, isoflavones	Prevents cardiovascular problems, promotes weight loss

Common Foods	Active Ingredients	Ailments and Diseases Helped
Flaxseed, walnuts; EPA and DHA omega-3 fatty acids	Anti-inflammatory	Reduce cardiac arrest, reduce triglycerides, lower blood pressure, prevent prostate cancer, promote kidney health
Fermented soy products	Nattokinase	Break down fibrin, the blood-clotting protein
Flaxseed, walnuts (160 mg alpha-linoleic acid/day)	B_6, copper, manganese, molybdenum, lignan, alpha linolenic omega-3 fatty acids	Reduce cholesterol, anti-inflammatory, build immune system, improve cognition
Fruits and vegetables	Vitamins A, C, coenzyme Q, and selenium, anti-inflammatory antioxidants and flavonoids	Build arterial elasticity, anti-inflammatory, reduce blood pressure, improve glucose and fat metabolism, reduce homocysteine

Table 13.02 Heart-Healthy Methylation Foods

Common Foods	Active Ingredients	Ailments and Diseases Helped
Garlic (activate enzyme at room temp) and beetroot juice, cloves	Phytonutrients, antiplatelet enzyme allinase, allicin, vitamin C, diallyldisulfide, chromium	Reduces blood pressure, protects arteries and veins, prevents clotting, anticarcinogenic, antidiabetic
Ginger	Gingerol	Reduces blood pressure, antiarthritic, blocks prostaglandins
Green tea	Antioxidants	Promotes heart health
Grapes and wine	Resveratrol is a nitrous oxide vessel relaxant.	Antiaging, heart health, anticarcinogenic

Common Foods	Active Ingredients	Ailments and Diseases Helped
High-protein foods—garbanzo beans, salmon, cocoa powder	Arginine	Produces vasodilator nitrous oxide, helps reduce blood pressure
Legumes: peas, lentils, beans	Folic acid and fiber	Reduces cholesterol, lowers homocysteine, lowers risk of heart attack
Licorice	Anethol, glycirryzin, isoflavone	Lowers cholesterol, protects liver
Oatmeal	Beta-glucan soluble fiber	Reduces cholesterol, reduces blood pressure
Olive oil	Squaline, a triterpine	Anticarcinogenic and heart-healthy
Citrus peel	Bioflavonoid mobiletin and tangeretin	Reduces LDL, apoloprotein B, and triglycerides
Peanut butter	Unsaturated fat, fiber, and vitamins	Heart-healthy
Turmeric, pistachio, rice bran oryzinol, and curry leaves	Leutin, cucurmin, phytosterol (stanol)	Heart-healthy foods, anticarcinogenic
Pine bark extract	Pycnogenol, antioxidant	Lowers blood pressure, relaxes vessels by nitrous oxide
Red chili	Capsaicin, carotenoids, vitamin C, B vitamins, potassium, magnesium, iron	Relieves pain, heart-healthy food
Red yeast (monascus pupureous)	Statin-like effect	Reduces cholesterol
Rice bran, soy products	Phytosterol, tocotrienol	0.8 mg per day reduces cholesterol (FDA[305])

---

305 http://www.efsa.europa.eu/en/efsajournal/doc/640.pdf

Common Foods	Active Ingredients	Ailments and Diseases Helped
Tarragon	Charvicol, polyphenol	Heart-healthy spice, great source of trace minerals, promotes appetite
Tomato	Lycopene, chromium	Antioxidant, protects against coronary heart disease, breast cancer, lung cancer
Seafood, milk, egg	L-taurine	Crosses blood-brain barrier, lowers blood pressure, reduces cholesterol
Spinach (steamed), kale, collard greens, Swiss chord, brussels sprouts, turnip, broccoli	Vitamins A, K, folic acid, and manganese	Helps blood clotting, prevents calcification of arteries, promotes bone health, fights Hodgkin's lymphoma
Walnuts	Polyphenol, omega-3 fatty acids, manganese, copper, tryptophan	Improves cognition and memory, boosts immune system, improves cardiovascular health
Watermelon	Lycopene	Promotes eye health
Hydrolyzed whey protein	Ace inhibitor like polypeptide	Controls blood pressure

13.08 Examples of Improvement of Cardiovascular Disease

We can improve our cardiovascular health by methylation diet, exercise, weight control, meditation, and reduction of day to day stress.

1. Weight reduction and muscle exercises (push-ups, pull-ups, and dips) for twenty minutes a day reduce anxiety and stress. Aerobic exercises for up to ten times more oxygen uptake are always better. Exercise and breathing during yoga reduce anxiety and stress also. Exercise reduces blood pressure, cholesterol, and weight gain. It promotes sleep.

2. Meditation or relaxed awareness by concentration and sound effects prevents heart attack by engaging the body's natural pharmacy for critical repair work.

3. Good diet and exercise help maintain a 120/70 blood pressure. Anger and depression are bad news for heart patients. One hour of exercise twice a week can help improve problems of high blood pressure, high C-reactive protein, high cholesterol, high blood glucose, high emotional stress, overweight condition, and constricted arteries.

4. Stress reduction lowers blood pressure and brings about better sleep.

5. People with robust neurons (CA3 region of hippocampus) are less prone to stress. They can better tolerate oxygen and glucose deprivation because of poor blood flow. They have more tumor-suppressor protein. This can happen with a sustained exercise program in order to decrease dense LDL and reduce total LDL. A combined program of diet, exercise, and stress management can improve HDL to above 40 mg/dl and reduce LDL cholesterol to around 110 mg/dl, with triglycerides below 100 mg/dl. Such a program can help reduce weight and remove signs of diabetes.[306]

13.09 Major Reference Institutions

Cardiovascular diseases are major health problem in the US. The following lead institutions conduct cutting edge research on dietary relations with and connections to heart health problems.

- Cagnon Cardiovascular Institute, New Jersey
- Cardiovascular Research Institute at University of Vermont
- Johns Hopkins Womens Cardiovascular Health Center
- Knight Cardiovascular Institute at Oregon Health and Science University
- Stanford Heart Center

---

306 http://jp.physoc.org/content/588/18/3347.full.pdf.

.

CHAPTER 14

Foods for Fighting Cancer

"A bad diet can corrupt our genes, their expression,
and their products—hormones, receptors, enzymes,
and growth inhibitors; it can kill us by cancer."
—Triveni P. Shukla

Cancer is a disease of the immune system, and it is very individual.[307] Its origin is either in a mutated gene or an altered epigenome.[308] As an epigenetic disease caused by environment, lifestyle, and diet, cancer affects one out of two men and one out of three women.[309] Genetic profiling of mutations that cause various malignancies may soon help remedy cancers. Research indicates that mitochondrial proteins can be used to kill cancer cells because cancer cells feed on functioning mitochondria.[310,311]

Food, alcohol, stress, overweight and obesity, sunlight, and occupational environment are controllable risk factors. Cancer can also be caused by HIV, hepatitis B, and H. pylori bacteria that cause stomach ulcers. Current statistics show that lung, colon, prostate, and breast cancer are most common. In view of many tests (such as yearly mammograms and prostate-specific antigen or PSA) having gone too far, a second opinion is almost a health-care necessity.

Foods that enhance the health of one million immune cells in our body can help us prevent cancer. *Even good old aspirin is found to reduce skin*

---

307 http://www.cancer.gov/cancertopics/understandingcancer/immunesystem/immune.pdf.
308 http://www.ncbi.nlm.nih.gov/pubmed/22789535.
309 http://www.cancer.org/research/cancerfactsstatistics/index.
310 http://www.nature.com/nrc/journal/v12/n10/full/nrc3365.html.
311 http://www.ncbi.nlm.nih.gov/pubmed/22399428.

cancer with no side effects.[312] Our common everyday foods have adaptogens, antioxidants, anticarcinogens, and cancer-fighting phytochemicals. A variety of such foods designed into our daily breakfast, lunch, snack, and dinner can deliver cancer-preventing apigenin (4,5,7 trihydroxyflavone), betaine, choline, and vitamins D and K. To be effective, the designed dish of the methylation diet should contain restricted calories, though. The mother's uterus is known to subdue response by immune cells of the fetus.[313] Without this privileged immunity, the antigens from the father and mother would not allow the growth of the fetus. Biology thus begins with this master defense of privileged immunity.

Every one of us has potential cancer cells in our body, and as such, all of us are prone to getting cancer. Although some of us may be more genetically disposed, cancer can also be caused by the stomach ulcer bacteria H. pylori and viruses like HIV and hepatitis B. Almost any body tissue can get cancerous and kill us. The elderly are more vulnerable to cancer.

Cancer is very individual, and it results from a normal cell gone out of control because of DNA mutation or change in our epigenome. The cells start out normal and then become terribly abnormal. They become enemies of our normal cells. If unchecked, they can enter, overtake, and eventually shut down a normal tissue and organ.

The medical cost of cancer has risen to $77.4 billion, not counting the productivity loss of $124.00 billion per year, and the cost is on the rise. In 2008 there were only 1.7 million new diagnoses. In 2013 there were 580,350 deaths, and diagnoses had risen dramatically. Current statistics make the point by revealing 238,590 cases of prostate cancer, 234,589 of breast cancer; 228 of lung cancer, 480 of colon cancer, and 36,800 cases of pancreatic cancer,[314] not including cancers such as leukemia, lymphoma, myeloma, bladder, esophagus, and liver and bile duct cancers.

14.01 Risk Factors

While our genes do determine our propensity to cancers, there are other risk factors that we can control to avoid them.

---

312 http://www.ucsf.edu/news/2013/06/106821/aspirin-may-fight-cancer-slowing-dna-damage.

313 http://www.cirm.ca.gov/our-progress/awards/maternal-and-fetal-immune-responses-utero-hematopoietic-stem-cell.

314 http://www.cancer.org/research/cancerfactsstatistics/cancerfactsfigures2013/index.

- genes and age
- weight, obesity, physical inactivity, and stress
- too much alcohol
- hormone diethylstilbestrol
- radon and other radiation
- too much sunlight
- HIV, H. pylori (the stomach ulcer bacteria)
- food: acrylamide in deep-fried foods, artificial sweeteners, fluoridated water, and grilled foods
- environment: asbestos, pesticides, herbicides, formaldehyde, bisphenol, phthalate, hair dyes

14.02 Genetics

Cancer is a disease of aging, and it results from a disordered and unstable genome.[315] The integrity of our genome is essential to proper functioning of 7.5 trillion cells in our body and even 2.5 trillion red blood cells. Genome instability is dictated by the foods we eat, the lifestyle we choose to live by, and the environment we live in. We know that environmental factors cause mutations responsible for lung, skin, and cervical cancers.[316] We need to control our exposure to food chemicals and additives, air and water pollution, and agricultural chemicals.

The key is controlling expression of genes. The cell cycle regulating gene P53, the guardian of the genome, is involved in most cancers. BRCA1 and BRCA2 genes are involved in breast and ovarian cancer. Ovarian cancer involves phosphoinositide 3 kinase mutation, the P13K gene. Also, the PARP inhibitor gene responsible for DNA repair is involved in many cancer cases. Acute lymphoblast leukemia manifests due to RNA that turns DNA "off" or "on" for production of an essential protein.[317]

An urgent medical need today is to decode the DNA of cancer cells and understand their clockwork. The best remedy may come from turning a cancer cell's DNA against itself. This will require an understanding of the driver mutations for each cancer and then setting right the immune system. Unfortunately we don't have this knowledge today. For now, the health-care

---

315 http://www.sciencemag.org/content/297/5581/543.
316 http://www.cancer.gov/cancertopics/causes.
317 http://cancer.sanger.ac.uk/cancergenome/projects/census/.

system depends on diagnostic tests and hopefully on possible vaccines in the near future. Our daily food is the best maintenance therapy today.

14.03 Lung Cancer

Lung cancer or carcinoma is the most common cause of death in the United States.[318] It is cancer of the lung tissue's epithelial cells, with no more than a 15 percent survival rate. The major cause is *DNA damage* and its poor repair. Small cell carcinoma can easily spread to the breast. Common dietary carcinogens are nitrosamines, nicotine via smoking, and benzopyrenes that come from smoking. Well-established major cancer risk factors are smoking, radon gas, air pollution, and occupational hazards like asbestos, which causes mesothelioma, the cancer of mesothelial cells.[319] Common symptoms are coughing up blood, weight loss, shortness of breath, bone pain, and fatigue. CT scan and radiographs give confirmatory diagnosis. Treatments include surgery, chemotherapy, and radiation. In the not-too-distant future, though, routine screening tests will include biomarkers in blood, a breath test for chemical profiling, and a saliva test for changes in airway tissue. A low-sugar and low-calorie diet with good lutein and lycopene intake is good for lung cancer patients.[320] Natural vitamin E supplements are very therapeutic.

14.04 Prostate Cancer

Prostate cancer is a slow-growing cancer of the prostate gland that produces seminal fluid. The cancer can spread to bone and lymph nodes. Men over age fifty are more prone to it. Mutations of BRCA1, BRCA2, HPC1, and vitamin D receptor gene are the common causes. Chromosomes involved are 8p, 13q, 16q, and p53. *Loss of the prostate cancer-suppressor genes is epigenetic in origin.* Major risk factors are genetics, obesity, statin drugs, high calcium, and lack of exercise. Common symptoms are difficulty in urination and erectile dysfunction. The EN2 gene that is involved in development can be detected in urine in order to diagnose prostate cancer. Cancer cells are devoid of zinc-restricting citrate production. Cancer cells thus produce energy for themselves and not for seminal fluid production. Biopsy offers a sure

---

318 http://www.cancer.org/cancer/lungcancer-non-smallcell/detailedguide/non-small-cell-lung-cancer-key-statistics.

319 http://www.cdc.gov/cancer/lung/basic_info/risk_factors.htm.

320 http://aje.oxfordjournals.org/content/169/7/815.full.

diagnosis. MRI and ultrasound are also used for diagnosis. Key treatments are surgery, radiation, and chemotherapy.

A plant-based vegetarian diet is a good preventive measure. The diet should avoid trans and saturated fats and be high in omega-3 fatty acids, vitamin D, lycopene, and antioxidant mineral selenium.[321]

14.05 Melanoma/Skin Cancer

Skin cancer or cancer of the lowest-level epidermis, called the neoplasm, is most common. This is nonmelanoma, known as basal cell carcinoma, and it accounts for 90 percent of skin cancer. Cancer of cells that produce color, melanocytes, is called melanoma. Melanoma is less common. Its diagnosis is done by biopsy. Mutations due to exposure to ultraviolet light and free radicals are major genetic causes. *DNA damage and poor DNA repair* play a huge role in onset of melanoma. Mutation of the BRAF gene is known to be involved in melanoma. Other genes implicated in melanoma are CDKN2A and CDK4 inhibitor. Major risk factors are smoking, ionizing radiation, and immune-suppressing agents. Common treatments are surgery, radiation therapy, and avoidance of UV radiation. Nonmelanoma skin cancer is much easier to cure. Gene therapy in the case of melanoma may become available in the near future.

Dietary prescriptions for skin cancer should include a high dose of antioxidants, low-dose daily aspirin, and minerals zinc and titanium oxide.

14.06 Breast Cancer

Breast cancer is the cancer of the inner lining of the milk duct. It represents 29 percent of all cancer cases, with a good survival rate of a high 80 percent. Genetic mutations involved in breast and ovarian cancers are those of cancer-suppressor genes BRCA1 and BRCA2. Gene p53 is also implicated. Major risk factors are age, female sex, and smoking. Treatments include hormone therapy, surgery, chemotherapy, radiation, and immunotherapy.

Diet control can be very helpful in treating breast cancer. A good diet for a breast cancer patient should be rich in dietary fiber, antioxidants, isoflavones, and fats and oils from seeds and nuts. Maintaining weight,

---

321 http://prostatecanceruk.org/information/living-with-prostate-cancer/healthy-living.

following a regular exercise program, and avoiding drinking and smoking can help prevent breast cancer.

14.07 Colon Cancer

Colon cancer is malignancy of epithelial cells of the colon and appendix. Irritable bowel syndrome often leads to colon cancer. Although mostly nongenetic, we now know that genes MSH2 and MSH6 on chromosome 6 are involved in colon cancer. Other gene mutation include those of the APC gene and p53. Common symptoms include rectal bleeding, anemia, weight loss, and constipation. Major risk factors are high red meat consumption, smoking, alcohol, and lack of exercise. Biopsy is the main diagnostic method. Treatments include surgery, chemotherapy, and radiation.

A good diet for colon cancer patients should include high dietary fiber, fruits, vegetables, whole grains, and an intake of 100 percent RDA of vitamin D.

14.08 Pancreatic Cancer

Pancreatic cancer is malignancy of the neoplasm of pancreatic tissue. This is the fourth largest cause of cancer death. Major risk factors are age, smoking, a diet rich in red meat and sugar, and occurrence of H. pylori stomach ulcer bacteria. Mutations of genes BRCA2 and PALB2 have been reported as causes of pancreatic cancer. Common symptoms are pain in the upper abdomen, heartburn, poor appetite, diarrhea, and weight loss. Treatments include radiation, surgery, and chemotherapy. Pancreatic cancer is very difficult to treat because immune cells are simply not allowed to do their work. Pancreatic cancer resists chemotherapy.

Diets for pancreatic cancer patients should be based on fruits, vegetables, and whole grains such that the patient gets the recommended daily allowance of vitamins D, B_{12}, B_6, and folate.

14.09 Ovarian Cancer

Ovarian cancer includes malignancies of the ovaries and fallopian tube. Common symptoms are pelvic pain, frequent urination, and bloating. Mutations of genes BRCA1 and BRCA2 are involved in ovarian cancer. Risk factors may include low-dose hormone contraception, high milk consumption, and high dietary fat consumption. Tube ligation and hysterectomy reduce chances of ovarian cancer. Diagnostic methods include a blood test for elevated levels of biomarker protein CA 125, pelvic examination, and transvaginal ultrasound. Treatments include surgery, radiation, and chemotherapy.

Diets for ovarian cancer patients should include antiangiogenic foods that contain kaempferol (endive, spinach, broccoli, kale), red onion, and tomato, which supplies lycopene.

14.10 Blood Cancers: Multiple Myeloma and Others

Blood cancers include cancers of red blood cells, white blood cells, platelets, and plasma cells that are produced in the bone marrow.

Leukemia is cancer of the white blood cells, which deprives the patient of the power of fighting infection. Lymphoma is cancer of the lymphocytes, a kind of white blood cell; it impairs the immune system. Myeloma is cancer of plasma cells. Cancerous plasma cells prevent production of antibodies, causing a poor immune system.

Diets for all blood cancer patients should provide enough protein and calories along with vitamins, minerals, phytonutrients, and antioxidants. Fruits and vegetables should be part of the diet. Diet design with proteins, antioxidants, and vitamin D_3 in the case of multiple myeloma, especially after stem-cell transplant, has become a preoccupation of both my wife, who is the patient, and myself as a caregiver. She has recovered greatly, and we know that diet is a great maintenance therapy.[322,323]

14.11 Stomach Cancer

Stomach cancer is often caused by H. pylori infection. It is a difficult-to-cure cancer and accounts for the second most cancer deaths. Genetic mutations of the AP gene on chromosome 5q and the CDH1 gene on chromosome 16 are frequently reported. Other risk factors are smoking, smoked meats and fish, salted fish and meat, and pickled vegetables. Common symptoms are absent in early stages. However, indigestion, weakness, fatigue, bloating,

---

322 http://blog.dana-farber.org/insight/2013/02/seven-tips-for-life-after-stem-cell-transplant/.

323 http://www.cancercenter.com/multiple-myeloma-cancer/nutrition-therapy/.

abdominal pain, and weight loss may be symptoms to be confirmed by the physician. Common diagnostic procedures are gastroscopic examination, CT scan, and biopsy. Treatments include surgery, chemotherapy, and radiation.

Diets based on fruits and vegetables allowing for the full recommended daily allowance of vitamins A and C, antioxidants, and high omega-3 fatty acids, represented by the Mediterranean diet, are good for stomach cancer patients.

14.12 Esophageal Cancer

Esophageal cancer is cancer of the inner linings of the esophagus. The cancer can spread to other layers of the esophagus. Afro-Americans suffer from a high incidence rate. *23andMe provides genetic risk factor analysis of esophageal cancer.* There may be up to twenty-six genes involved. Risk factors to avoid are smoking, high alcohol consumption, and gastrointestinal reflux (GRD). Common symptoms include chest pain, weight loss, pain in swallowing, persistent cough, heartburn, and hiccups. Diagnosis is based on barium-swallow X-ray, CT scan, PET (positron emission tomography), and endoscopy.

Patients need high protein and a balanced-calorie diet, but all this has to be in the form of liquid and semisolid foods such as ice cream, smoothies, pureed fruits and vegetables, scrambled eggs, soup, and Jell-O. High-protein versions of Boost and Ensure are great liquid drinks that can be consumed with straws. Puddings and milkshakes can be prepared with added protein powder.

14.13 Cancer of the Bile Duct

Cancer of the epithelial cells of the bile duct is low in incidence but hard to cure. Well-known risk factors are ulcerative colitis, hepatitis B and C, and parasites. Common symptoms are abdominal pain, weight loss, jaundice, itching, and changes in color of urine and stool. Diagnosis is based on ultrasound imaging, blood tests for alkaline phosphatase, bilirubin, and gamma glutamyl transferase, liver functions, biopsy, and tests for antigens CA-19-9 and CEA. Treatments include chemotherapy, radiation, and surgery.

A high-protein and balanced-calorie diet is necessary for the patient. Foods of liking should be selected for higher appetite. Often vitamin C and E supplements and selenium from Brazil nuts are necessary. Wine and beer can be included in the daily diet. Most important of all is to keep track of and maintain weight.

14.14 Testicular Cancer

Testicular cancer is the cancer of germ cells where not all tumors are malignant. Fortunately, the incidence is very low, 0.4 percent, and it is an easy cancer to cure. Too many chromosomes, triploid and tatraploid, are implicated in germ cell cancer. Isochromosome 12p is common in 80 percent of the cases. Common symptoms are lumps in the testicles, swelling of the testicles, pain, dull ache, low back pain, and heaviness. Diagnosis doesn't employ biopsy for fear of spreading it further. Instead, examination of the tissue is the common diagnostic procedure. Known risk factors in testicular cancer are high-fat and saturated-fat diet, high-cholesterol diet, high intake of milk and red meat, and high galactose from lactose in milk. A high-fat diet poses the highest risk for incidence of germ cell cancer.

Testicular cancer patients should select fruits and vegetables, poultry meat, unsaturated fats, and whole-grain foods in their daily diet. They should limit high-saturated fat foods.

14.15 Food and Cancer

Breast, colon, and prostate cancer patients can maintain a good life by healthy food and a physically active lifestyle. Twelve million people over sixty years of age survive cancer in the United States, and close to 14 percent of them live after the cancer diagnosis by cancer fighting foods

Findings in nutritional oncology suggest that dietary fiber, carotenoids, vitamin A, vitamin D, folic acid, vitamin B_{12}, and minerals selenium and zinc should be present in the daily diet in order to fight cancer. Dietary fiber not only acts as a prebiotic but also helps in transport of antioxidants through the gastrointestinal tract. Antioxidants should deliver 10,000 ORACS (oxygen radical absorbance capacity, expressed as micromole per 100 gram of Trolox or vitamin E equivalent) per day for protection of DNA. High fiber consumption along with fruits and vegetables is very healthful in supplying vitamins and minerals. Whole-grain consumption lowers visceral fat and reduces cancer risk. A good way to prevent triggering a latent cell to become cancerous is to (1) consume vitamin K2 from hard Gouda, Edam, and Emmentaler cheeses; (2) avoid grilled and pan-fried meats and other foods; and (3) exercise routinely every day at least five minutes vigorously for a high heart rate. Most importantly, living on a restricted-calorie methylation diet can be a boon.

14.15.01 Calorie-Restriction Diet for Energy Balance: A calorie-restricted diet leads to marked reduction in cancer risk (University of Texas, M. D. Anderson Cancer Center[324]). The benefits are believed to come in three different but related ways: (1) adequate leptin level; (2) reduced inflammation; and (3) low level of growth factor IGF-1, which regulates cell growth and death, reduced energy uptake, and reduced immunological defects.

14.15.02 Avoidance of Grilled and Charred High-Protein Foods (Meats): Carcinogens from protein-degradation products are known to be carcinogenic.

14.15.03 Minerals and Vitamins in Cancer Development: RDAs of niacin, folic acid, and vitamin C should be doubled for healthy DNA.

14.15.04 Vitamin D in Breast Cancer: Vitamin D induces cancer cell death, and up to 800 IU a day should be part of a proper diet. There are diverse metabolic, inflammatory, and immune system genes induced by vitamin D.

14.15.05 Vitamin K2 in Cancer Prevention: Menaquinone and menatetrenone are vitamin K2 chemicals available in hard Gouda, Edam, and Jarlsberg cheeses. Consuming two slices of Gouda cheese per day is a great help in fighting cancer, cardiovascular, diabetes, and osteoporosis diseases.

14.15.06 Caffeine and Cancer: Caffeine protects from cancer of the oral cavity, pharynx, endometrium, and liver. Coffee contains antioxidants and anti-inflammatory compounds such as caffeine, cafestol, and caffeic acid. Caffeine passes through the blood-brain barrier; it affects breast tumors, brain tumors, nerve cells, and neurons. However, one should keep daily intake to no more than 40 mg/day.

14.15.07 Saffron and Cancer: Saffron components have strong antioxidant, anti-inflammatory, cytotoxic, anticarcinogenic, and antitumor properties. The list of components includes crocetin, various crocins (such as picrocrocin), zeaxanthin, lycopene, beta-carotene, and safranal (the main component of saffron's fragrant essential oil), which is very good against breast cancer.

---

324 http://www.cancerandmetabolism.com/content/1/1/10

14.15.08 Betaine and Cancer: High intake of betaine (from sugar beets) and choline (eggs, almonds, garbanzo beans, seeds of pumpkin and sunflower) prevents breast cancer [325].

14.15.09 Exercise and Cancer: At least thirty to forty-five minutes five days every week is a highly healthy exercise program.

Avoiding grilled foods, consuming calorie-restricted diets, getting enough of vitamins D and K2, and enjoying the benefits of betaine, caffeine, choline, and saffron can be great nutritional tools in fighting cancers. Table 14.01 lists cancer-fighting phytochemicals from everyday common foods that can be used for variety in the daily design of foods and their preparation.

Table 14.01 Common Foods for Cancer-Fighting Phytochemicals

Common Foods	Phytochemicals	Mode of Action
Fruits & vegetables	Vitamin C, carotenoids	Effective against pancreatic cancer (Swedish study[326])
Asparagus, papaya, cabbage	Folic acid, B$_{12}$	Uracil incorporation in DNA, folate recycle, low methylation, a natural vaccine
Apple, berries, tomato, red grapes, and pea pods	Quircetin, delphinidin	Prevents H. pylori and stomach ulcer
Blueberries, strawberries	Elagic acid	Anthocyanins boost the immune system and protect capillaries from oxidative damage.
Bok choy	Isothiocyanates, brassicin	Anticarcinogens, reduced risk of cancer, reduce estrogen
Black pepper	Piperine, Chromium	Interferes with self-renewal of cancer stem cells

---

325 http://www.nature.com/bjc/journal/v102/n3/full/6605510a.html

326 http://www.swedish.org/services/cancer-institute/cancer-types/pancreatic-cancer

Common Foods	Phytochemicals	Mode of Action
Broccoli, brussels sprouts, cauliflower, and kale	Sulphoraphane and indol-3-carbinol, 3,3' diindolylmethane	Help produce cancer-killing detoxification enzymes. Quench free radicals. Indol-3-carbinol inhibits growth of estrogen-responsive cancer.
Cabbage and cauliflower	Isocyanate, vitamin K, dietary fiber	Anticarcinogenic, protects against prostate cancer, lowers cholesterol
Carrots, red, yellow	Beta carotene	Critical to DNA repair
Cheese, eggs, beef, lamb for conjugated linoleic acid	Modified omega-6 fatty acids	Anticarcinogenic, prevents atherosclerosis, builds immune system
Chicken and broccoli	Sulphoraphane, antioxidant	Fight cancer and prevent cellular damage
Citrus rind, fruits	Lemonine, vitamin C	Inhibits key protein that promotes cancer-cell growth, reduces skin cancer; a natural antioxidant
Garlic	Allinase	Detoxifies chemical carcinogens and stimulates immunity, protects red blood cell membranes, fights cancer
Ginger	Gingerone	Scavenges free radicals and protects DNA
Green tea	Epigallocatechin, gallate	Kills cancer cells
Gouda, Emmentaler, and Edam cheeses	Menaquinone and Menatertrenone	Controls cancer and prevents diabetes, heart-healthy.
Inulin/ Jerusalem Artichoke	Fructooligosaccharide	Colon health

Common Foods	Phytochemicals	Mode of Action
Horseradish, 10X better than broccoli	Glucosinolates	Improves carcinogen detoxification by liver
Leafy dark green vegetables, orange pepper	Zeaxanthin	Prevents skin cancer; a powerful antioxidant
Legumes, soy, tofu	Indol-3-carbinol, genesteine	Protects from breast, ovarian, and colon cancer
Mustard, garden greens, kale	Indol-3-carbinol	Opens tumor cell receptors to chemotherapy drugs for effect and efficiency
Watercress	Isothiocyanates	Triggers enzymes that stop cancer-cell growth
Shiitake mushroom	Lentinan	Boosts immune response
Soybeans	Saponins	Antiprotozoal (giardia) agent
Soy, barley, wheat, rye; soy contains highest level.	Lunacil, isoflavones	Named after Filipino word *lunas* for cure,; kills cancer cells
Soy foods and fiber	Isoflavone-foods, iron, molybdenum, manganese, potassium, zinc, copper, niacin, folate, protein	Exhibit estrogen-like effects, prevent breast and prostate cancer, reduce risk of heart attack
Supplements	Stevioside	Natural artificial sweetener
Tomato	Lycopene	Free radical scavenger protects against breast and prostate cancer
Turmeric	Curcumin, phenolics	A tonic for the stomach and liver and a blood purifier; anti-inflammatory and antioxidant properties

14.16 Key Anticancer Foods

Antioxidant *delphinidin from berries* and *flavonoids from broccoli and kale* are great curatives against cancer. Blueberries with anthocyanins may reduce tumor growth, spinach with folate guards against DNA damage, broccoli-like cruciferous vegetables reduce stomach cancer, and tomatoes have natural sunblock that protects against skin cancer. The antioxidant intake should be doubled for cancer patients. Rice cooked in a puree of broccoli, brussels sprouts, cauliflower (Brassica in general) for good daily dose of 3, 3' diindolylmethane is a great addition to our diet. Tables 14.01 and 14.02 have an extensive list of anticarcinogenic foods.

Table 14.02 Common Foods for Cancer-Fighting Minerals and Vitamins

Nutrient Category	Nutrient	Function
Vitamins	Vitamins A, C, and E, niacin	Repair DNA, manage telomerase length
Minerals		
Almonds, peanut butter	Magnesium	Colon cancer
Brazil nuts	Selenium	Prevent cancer
Zinc and iron	Zinc and iron	Endonuclease and superoxide dismutase, poor synthesis of catalase and hemoglobin, prevent DNA oxidation
Functional Foods		
Arginine	Amino acid	Increased lean mass and bone density by growth hormones
Ginseng	Adaptogen, ginsenosides	Herbal tonic in Europe and Russia, offer nonspecific resistance to our body against stress factors, adaptogens switch from stimulating to sedating mode

Nutrient Category	Nutrient	Function
L-glutamine	Sustamine in Japan, glutamine peptone	Increases hydration
Milk thistle	Silymarin	Protects liver; well-researched and used in Europe
Phospholipid	Phosphatidylserine	Improves cognitive function; good sources are meat products and vegetable oils
Picrorhiza (kutaki)	Glycosides, picroside 1, an antioxidant	Perennial herb grown in India and used in ayurvedic medicine in order to treat bronchial and liver problems
Rosemary	Anticarcinogens	Prevents carcinogen buildup during cooking/grilling
Salsa and avocado	Lycopene, healthy fat	Potent anticarcinogens, prevent lung and bladder cancer
Salmon and watercress is a great combination.	Omega-3 fatty acids from salmon, phenylethyl isocyanate from watercress	Prevents leukemia and kidney cancer
Saffron	Crocin	Anticarcinogenic, antioxidant, lowers cholesterol and triglycerides, improves memory
Schisandra Chinese berries	An adaptogen	Used as medicine in China for centuries, it helps liver functions.

Nutrient Category	Nutrient	Function
Thyme, tomato sauce, peppermint, red wine	Apigenin, flavonoid	Anti-inflammatory, antioxidant, anticarcinogens, gout protection by blocking uric acid
Whey protein products	Lactoferrin, glycopeptides	Promote eye health, prevent tumor growth, improve immune system, helps iron absorption
Yellow onion and turmeric	Cucurmin from turmeric, quircetin from onion	Reduce precancerous polyps, anticarcinogenic
Antiangiogenic food such as Apples, artichoke, blackberries, blueberries, broccoli, brussels sprouts, bok choy, dark chocolate, cranberries, cauliflower, cherries, garlic, ginger, ginseng, grape-seed oil, green tea, lemon, kale, licorice, parsley, pineapple, grapefruit, lavender, mushroom, nutmeg, olive oil, oranges, raspberries, pumpkin, soybean, strawberries, red grapes, red wine, tomatoes, turmeric	These foods containing many phytochemicals prevent formation of blood vessels tfor cancer cells and starve them to death.	

The key to designing foods for cancer patients is variety. The variety comes from a list of berries, beans, broccoli, cabbage, carrots, cauliflower, chili pepper, cruciferous vegetables, dark chocolate, dark green vegetables for folic acid, figs, flaxseed, garlic, ginger, grapes, grapefruit, green tea, lentils, kale, licorice, milk for vitamin D, mushroom, oranges, papaya, nuts, rosemary, seaweed, smoothies, soy foods, sweet potato, wine for resveratrol, turmeric, tomatoes, turnips, and whole grains. We can select a combination that can easily deliver vitamin and mineral antioxidants, omega-3 fatty acids, and other heart-healthy phytochemicals.

14.17 Major Reference Institutions

The following are major research institutions that lead in cancer research, particularly diet-cancer relationships.

- Cancer Research Institute
- Center for Cancer Research
- Fred Hutchinson Cancer Research Center
- Dana Farber/Harvard Cancer Institute
- Geoffrey Beene Cancer Research Center at Memorial Sloan-Kettering Cancer Center
- Mayo Clinic
- M. D. Anderson Cancer Center
- MIT Center for Cancer Research

CHAPTER 15

Foods for Long Life

"Life expectancy would grow by leaps and bounds if
green vegetables smelled as good as bacon."
—Doug Larson

Our evolution with oxygen is paradoxical because our metabolism and physiology require close to 1.5 Kg highly electronegative oxygen daily. Its utilization has to be efficient, and none of the oxygen should go amok. The fact is that less than 0.25 percent of it does go amok, as reactive oxygen species or free radicals that can damage our DNA, the brain tissue, and our arteries. This happens more as we age. But father evolution has been extremely careful in installing electron-donating antioxidants like vitamins A, C, and E; endogenous antioxidants like glutathione and melatonin; and antioxidative enzymes like superoxide dismutase, peroxidase, catalase, and $CoEnzQ_{10}$. Therefore, we need to manage dietary antioxidants during our life, much more so during our senior years when mitochondria don't function well and more electrons go amok.

Thus the daily foods listed in Table 3.04 become much more critical as we age. Food selection during our senior years should target maintaining a healthy brain and healthy vital systems. High-antioxidant foods can prolong the life of each and every one of our body systems because they help reduce stress and manage electrons and protons gone amok. Exercise, meditation, and yoga can augment the health value of such foods and thus enhance longevity and quality of life. We have mapped our genome, but reengineering it for a long life is not yet possible. The means of defeating death may be a kind of dream at the moment. However, with the right foods and right epigenome, we can get rid of inflammations, infections, cancer, autoimmune diseases, and toxins. We can thus become less vulnerable to old-age physiology. We can maintain our epigenome by use of properly selected food, lifestyle, and environment around us. Our need for daily intake of antioxidants, vitamins D, B$_6$, folic acid, B$_{12}$, and K2 (arginine, tryptophan, and tyrosine), and minerals magnesium, manganese, copper, cobalt, selenium, and zinc—just call them methylation or gene-expression nutrients—becomes critical. Actually senior citizens should double their daily antioxidant intake and get enough vitamin B$_{12}$. The latter fights dementia or Alzheimer's disease.

Lifestyle and daily nutrition can slow down shrinkage and wrinkles, dopamine depletion, failure of sense organs, and neuronal death. Memory and cognition can thus be preserved for a relatively longer time. Lifestyle and nutrition can improve gene expression and neuronal connections. This chapter has practical information on foods for DNA protection and for memory and cognition. It emphasizes exercise, sleep, low-stress living, and living in interaction with people. We should fashion a lifestyle and preserve it for the long years that we intend to live. For a long life, we need to keep our mitochondria stable and strong.[327,328]

Centenarians in Loma Linda, California (Seventh-Day Adventists), Nicoya in Costa Rica, Okinawa in Japan, and Sardinia in Italy believe that plant-based foods, family, and a socially ordained lifestyle give them valid reasons to live and keep living.[329,330]

By design, aging is a diet-dependent slow process, and stress is a dominant factor that accelerates aging. Actually 70 percent of aging is due to environment. We need to have our body and mind connected in order to slow down aging because health is an adaptation of our 200-million-year-old physical or reptilian brain, which can be conditioned for a longer life.[331] The adaptation has evolved via the neocortex, which controls emotion and thought. The thinking brain (the neocortex) can be made extrafunctional by exercise, faith, friends, and food, and aging can be delayed.[332] This is a great

327 ht http://www.ellisonfoundation.org/research/mitochondrial-damage.

328 http://www.ncbi.nlm.nih.gov/pubmed/22399429.

329 http://ngm.nationalgeographic.com/ngm/0511/feature1/.

330 http://www.nbclosangeles.com/news/health/Loma-Linda-Leads-in-Longevity-123957424.html.

331 https://www.lef.org/magazine/mag2012/feb2012_Novel-Magnesium-Compound-Reverses-Neurodegeneration_01.htm.

332 http://www.canada.com/health/Exercise+best+anti+aging+treatment+study+suggests/4321718/story.html.

news for the fast-expanding senior citizens in America, estimated to be 70 million by 2030, almost double what it is today.

15.01 Genetics and Our Genes

Progeria children age early.[333] There are forty-four genes that slow down as we age. Less than a dozen that are involved in metabolism, growth, and fat and cholesterol processing, if slowed down, accelerate aging. We know that the IGF-1 gene, which produces an insulin-like growth-promoting protein, if defective, causes poor growth and shorter life.[334] The most important is the gene for telomerase, which prevents shortening of DNA. An example is the FOXO3 gene variant, which adds years to long-living centenarians.[335] Equally important is the daf2 gene. But epigenetics and stress-free social bonds are more powerful in extending our lives and preserving our immune system.[336] Antioxidants, vitamins, and anti-inflammatory foods keep the human genome stable for a disease-free longer life.

Oxygen, which keeps us alive, becomes a bit difficult to manage as we age. Our mitochondria begin to leak more electrons that produce free radicals of superoxide and hydrogen peroxide and cause oxidation beyond what our body's antioxidant system can effectively quench.[337] Sustained stress accentuates this free-radical problem. There are five practical ways to extend life.

1. Good supply of energy to our cells.
2. Good delivery of exogenous antioxidants via our daily food.
3. Good delivery of a daily dose of nutrients that can enhance the supply of endogenous antioxidants like glutathione.
4. Fortification of defenses by intake of coenzyme Q_{10} from canola oils and nuts. Fruits, vegetables, eggs, and dairy products are other sources.
5. Reduction of the risk of diabetes, an ailment that is connected to all other chronic diseases of the heart, eyes, cancer, and liver, because glucose metabolism in our glucose factory is key.

---

333 http://www.progeriaresearch.org/meet_the_kids/.
334 http://www.nature.com/ejhg/journal/v21/n6/full/ejhg2012223a.html.
335 http://www.stanford.edu/group/brunet/Carter%20ME%202007.pdf.
336 http://www.cossa.org/caht-bssr/linda%20george.pdf.
337 http://altmedicine.about.com/od/antiagingdiets/a/antiaging_antioxidants.htm.

Signs of aging include wrinkles, degeneration of bones, organs, and glands, and weak memory. Such signs can be delayed by eating foods that repair DNA and chromosomes and boost energy-producing activity of mitochondria in our 7.5 trillion cells. Such foods must provide us daily doses of vitamins B_6 and B_{12}, anti-inflammatory vitamin D, alpha-lipoic acid, taurine, and creatine. No doubt time weakens our organs, brain, and muscles. We become less functional as we age because the gene-expression system becomes faulty.

Good health can be diagnosed simply by observing that there are no problems on the skin below the toes, no signs of skin cancer, no bleeding of gums, minimal change in the smell of breath, a constant weight without sudden loss, no fainting, regular good nights' long sleep, normal sex life, good memory and cognition, and regular bowel movements without frequent constipation.

15.02 Longevity by Managing the Daily Food

The world over, 35.6 million people suffer from dementia. In the United States, Alzheimer's affects 5 million,[338] and another 1.8 million have other forms of dementia. This can be avoided by choosing foods that provide vitamin B_{12} or using B_{12} supplements.

Alzheimer's is caused by beta-amyloid deposits and narrow blood vessels to the brain (vascular dementia). Dementia is caused by alpha-synuclein protein deposit.[339] There exists an enzyme that causes overproduction of beta-amyloid. The brain gets atrophied and loses synapses. Its fatty tissue suffers from oxidative damage. DHA in brain cells and phospholipids in the brain-cell membranes add extra years to our life. A healthy diet can prevent 80 percent of heart-disease problems and 40 percent of cancer as we age (World Health Organization[340], [341],[342]). Foods during senior years should maintain cellular function and long life of telomere.

Furthermore, foods should maintain intestinal morphology with TGF (transforming growth factor), sialic acid, and phosphatylcholine; regulate metabolism with minerals calcium, iron, and vitamin E; bind and eliminate

338 http://www.ninds.nih.gov/disorders/dementias/detail_dementia.htm.
339 http://www.ninds.nih.gov/disorders/dementias/detail_dementia.htm.
340 http://whqlibdoc.who.int/trs/who_trs_916.pdf
341 http://www.who.int/dietphysicalactivity/en/
342 http://www.who.int/nmh/publications/fact_sheet_diet_en.pdf

toxins and pollutants; help manage pathogens by lactoferrin and lysozyme; boost immunity via prostaglandins and cytokines; help manage hormones and neuropeptides; and promote cognition and memory. Foods should include choline from eggs, cinnamon, brussels sprouts, phytonutrients from foods of bright colors, and probiotic and prebiotic dietary fiber.

Researchers all over the world are investigating the genetics of personal diseases and the molecular basis of appetite and metabolic regulation. Researchers are constantly in search of foods for health and long life, including those that reduce exogenous stress, kill pathogens, enhance immunity, and promote optimal growth and development.

Our daily food should include 3-phenyl indol 3-acetamide, which reduces anxiety and boosts memory, and ATP manufactured with ribose [343], [344]. Toasted vegetables like asparagus add phytochemicals and prebiotic dietary fiber. Ubiquinol from yeast extract and avocado, a precursor of CoQ10, are good for diets for seniors. Foods for the elderly should include broccoli-type cruciferacea vegetables and high-quality protein containing arginine, tyrosine, and tryptophan.

Proteins for neurotransmitters—brain cell-to-brain cell communication—are needed daily. Tryptophan conversion to serotonin is necessary not just for sleeping but also for learning and memory. Dietary carbohydrates move tryptophan across the blood-brain barrier. Omega-3 fatty acids such as EPA and DHA are a must in foods for brain health of the elderly. Antioxidants of polyphenols (catechins, epigallocatechins, thearubigins, theaflavins) can prevent DNA damage and preserve cognition.

Cold-pressed plant oils, hemp oil for omega-3, avocado oil for lycopene and carotene, and buckthorn oil-based lotions are life-sustaining and life-extending food nutrients. We should use buckthorn and vegetable oil lotions. Tables 15.01, 15.02, and 15.03 represent a compilation of data respectively on foods that protect DNA, foods for memory and cognition, and foods for longevity. A close examination of the information in these tables reiterates and suggests regular consumption of foods containing vitamins A, beta-carotene, C, D, and E, with higher doses of B_1, $B2_2$, B_6, and B_{12}; minerals calcium, chromium, copper, iodine, iron, selenium, and zinc; a variety of antioxidants, including polyphenols and flavonoids; and high-quality proteins with respect to amino-acid balance. I extend the advice to

---

343 http://www.mdpi.com/1420-3049/18/6/6620
344 file:///C:/Users/Owner/Downloads/molecules-18-06620.pdf

taking supplements of vitamin D$_3$ and omega-3 fatty acids. This then is the methylation diet. A lot more nutrients can come from a selection of fruits, vegetables, tree nuts, seeds, and dietary fiber. Yellow, orange, and green color in the case of fruits and vegetables is a good guide for sufficiency of antioxidants, phytochemicals, and phytonutrients. Tree nuts and seeds of sunflower and pumpkin are great sources of fiber, minerals, good fat, protein, and vitamin E. Occasional consumption of red wine in a social setting, while snacking on nuts and seeds, is very good for relaxation and stress control.

Table 15.01 Foods for Protecting DNA

Common Foods	Active Ingredient	Effect
Chicken soup	Cysteine amino acid	Anti-inflammatory
Astragals	Cycloastrogenol, fiber, telomerase activator; an adaptogen	DNA damage control, antiaging, anti-inflammatory, hepatoprotective, anti-HIV
Basil, great for salads, eggplant, tomato preparations	Vitamin K, eugenol, lemonine, flavonoid, orientin, vicerin	DNA damage control
Oranges	Vitamin C, hesperidins	DNA damage control, prevent cancer and varicose veins, improve digestion
Pomegranates	Polyphenols	DNA damage control, prevent cancer, heart-healthy
Black and green tea	Catechins	DNA damage control
Gingseng, ashwagantha,	Adaptogenic foods	Epigenome stability and homeostasis, stress control
Licorice	Glycirrhizin	Promotes longevity
Shiitake & maitake mushrooms	Vitamin D and B vitamins	Methylation nutrients

Table 15.02 Foods for Memory and Cognition

Common Foods	Active Ingredients	Ailments and Diseases Helped
Salmon, olive oil	Omega-3 fatty acids (EPA, DHA)	Anti-inflammatory and neuroprotective, brain-cell membrane function, improved cognition
Blueberries	Anthocyanins	Boost memory and protect from urinary infection
Cacao	Theobromine	Boosts mental capacity
Chocolate	Flavonoids and catechins, anandamide, phenyl ethylamine	Antioxidants, reduced heart problems; anandamide brings relaxation, a natural opiate
A mononucleotide	Citicholine	Neuroprotective effect, improved cognition, choline-like effect
Tea	Theanine, catechin, flavonoids	Improves cognition
Mushrooms	Vitamin D	Improve cognition
Walnuts	Polyphenols, omega-3 fatty acids, copper, manganese, tryptophan	Boost cognition, improve neural connections, boost immune system

Table 15.03 summarizes features of diet and supplements during our senior years.

Table 15.03 Managing Diet and Food Supplements for Longevity

Food	Intake
Water	Drink at least two liters a day.
	Avoid refined carbohydrates (sugar, cola, sweets). Avoid pesticides, herbicides, pollutants, carcinogens, and chemicals in daily food by way of additives.

Food	Intake
Major nutrients	Proteins with balance between essential and nonessential amino acids as snack foods (up to 30% of total calories per day). Take high-arginine proteins like nitrous oxide, good for blood circulation. Complex carbohydrates reduce stress. Good fats for 25% of total calories can come from pistachios, almonds, olive oil, and avocado, but no more than 25% of calories should come from even these good fat sources. A very good fat is from avocado, which also provides mannoheptulose, a natural longevity nutrient. Omega-3 fatty acids are good against inflammation. They are the key to electrical signaling and messaging (DHA and EPA make up 20% of our brain-cell membrane). Use salmons and walnuts for omega-3 fatty acids. Increase dietary-fiber consumption by way of lentils, cooked chickpeas, apples, and raspberries, as well as probiotics like yogurt and sauerkraut. Flaxseed lignans are excellent dietary fiber. Egg yolk, cabbage, and cauliflower for at least 550 mg per day of brain-food choline, a B vitamin. It is a memory food because it helps makes acetylcholine, a neurotransmitter. Soy nuts for phospatidylserine
Vitamins and minerals	Vitamins A, beta-carotene, C, D, E, and K Higher dose through foods of B_1, $B2_2$, B_6, and B_{12} Calcium, chromium, copper, iodine, iron, selenium, zinc
Phytonutrients	Coffee and chocolate: Caffeine, theophylline, and theobromine are also present in chocolate. Coffee in moderation is good, but none after 3:00 p.m. Caffeine is known to have antiaging benefits. One cup of coffee gives 150 mg caffeine. 200 mg (1.5 cups) can reduce stress. Carbonated 12-oz cola has only 15 mg of caffeine (Absorbs in 45 minutes. A 70-Kg person should not consume more than 280 mg/day). Green tea polyphenolic antioxidants, in particular epigallocatechin-3-gallate sold by Unilever, United Kingdom); amino acid theanine in black tea for relaxation

Food	Intake
Antiaging foods	Almonds, leafy vegetables, astragals, avocado, chocolate, beans, berries, green tea, melon, seaweed, turmeric, colorful vegetables and fruits, wine, curcumin from turmeric, ferulic acid from cereal bran, mericetin flavonoid from walnuts, rosemarinic acid from sage, mints, and rosemary herbs, red and yellow peppers, resveratrol from red wine, picnogenol, and lipoic acid. Use a combination of ground cloves, ground allspice, sage, marjoram, cinnamon, oregano, and thyme in your recipes of choice. These provide antioxidants, antimicrobials, and anticarcinogenic nutrients.
Food supplements	Multivitamin tablets are a good thing to take. Take thyroid hormones under prescription and *Antioxidants four times a day* by way of tea, coffee, and chocolate. **Gamma Amino Butyric Acid (GABA)** can cause insomnia and mental disorders. Use GABA supplements only under a physician's advice. CoEnzQ10 ia an other good supplement
Good foods	A glass of wine or grape juice for resveratrol, which has a calorie restriction-like effect. Dark chocolate with tryptophan induces serotonin and endomorphins for sleep and painkilling; it has high flavonoids, which reduce platelet clumping.

Quality in day-to-day life improves by cultivating a healthy lifestyle, reducing weight, avoiding blood glucose, reducing the stress hormone cortisol, getting at least five to ten minutes of sunshine daily, and restricting calories to 80 percent by fasting once a week, and improving cellular function by routine exercise. Table 15.04 prescribes thirty minutes of exercise, a minimum of seven hours' sleep, use of antioxidant foods to quench reactive oxygen-free radicals that regularly escape energy-producing mitochondria in the cell, and a routine practice of one's chosen faith.

Table 15.04 Managing Longevity by a Designed Lifestyle

Exercise	At least thirty minutes daily of walking and moving around for good health and cognition. Commit to breathing exercises at least one hundred times a day, fifty times each nostril.
Sleep	Sleep for at least seven hours a day. This is critical for growth hormones.
Manage stress	Cultivate mental powers, be an extrovert, be purposeful, and stay calm. Consume food antioxidants and avoid oxidative stress.
	Avoid Parkinson's, Alzheimer's, diabetes, dementia, and depression by routine methylation diet.
Faith, friends, and social activity	Cultivate relationships, connect with people, and be on the go to creating something new.

Serious research is underway to slow down or even relatively reverse aging. The best proof to this general effect is increasing life expectancy. A tremendous amount of research data and the speed at which they can be meta-analyzed give hope for a sustainable longevity with quality of life. We must understand the risk of too many medications, urinary infections, normal pressure hydrocephalus (spinal fluid in the brain), and an unhealthy thyroid. We must seek advice from our physicians punctually and periodically.

15.03 Foods for Routine Therapy

Seniors should double daily intake of antioxidants and vitamin D_3, consume prebiotic dietary fiber and probiotic foods, get digestive enzymes from grains and legume sprouts, and have one-half ounce of walnuts and flaxseed per day for anti-inflammatory omega-3 fatty acids.

15.04 Major Reference Institutions

The following research institutions perform cutting edge research on anti-aging and longevity.

Buck Institute, Swiss Federal Institute of Technology, National Institute on Aging, and University of Michigan provide us confirmed knowledge about the aging process and prolonging age by proper diet. The studies on the Okinawa diet have been very revealing.

CHAPTER 16

Antioxidants, Vitamin D, Enzymes, and Gene Therapy

> "Fighting electrons gone amok, preventing inflammation by
> sufficient vitamin D_3 and omega-3 fatty acids, and helping
> daily digestion with enzymes in sprouts and vegetables
> can prevent chronic diseases and extend life."
> —Triveni P. Shukla

We live by our genes and gene products: enzymes, cofactors, and antioxidant power of vitamins and minerals. Our diets and food habits have evolved as an adaptation to oxidative stress because of a miniscule amount of oxygen we breathe always going amok. Actually, the story of electrons and antioxidants as it relates to our health is tied to our evolution.[345,346] As oxygen-dependent creatures, we live by electrons hopping over proteins and by extruding protons across the mitochondrial membrane[347] for energy production as ATP. Toxic reactive oxygen species (ROS) or free radicals are byproducts of our daily metabolism, which consumes 1 kg pure oxygen. Electrons jump off oxygen, or as they say, "leak away" and produce reactive oxygen species. Antioxidants like endogenous glutathione, carotenoids, and vitamins A, C, and E, and minerals selenium and zinc as part of our enzymes keep a balance by quenching or preventing formation of ROSs. Melatonin, as an indolamine-type terminal antioxidant, is better than vitamins A, C, and E in quenching even more dangerous peroxinitrite free radicals of nitric

---

345 http://www.ncbi.nlm.nih.gov/pmc/articles/PMC3249911/.
346 http://www.ncbi.nlm.nih.gov/pubmed/14527634.
347 http://nhscience.lonestar.edu/biol/etc/respirat.html.

oxide.[348,349] Although noncaloric, antioxidants are necessary for the balance and maintenance of the reduction-oxidation equilibrium, a precondition for homeostasis and a balanced body chemistry. Antioxidants are part of our grand defense mechanism. ROSs are not all bad, though. Needed on demand by our immune system to kill pathogens, they must be managed and managed well by dietary antioxidants in order to avoid cells' protein, lipid, and DNA damage.[350]

Oxygen metabolism for production of energy molecule ATP in our cells is regular and almost 99 percent efficient. But one percent of electrons do leak away from stepwise molecule-to-molecule electron transfer in our cells' mitochondria. The leakage increases as we age and ROSs with unpaired electrons are produced. They damage DNA, kill cells, prevent appropriate nzyme production, cause inflammation, and alter structures of cholesterol and cell membranes. The result is oxidative stress. Antioxidants in our plant-based foods neutralize these free radicals in our body just as they do in the plant systems. They either prevent formation of free radicals or outright remove them. As reducing agents, antioxidants work as water-soluble molecules (vitamin C, glutathione, lipoic acid) in the cell or as lipid-soluble molecules (vitamins A, D, E, and K) in the cell membranes. ROSs and electron-managing antioxidants permeate our whole life, and they must be kept in balance.

As anti-inflammatory agents, antioxidants protect against many chronic diseases, such as diabetes, obesity, metabolic syndrome, infections, cardiovascular diseases, eye disorders, joint pain, cancers, ulcerative colitis, bowel diseases, gingivitis, asthma, and allergies, by preventing inflammation. The most powerful antioxidant is natural glutathione produced in our body.[351] Fresh fruits and vegetables and meat are good sources of glutathione, which must be complemented with the minerals selenium and zinc. In order to avoid bad health complications of meat eating, it is wiser to get glutathione from six daily servings of fruits and vegetables, which also provide vitamin A, carotenes, and vitamin C. Vegetable oils provide vitamin E, the most important lipid-soluble antioxidant. We can get polyphenol antioxidants from coffee, soy products, grapes, wine, chocolate, and

348 http://www.menshealth.com/health/power-melatonin.
349 http://www.lef.org/magazine/mag2012/sep2012_7-Ways-Melatonin-Attacks-Aging-Factors_01.htm.
350 http://www.ncbi.nlm.nih.gov/pubmed/21356165.
351 http://www.glutathionescience.com/glutathione_science_002.htm.

cinnamon. Pyrroloquinonline quinone in our diet (fermented soy products, edamame, spinach, green peppers) is a very powerful antioxidant.[352,353,354] Micronutrient minerals selenium and zinc are necessary for the functioning of antioxidative enzymes.

Common foods such as cereal brans, herbs (basil, dill, oregano, peppermint, sage, savory, tarragon, and thyme), spices (cardamom, cinnamon, clove, coriander, cumin, garlic, ginger, mustard seed, onion, oregano, parsley, paprika, pepper, and turmeric), dried fruits (apples, dates, figs, pears, plums, and raisins), seeds, and nuts can provide sufficient level and variety of antioxidants for quenching free radicals, boosting our immune system, preventing mitochondrial DNA damage, and preventing damage of cell membranes. Also, antioxidants ward off infections and flu.[355] Even daily exercise produces free radicals that the antioxidants can quench. The brain is most vulnerable to oxidative stress, which causes neuron cell damage in the case of Alzheimer's and Parkinson's diseases. Antioxidants eliminate this serious vulnerability of old age.

Enzyme therapy includes consumption of a number of enzymes. Superoxide dismutase quenches superoxide free radicals to oxygen and hydrogen peroxide, catalase quenches hydrogen peroxide to oxygen and water, and peroxidase enzymes do similar quenching. Vitamin D and targeted gene products are yet other therapies for chronic disease management.

16.01 Antioxidant Therapy

We live well because our cells always work well with oxygen. Our DNA punctually keeps giving right instructions for making right proteins, enzymes, and molecules for signaling. Our body, the molecular factory, keeps running efficiently. However, inefficiency and imbalance in the mitochondria of our cells can convert vital oxygen into free radicals or reactive oxygen species (ROSs) with very reactive unpaired electrons. Examples of such free radicals are hydrogen peroxide, singlet oxygen, and hydroxyl ion. They steal electrons from our molecules of life, such as DNA and RNA, and damage them; they damage our cell membranes. Free radicals

---

352 http://www.ncbi.nlm.nih.gov/pmc/articles/PMC1136652/.

353 http://www.swansonvitamins.com/health-library/products/pyrroloquinoline-quinone-brain-health-anti-aging-supplement.html.

354 http://examine.com/supplements/Pyrroloquinoline+quinone/.

355 http://umm.edu/health/medical/reports/articles/ear-infections.

can even kill cells. They oxidize LDL in our bloodstream. Toxic metals in our foods accelerate this stealing of electrons, which is behind stress and 60 percent of all chronic diseases.

Free radicals oxidize fatty materials and harden them, causing loss of cell-to-cell communication, reduction in supply of key life-sustaining molecules, and mitochondrial DNA damage, with loss of energy production. The chain reaction produces more and more free radicals. They must be stopped by antioxidants (see tables 16.01 and 16.02) because extensive damage simply means death of our cells, causing ailments and diseases. Table 16.01 has indications of daily intake for foods on which reliable data are available.

Antioxidants are reducing agents. They can donate electrons to free radicals before they begin to steal from DNA and RNA. Colored plant foods loaded with antioxidants have been an evolutionary food-selection criteria for antioxidants. They were needed to quench or sequester free radicals in plant leaves.[356] When we consume them, antioxidants help improve heart and eye health, the immune system, and cognition. They improve our vision, extend our life span, help gene expression, help kill cancer cells, prevent inflammation, and promote signal transduction. They manage electrons gone amok.

We can pick daily fruits and vegetables for flavonoid and other antioxidants (see table 16.02) for an effective dose with a variety of taste and flavor. We can use regular oranges for flavonones; other fruits and vegetables for flavonols; tofu for isoflavones; and tea, coffee, whole grains, and chocolate for a variety of polyphenols.

In general, antioxidants reduce oxidative stress. A combination of antioxidants, fiber, phytochemicals, and antioxidant minerals selenium and zinc in our daily diet can help maintain a healthy body and a healthy mind. Vitamins C and E as antioxidant vitamins slow down aging and extend life. Vitamin C or ascorbic acid neutralizes free radical hydrogen peroxide, and vitamin E protects the cell membranes. A single molecule of vitamin E neutralizes 120 free radicals.

---

356 http://www.ncbi.nlm.nih.gov/pubmed/16217560.

Table 16.01 Common Antioxidants in Our Foods

Antioxidant	Daily Intake	Function
Alpha-carotene/ lutein	2 mg	Found in carrots, tomatoes, winter squash; great for eye health
Beta-carotene	6 mg	Carrots, winter squash, bell peppers, and pumpkin are good sources. It is pre-vitamin A carotenoids.
Alpha-lipoic acid	0.5 gram	An antioxidant from broccoli, spinach, collard greens, and chord
Anthocyanins	12.5 mg	Cyanidin, delphinidin, malvidin, pelargonidin, peonidin, petunidin, pterstibene (resveratrol-like), stilbenoids, resveratrol. Use bright-colored berries, red onions, apples, cocoa, grapes, and eggplant.
Proanthocyanidins	100 mg	Tea, black currants, cranberry, grape seed and skin
Astaxanthin	4 mg	An approved food color as terpene carotenoids. Good sources of this eye-health antioxidant are salmon, trout, and shrimp.
Canthaxanthine	150 mg	A powerful antioxidant
Capsaicin	0.5 mg	A capsaicinoid from chili pepper against pain, cancer, and psoriasis. It increases energy metabolism.
Cucurmin	75 mg	As a natural phenol from turmeric, it is a great antioxidant; anti-inflammatory.
C3G glucoside		Anthocyanin flavonoid antioxidant from purple corn, popular in Mexico, blackberries, and black currants; good for eye health
Chalcones/quircetin		Great antioxidant in beer (hops)
Chicoric acid		Basil leaf and chicory root antioxidant

Antioxidant	Daily Intake	Function
Coenzyme Q_{10}	150 mg	Natural electron transport chain antioxidant found in meat, oils, fish, nuts, beans, seeds, eggs, and chicken
Pyrroloquiniline quinone	2 mg	Fermented soy products, edamame, spinach, green peppers
Selenium, a mineral	100 mcg	Works with vitamin E; an antioxidant that protects cells
Superoxide dismutase		An iron and manganese enzyme that quenches free radicals. Cabbage and broccoli are good sources.
Ubiquinol	60 mg	Reduced electron-rich form of powerful antioxidant; donates electrons and quenches free radicals. Like coenzyme Q_{10}. Chicken, beef liver, and broccoli are good sources.
Vitamin C	120 mg	Donates electrons and quenches free radicals
Vitamin D_3	1000 IU	Acts like cancer drug transferrin; inhibits iron-catalyzed oxidation
Vitamin E, tocopherol	30 mg	Breaks free radical chain reaction; present in most virgin oils
Zinc	10 mg	It is an electron-donating antioxidant to OH (hydroxyl) radicals. Helps 100 enzymes, immune system, taste & smell, and growth.

Note: Beta-carotene, lycopene, beta-cryptoxanthin, zeaxanthin, astaxanthin, and lutein belong to the carotenoid family of antioxidants. Flavonoid antioxidants include quircetin, rutin, hesperidins, apigenin, and luteolin, and then there is the famous epigallocatechin-3-gallate from green tea. Isothiocyanates, resveratrol, and tannins are other antioxidants.

Whereas pH is an indicator of proton activity in our body, oxidation-reduction potential—call it electromotive force—is an indicator of electron activity. Antioxidants modify cellular "oxidation-reduction" state.[357] Oxygen

---

357 http://www.ncbi.nlm.nih.gov/pmc/articles/PMC2952083/.

radical absorbance capacity (ORAC) values and the power of antioxidants in common foods are listed in table 16.03. *Vegetables* such as broccoli, bok choy, cauliflower, cabbage, collards, carrots, garlic, horseradish, kale, onions, tomatoes, and spinach contain powerful antioxidants (see table 16.02). So do *fruits* such as apples, apricots, berries, citrus, grapes, and pears. *Herbs and spices,* including cinnamon, chocolate and cocoa, tea, and turmeric, are also great sources of antioxidants. Our daily food for antioxidants simply means vitamin C, vitamin E, beta-carotene, lutein, lycopene, coenzyme Q_{10}, pyrroquinoline quinone, and polyphenols from a variety of foods. *Phenolic* antioxidants from whole grains, fruits and vegetables, cocoa powder, and green tea are also anti-inflammatory, cardioprotective, and anticarcinogenic.[358] Taken with (1) an omega-3 and omega-6 blend of fatty acids in a ratio of 1:2 and (2) foods rich in fiber and low-sugar complex carbohydrates, antioxidants can prevent chronic disease and enhance the quality of life for a long time. Actually the underlying science is the basis of health claims approved by the FDA. Good antioxidants are the ones with lower oxidation-reduction potential. Blackberries, walnuts, strawberries, artichokes, cranberries, brewed coffee, raspberries, pecans, blueberries, ground cloves, grape juice, and unsweetened baking chocolate are the major antioxidant-bearing common foods.[359] Concentration (mMole/100 g) of antioxidants in foods[360] and their oxdn-redn potential[361] have been variously reported.

Table 16.02 Antioxidants from Common Fruits and Vegetables

Antioxidant	Average Daily Intake	Function
Flavonoids	190–200 mg	Cancer prevention and protection from inflammation
Flavanones	15 mg	Antioxidants
Hesperidins		Anti-inflammatory and immune-boosting citrus antioxidants

358 http://www.ncbi.nlm.nih.gov/pubmed/22070680.
359 http://www.ncbi.nlm.nih.gov/pubmed/16825686.
360 http://voh.chem.ucla.edu/vohtar/fall06/classes/153C/pdf/Best%20antioxidant%20foods.pdf.
361 http://www.sfrbm.org/frs/Frei.pdf.

Antioxidant	Average Daily Intake	Function
Naringinin		Citrus and grapefruit fruit antioxidant
Flavonols	160 mg	Glycosilated flavonoids from black tea, buckwheat
Catechins, epicatechins, gallocatechins, kaempferol; quircetin, theaflavin	A daily onion and fruits and vegetables can provide cancer-fighting flavonoids in sufficient amounts.	Antioxidants in cocoa products and chocolate, peaches, even vinegar. Kaempherol is a glucoside antioxidant against cancer and cardiovascular disease (found in apples, broccoli, kale, cabbage, beans, tea, capers); quircetin is a flavonoid antioxidant found in capers, kale, dill, and sweet potatoes; and theaflavins are polyphenol antioxidants from tea against HIV and cancer.
Flavones, apigenin, luteolin, & tangeretin	20–50 mg	Beneficial effects against atherosclerosis, osteoporosis, diabetes mellitus, and certain cancers. Cereals are a good source.
Eugenol	Up to 150 mg	Very potent antioxidant from cloves
Pycnogenol	20 mg	A proanthocyanindin polyphenol antioxidant
Glutathione	500 mg	Our body produces most of our daily need for this powerful antioxidant
Isoflavones, daidzein, genistein, glycetin	25 mg	Anticarcinogenic isoflavones natural to soy foods provide up to 30 mg/day in Asiatic countries.
Lutein & zeaxanthin	2.5 mg	Yellow xanthophyl carotenoid of green vegetables, broccoli, kale, turnips, and Swiss chord
Lycopene	20 mg	Tomatoes and watermelon
Melatonin	5 mg	A terminal antioxidant and sleep-wake rhythm hormone from olive oil, beer, wine, grape skin, and walnuts

Antioxidant	Average Daily Intake	Function
Phenolic Acid Esters	800–1000 mg	Antioxidants from daily fruits and vegetables
Chicoic, chlorogenic		Coffee, cinnamon, cranberry
Ellagic, ellagitannins, garlic, and gallotannins		Fruits and vegetables
Rosemarinic acid		Rosemary
Salicylic acid		Blackberries and blueberries

Table 16.03 Oxygen Radical Absorbance Capacity (ORAC) Values of Antioxidants Present in Our Daily Foods

Common Foods	Micromole in ORAC = Trolox Equivalent, TE per 100 grams
Spices, cloves, ground	314,446
Cinnamon, ground	267,536
Oregano, dried	200,129
Turmeric, ground	159,277
Cocoa, dry powder, unsweetened	80,993
Unprocessed cocoa beans	28,000
Black pepper	27,168
Cinnamon, 1 tablespoonful	21,402
Oregano, 1 tablespoonful	16,010
Elderberries, 1 cup	14,697
Lentils, 1 cup	13,981
Turmeric, 1 tablespoonful	12,742
Chocolate, 30 grams	12,060
Pinto beans	11,864

Common Foods	Micromole in ORAC = Trolox Equivalent, TE per 100 grams
Artichoke, cooked, 1 cup	7,904
Pomegranate juice 100%	5,923
Prunes, ½ cup	7,291
Plums, dried	5,700
Black plums, 1	4,844
Red wine	5,693
Red kidney beans	13,727
Cranberry, raw, 1 cup	9,584
Blueberries, raw and anthocyanins, 1 cup	9,019
Cherries	4,620
Sweet cherries, 1 cup	4,873
Apples, red, quircetin, and other polyphenols	5,900
Granny Smith, 1 apple	5,381
Watermelon, a great source of lycopene	3,800
Raisins	2,830
Blackberries, 1 cup	7,701
Raspberries, 1 cup	16,058
Strawberries, 1 cup	5,938
Pears, 1 medium	5,255
Pecans, 1 oz	5,095
Kale	1,700
Spinach	1,260
Brussels sprouts	980
Broccoli	890
USDA: ORAC values as trolox equivalent (micromole TE per 100 gram). Trolox is a vitamin E analogue. Freeze-dried fruits and vegetables have high ORAC values.	

The most powerful in-vivo antioxidant in our body is *glutathione*. This is the food for white blood cells, T-cells, and B-cells. Glutathione is meant for our fighting cellular military. Sulfur-containing foods like broccoli, cauliflower, onions, and garlic boost its production, and so do selenium and vitamin B_6. Intake of wheat germ, whole grains, and whey protein is good for boosting glutathione. Glutathione converts to the enzyme glutathione peroxidase, which sequesters free radicals and protects our cells from damage. Vitamin E, good against heart and skin diseases, is an antioxidant that fights inflammation. Anticancer selenium is part of the enzyme superoxide dismutase that breaks down free radical peroxides. The electron-donating power of antioxidants must fit the reduction-oxidation potential of a free radical. Vitamin C, flavonoids, and lipoic acid quench hydroxy free radicals; superoxide dismutase quenches superoxide; vitamin C, beta-carotene, and flavones quench hydrogen peroxide; and ubiquinol and beta-carotene quench lipid hydroperoxides.[362] Our approach to selecting daily foods should focus on preventing inflammation by proper intake of antioxidants. Dietary antioxidants below a redox potential of 0.45 millivolt can inhibit in-vivo oxidation.

16.02 Major Antioxidants

Anthocyanins, beta-carotene, pyrroquinolin, catechins, cryptoxanthin, flavonoids, indoles, isoflavones, lignans, leutin, lycopene, polyphenol, vitamin C, ubiquinone, and vitamin E are major organic antioxidants. Copper, manganese, selenium, and zinc are the inorganic mineral antioxidants.

The most powerful antioxidant is glutathione, the mother of all antioxidants that our body produces. There are 500 milliMoles/l or 1.00 to 1.5 grams of glutathione in our entire blood. Garlic, onion, cabbage, cauliflower, and DNA methylation nutrients like vitamins B_6, B_{12}, and folic acid help maintain a proper glutathione level in our body.

16.03 Strategy for Antioxidant Intake

Berries, broccoli, grapefruit, garlic, lemon, oranges, red grapes, and tomatoes are common produce items. We need to create a culture of eating fruits and vegetables. We need to get in the habit of garnishing our dishes, creating fruit and vegetable snacks for health, and having chocolate, tea, and coffee for

---

362 http://www.intechopen.com/download/get/type/pdfs/id/39554.

after-dinner enjoyment. Their use by light cooking releases antioxidants. The idea of chocolate before bedtime has great nutritional merit. We have many others listed in table 16.03. In order to make them more effective, though, we should avoid exposing ourselves to carcinogens (charred and grilled foods and storage of foods in plastic containers), unfiltered tap water, radon gas, cell phones, TV, and medical devices, and prolonged exposure to UV light during sunbathing. We should avoid conditions that promote free radical generation.

16.04 Antioxidant Enzymes

There are three major antioxidant enzymes: superoxide dismutase, which requires copper, zinc, and manganese to break down superoxide free radicals; iron-dependent catalase, which breaks down hydrogen peroxide; and glutathione peroxidase, which maintains a low oxidative state.

Antioxidant minerals like copper, iron, and selenium work with enzymes as cofactors. They are involved in cell division and cell elongation, and they modify cell senescence and death. Antioxidants, nonenzymatic or enzymatic, are an integral part of human physiology, and there is a hierarchy and synergy of their work, meaning we need to consume a variety of antioxidants.

Free radical quenching is not a simple biochemical reaction. On the one hand, they defend us by helping our immune cells to engulf bacteria; and on the other, they cause DNA damage, lipid oxidation, and stress. Just to quantify, the redox potential of a hydroxyl free radical is + 0.23 volt, of hydrogen peroxide is + 0.36, and of superoxide is + 0.07 volt. Thus, different antioxidants are suited for dealing with different free radicals. For instance, vitamin C cooperates with vitamin E, and it can repair and recycle vitamin E. Whereas enzyme antioxidants such as superoxide dismutase, catalase, and glutathione peroxidase prevent initial free-radical attack, vitamin C, vitamin E, glutathione, and coenzyme Q_{10} repair oxidizing radicals.

16.05 Vitamin D Therapy

Stored in body fat, vitamin D is a steroidal, hormonelike, super vitamin.[363,364] We know that most tissues and cells in our body have vitamin D receptors

---

363 http://www.pnas.org/content/109/46/18827.full.pdf+html.

364 http://www.mynaturalawakenings.com/BREV/September-2010/Vitamin-D-ndash-The-Master-Hormone-Document-Actions/.

and the necessary enzyme system to convert D2 to the active form of vitamin D_3, which plays a critical role in preventing chronic diseases of cancer and the autoimmune system. It fights infections and reduces risk of cardiovascular diseases. It is involved in bone health, weight loss, and reducing depression. Vitamin D_3 is critical to calcium equilibrium in our body. Other roles of vitamin D include its involvement in nerve-cell proliferation, gene expression, insulin release, and glucose tolerance. It is also involved in infections, inflammations, heart diseases, and skeletal and musculoskeletal health problems. This is why the Institute of Medicine has revised daily intake to be 600 IU per day.[365]

Vitamins A and D and minerals calcium and phosphorus work in an orchestrated metabolic synergy. Vitamin D unlocks DNA blueprints for cells to produce various products. As such, it is critical to the biochemical machinery of all cells.

Vitamin D therapy, therefore, goes beyond conquering rickets, characterized by imperfect calcification, softening, and distortion of the bones. Unfortunately, vitamin D deficiency is now common in children, adults, and senior citizens alike.[366] It is known to retard growth and cause growth deformities in utero and during childhood. Later in life, the deficiency is responsible for osteopenia, osteomalacia, muscle weakness, and increased risk of fracture. Thus, vitamin D is a predictor of bone health, major chronic diseases, and disorders such as schizophrenia, depression, and wheezing illness related to lung function. Vitamin D prevents multiple sclerosis.

Less than 20 nanograms/milliliter of Vitamin D_3 in serum is a sure sign of deficiency, and more than 1 billion people worldwide are vitamin D-deficient. At least 40 percent of the elderly in Europe, 52 percent of Hispanic and black adolescents in the United States, 48 percent of white preadolescent girls in the United States, and up to 30 percent of children even in tropical countries with good sun exposure are vitamin D-deficient.[367] The reason is junk and imbalanced food.

Vitamin D intake via our diet needs to increase from the old RDA of 400 IU per day to at least 600–1000 IU per day. Much of the daily need can be met by dietary upgrade with canned and fresh salmon, fresh and sun-dried shiitake mushrooms, exposure to sunlight, fortified milk, fortified orange

---

365 http://www.iom.edu/Reports/2010/Dietary-Reference-Intakes-for-calcium-and-vitamin-D.aspx.

366 http://www.scientificamerican.com/article/vitamin-d-deficiency-united-states/.

367 http://www.scientificamerican.com/article/vitamin-d-deficiency-united-states/.

juice, yogurt, milk, cultured milk, butter, and fortified breakfast cereals. The missing amount can come from consumption of over-the-counter vitamin D₃ supplements. Health problems are bound to multiply if the need remains unmet.

- **Osteoporosis:** The major cause of osteoporosis is deficiency of vitamin D and calcium. Deficiency of vitamin D in particular can cause fibromyalgia, chronic fatigue, peripheral neuropathy, hypertension, diabetes, and heart diseases. A daily intake of 600–800 IU of vitamin D₃ with a minimum of 600 mg of calcium can reduce the risk of bone fracture.

- **Cardiovascular function:** Vitamin D controls the release of stress hormones that lead to high blood pressure and inflammation. Vitamin D deficiency is now known to be linked to high blood pressure among the US black population in particular. The Institute of Medicine prescribes 30 nanograms/milliliter vitamin D in our blood, and there is serious risk below 15 micrograms/ml. The daily intake prescription is at 600 IU.

- **Cancer:** Vitamin D may prevent colorectal cancer by suppressing growth and blood-vessel formation (angiogenesis) for tumors. A trial including 1,200 postmenopausal women on 1,000 IU vitamin D and calcium demonstrated a 27 percent reduction in risk.[368,369]

- **Uterine fibroids:** Vitamin D deficiency is associated with uterine fibroids,[370] a common tumor of the female uterus.[371] The methylation diet containing vitamin D and reduced intake of fat and high-glycemic starchy foods can help manage this problem.

- **Antioxidant vitamin:** Vitamin D has potent anti-inflammatory properties. It is a more potent antioxidant than vitamin E. Also, it is an analgesic.[372]

---

368 http://www.cancer.gov/cancertopics/factsheet/prevention/vitamin-D.

369 http://www.ncbi.nlm.nih.gov/pmc/articles/PMC1470481/.

370 http://www.ncbi.nlm.nih.gov/pubmed/23493030.

371 http://www.ncbi.nlm.nih.gov/pubmed/18534913.

372 http://www.ncbi.nlm.nih.gov/pubmed/8325381.

- **Depression:** Vitamin D suppresses cytokine production in brain tissue. More sunlight exposure means more Vitamin D and a higher release of serotonin, which can control depression.[373]

- **Autoimmune diseases:** Vitamin D as an anti-inflammatory vitamin can protect against lupus and multiple sclerosis, a disease in which the body's immune system goes against itself.[374]

- **Chronic diseases**: Vitamin D can correct polycyclic ovarian syndrome (PCOS). Its deficiency is known to cause Sjogren's syndrome, multiple sclerosis, rheumatoid arthritis, thyoroiditis, and Crohn's disease.

- **Brain development:** Vitamin D is involved in brain development. It is the best remedy for mental illness by maintaining the health of neurons and the nervous system via production of nerve-growth factors.[375] Vitamin D protects brain cells from free radical damage. Vitamin D prevents autoimmune damage to brain cells, and its deficiency may be related to autism.

- **Muscle strength:** Vitamin D deficiency causes muscle weakness. Skeletal muscles require vitamin D for their routine performance and strength.

- **Fetal origin:** A low-vitamin D diet for the pregnant mother can give rise to schizophrenia and multiple sclerosis in the offspring.[376]

16.05.01 Nonskeletal Functions of Vitamin D: Brain, prostate, breast, immune cells, and colon tissues have vitamin D receptors. *Vitamin D controls the actions of some two hundred genes, including those involved in critical cell growth and*

---

373 http://www.ncbi.nlm.nih.gov/pubmed/23377209.

374 http://health.usnews.com/health-news/diet-fitness/diabetes/articles/2010/08/24/vitamin-d-may-influence-genes-for-cancer-autoimmune-disease.

375 http://www.foxnews.com/health/2013/12/04/vitamin-d-deficiency-may-damage-brain-study-finds/.

376 http://schizophreniabulletin.oxfordjournals.org/content/early/2012/12/10/schbul.sbs148.full.

functions.[377] Vitamin D$_3$ is used for treatment of psoriasis and tuberculosis, and it can be used as an immunomodulator. Also, it increases heart-muscle contractibility and insulin production.

Vitamin D function requires magnesium; magnesium and calcium control heart-muscle contraction for regular beating. Magnesium relaxes and calcium helps contraction of muscles. Critical to the energy process as a whole, magnesium is critical to type 2 diabetes in relation to conversion of sugar to energy. As a matter of fact, magnesium is next to potassium in concentration as positively charged ions in our cells. These two minerals are very critical to our health.

Vitamin D, calcium, vitamin A, phosphorus, and other minerals are essential for maintaining our immune system and muscle strength. Vitamin D controls calcium absorption and its transport to and across our cells.

Vitamin D is a good preventive against both type 1 and 2 diabetes and internal inflammation. Vitamin D deficiency causes poor production of and release of insulin, and glucose intolerance is inversely related to vitamin D in our blood supply. Vitamin D can double absorption of calcium and phosphorus, necessary for bone health. The absorption though intestine can increase up to 40 percent through small intestine.

16.06 Sources of Vitamin D

The human body can make vitamin D from cholesterol when skin is exposed to sunlight. Vitamin D$_3$ is cholecalciferol, and D2 is ergocalciferol. We get vitamin D from oily fish, exposure to sunlight, and only a few unfortified foods, such as mushrooms and yeast. Salmon, eggs, trout, tuna, and meat are good sources of vitamin D. Cod liver oil can supply up to 1000 IU. Tuna, mackerel, and sardines consumed in 3.5-ounce portions can supply up to 250 to 300 IU. Good sources for vegetarians are milk, shiitake mushrooms, fortified cereals, margarine, yogurt, fortified orange juice, and yeast. In order to fulfill our daily need, the best practice is to use over-the-counter vitamin D$_3$ tablets of 500 to 1000 IU daily. The following recommendation in regard to vitamin D intake should be followed:

The liver, kidney, and small intestine process and help transport vitamin D in our body. As a general rule, we need a minimum of 30 nanograms/milliliter active vitamin D$_3$ in our serum. Our skin has great capacity to make vitamin D$_3$. Full sunlight exposure of even twenty minutes a day provides an adequate

---

377 http://www.dana.org/News/Details.aspx?id=43118.

(equivalent to 20,000 IU of D₂) level of vitamin D₃, which is stored in body fat. Causes of deficiency are limited light exposure, poor bioavailability due to heritable disorders, hyperthyroidism, osteomalacia due to tumors, reduced fat absorption, metabolic loss, liver failure, and chronic kidney disease.

Daily health can get a boost in vitamin D by

1. Increasing daily intake to around 800 to 1000 IU of vitamin D₃ (4000 IU for lactating women).
2. Maintaining 400 IU of vitamin D₃ for children (Canadian standard).
3. Increasing intake to 800 IU for dialysis patients.
4. Increasing Vitamin D intake for liver patients with poor fat absorption.
5. Managing daily sunlight exposure for at least twenty minutes.

16.07 Enzyme Therapy

There are 10,000 enzymes needed to run our life processes. We exist and reproduce because of enzymes, and our molecular factory runs because of enzymes. A good source of enzymes for our digestive system is raw food itself. When we consume mostly processed foods, our body has no choice but to depend on enzymes made in the stomach or those delivered to the intestine from the pancreas. This enzyme supply depends on our genes, age, and foods we eat. Our digestive, nervous, and immune systems need to function very closely, and enzymes are the key for their unified and interrelated functioning. A mutation that changes the gene responsible for making these enzyme proteins results in a disorder, a disease, or an intolerance. Important types of enzymes from a therapeutic viewpoint are (1) gastric and pancreatic enzymes; (2) antioxidant enzymes; and (3) asparginase-like enzymes, used in cancer therapy. Gastric and pancreatic enzymes break down proteins and starch and also certain enzyme inhibitors.

Poorly digested proteins can act like an invading antigen and threaten our immune system. Digestive enzyme therapy thus can modify an ingested protein to become friendly to the immune system. Our digestive system produces twenty-two enzymes, including protease, amylase, lipase, and cellulase, which often work with minerals and cofactors.

Sprouted grains, seeds, and legumes can provide vitamin D in addition to bromelain, chymotrypsin, catalase, diastase, mycozyme, pancreatin, and pepsin for efficient digestion. In addition, sprouts are a good source of polyphenol antioxidants and vitamins. Antioxidant enzymes catalase, superoxide dismutase, and coenzyme Q_{10} can also be therapeutic. In addition, sprouts from broccoli, mustard, caraway, lentils, and small beans serve as a source of proteinase enzymes for digestion of high-protein foods for pancreatic cancer patients. They can boost the immune system, reduce inflammation, and improve metabolic efficiency.

A good enzyme product as food is a blend of multiple enzymes, antioxidants, prebiotics, and probiotics, including omega-3 fatty acids that can function synergistically. Actually therapeutic enzymes can be used in many ways:

1. Lysosomal diseases can be treated with lipids and glycoprotein enzymes.
2. Phenylketonurea is treated with enzymes and cofactors.
3. Protease can be used for breaking down fibrin, the clot-forming protein.
4. Enzymes can be used as curatives for digestive fluids affected by gene mutation in cystic fibrosis.
5. Lysosomal and glycogen storage diseases can be cured by enzymes.
6. Antioxidant enzymes glutathione peroxidase, superoxide dismutase, catalase, and coenzyme Q_{10} from sprouts can be routinely used to reduce oxidative stress.
7. Enzymes can be used in order to improve bioavailability of key nutrients and induce fat loss by lipase.

Today the enzyme lactase is used commercially in low-lactose milk production for lactose-intolerant populations all over the world. The enzyme bromelain, which is present in pineapple and kiwi, contains the bromelain-like actinidin enzyme. Papain from papaya is used as a meat tenderizer, and so is bromelain. Just think of the value of pineapple-enhanced easy-to-digest pork dishes of already partly tenderized protein. Fermented foods like yogurt, kefir, kimchi, and sauerkraut are good sources of enzymes. For maximum effect, such fruits and food preparations can be used between meals against inflammation, cardiovascular disorders, cancer, and colon health, and to improve immune response. The asparginase enzyme has potential for cancer treatment. Sprouts, for instance, have been used in Japan for centuries for facilitation of high-protein digestion, inactivating enzyme inhibitors in soybean foods, and triggering weight loss.[378,][379] However, enzyme replacement therapy in case of a serious disease should be undertaken only in care of a physician.

16.08 Gene Therapy

Genetic disorders result from impaired or absent genes. "No genes" simply means "no enzyme proteins." But we now have technology to correct a target gene. It is now possible to (1) insert a normal gene in the genome at a nonspecific location, (2) swap a normal gene for the abnormal gene, (3) correct an abnormal mutated gene by reverse mutation, and (4) alter the on/off regulation by a gene by adding large sections of DNA on specific genome sites. However, the process is extremely difficult through common virus vectors. Today, direct delivery of a therapeutic DNA into a target cell is possible only for certain tissues. DNA can be used as a pharmaceutical drug on the one hand, and genes can be corrected by designed mutation on the other. A gene that encodes a protein drug can be designed.

Molecular therapy with genes has great potential in the case of single-gene diseases like cystic fibrosis, hemophilia, muscular dystrophy, chronic lymphocytic leukemia, Parkinson's disease, multiple myeloma, lipoprotein lipase deficiency in pancreatitis, sickle cell anemia, retinal diseases, and PKU. Although there have been a few success stories during the 2010 to 2013 period dealing with blood disorders,[380] routine gene therapy in hospitals is still not common. Although somatic gene therapy for nonreproductive cells appears more successful,[381] successful gene deliveries have yet to be worked out, and the multiple gene effects of Alzheimer's disease, arthritis, and diabetes are too complex for gene therapy. Also, there are too many hurdles of encapsulation and release of genes in the case of lung cancer, reengineering of lymphocytes in the case of myeloma, treatment of myeloid leukemia, and design of a deafness gene. Potential of near-future success may involve Parkinson's disease, gene silencing in the case of Huntington's disease, and treatment of cancer, viral infections, allergies, and immune response.

---

378 http://umamimart.com/2011/09/japanify-namuru-spicy-bean-sprouts/.

379 http://wholegrainscouncil.org/whole-grains-101/health-benefits-of-sprouted-grains

380 http://www.ndsu.edu/pubweb/~mcclean/plsc431/students/brandi.htm.

381 http://www.walesgenepark.co.uk/all-about-somatic-gene-therapy-2/.

Although the FDA has not approved any gene therapy to date, germ-line gene therapy via sperm or egg cell and somatic gene therapy for individual patients are at the forefront of cell-based technologies for gene therapy.[382,383,384] Currently reported therapies are only experimental. Gene therapy seems to be possible in the case of choideremia eye disease[385] and the long-awaited HIV vaccine containing synthetic genes.[386] Other possible success stories include treating the genetic form of blindness by Spark Therapeutics, fixing the defective eye gene RPE65, treatment of bone marrow stem cells, and liver disease. The first gene therapy approval happened in Europe in 2012 for treatment by Unique.

16.09 Major Reference Institutionst

Antioxidants are key to our health. The following are lead institutions that excell in research on dietary antioxidants.

Antioxidant therapy is a growing field of clinical practice. Institutions of repute are the National Center for Complementary and Alternative Medicine; the Center for Integrative Cancer Treatment in Evanston, Illinois; and the St. Louis University School of Medicine. These are good sources of information on antioxidant and vitamin D therapy. A good proof of natural enzyme therapy is for-sale sprouted grains in the produce section of the retail store. Gene therapy has yet to evolve and mature.

382 http://www.qcc.cuny.edu/socialsciences/ppecorino/MEDICAL_ETHICS_TEXT/Chapter_7_Human_Experimentation/Case_Study_Cystic_Fibrosis_Gene_Therapy.htm.
383 http://www.cbc.ca/news/health/gene-therapy-leukemia-treatment-successful-1.1002779.
384 http://www.ama-assn.org//ama/pub/physician-resources/medical-science/genetics-molecular-medicine/current-topics/gene-therapy.page.
385 http://www.thehindu.com/sci-tech/choroideremia-gene-therapy-shows-promise/article5579162.ece.
386 http://www.newscientist.com/article/dn24842-synthetic-gene-helps-hiv-vaccine-hit-shapeshifting-foe.html.

PART FOUR
OUR CHOICES

Our choices as to exercise, meditation, relaxation,
yoga, and spirituality for stress reduction can
and do affect our health by altering nuclear and
mitochondrial DNA and gene expressions.

CHAPTER 17

Exercise Therapy

"Those who think they have no time for exercise will
sooner or later have to find time for illness."
—Edward Stanley

Exercise is part of being human, and it is the language of our physical brain, the body's master control system.[387] To exercise is to deliver oxygen to our body and to tone our musculoskeletal system at the same time. Exercise primes both our body and mind by structure and function.[388]

Oxygen, a very electronegative gas, became central to human evolution eons ago. We have evolved now with a normal cardiac output of 1.34 milliliters of oxygen per minute, toted on the hemoglobin molecule for our energy-producing cells.. This requires that we breathe five liters of air per minute, such that each gram of hemoglobin can bind 1.34 milliliters of oxygen. In the seven liters of blood in an average human body, there are 10×10^{21} molecules of hemoglobin, each with four oxygen molecules.[389] Just 100 ml of blood per minute is capable of delivering twenty milliliters of oxygen.

Oxygen is essential for the functioning of the heart and brain, and aerobic or breathing exercises are excellent means of improving the oxygen supply to the heart, brain, and other body cells. Exercise is a biological need and a message for growth. It doesn't cause any direct effect but works indirectly by strengthening muscles, improving cardiovascular and respiratory systems,

---

387 http://www.fi.edu/learn/brain/exercise.html.
388 http://health.howstuffworks.com/wellness/diet-fitness/information/mind-body-exercise-connection.htm.
389 http://www.austincc.edu/~emeyerth/hemoglob.htm.

and integrating neural, respiratory, cardiovascular, muscular, and skeletal systems.

As a stimulus, exercise helps manage the rate and depth of fatigue for normal body response. Deep breathing and aerobic exercises feed the brain, heart, and other muscle cells with oxygen. During exercise, we shake and move our body, sweat to remove toxins, and increase metabolism. Two to four minutes of anaerobic exercises of lifting, running, jumping, and uphill climbing for building muscle mass and increasing oxygen volume intake is of great health value because working muscles need three time more oxygen than resting muscles. Both aerobic and anaerobic exercises are necessary.[390] Whereas anaerobic exercise promotes strength, speed, and power by training muscles, aerobic exercise is good for delivery of oxygen to the cells. Our cells need twice the weight of oxygen compared to the weight of glucose for energy production. Exercise should be low-intensity for a longer time, involving walking, running, rowing, and bicycling. It is all about flow, motion, rhythm, and oxygenation, and it does affect gene expression.[391] Short-time intense exercises can be good also.

As a science of movement, exercise is interdisciplinary and linked closely to diet and nutrition. Weight-loss supplements per se don't work without exercise. Portion control and high-satiety diets for weight reduction and control work only when complemented with exercise. Exercise helps breaks down fat stores, balances hormones, replaces fuel stores, helps produce more ATP energy, helps repair cells, innervates, and helps speedy production of molecules of life for routine physiology because it increases cardiac output for increased oxygen-carrying capacity of the blood. Consumption of a large amount of oxygen during exercise helps reduce weight. Exercise trains muscles to use oxygen, increases metabolism, improves cognition, controls neuropeptide Y (NPY) production, and reduces oxidative stress and stress hormones. Exercise is directly linked to the health of mitochondria.[392,393]

We should think of exercise as *meditation in motion,* which results in relaxation, reduced anxiety, and reduced stress. It is the cheapest medicine for our overly stressed body and mind.

Physical inactivity became synonymous with modernity during the past sixty years. Exercise and physical activity simply walked away from our

---

390 http://www.who.int/dietphysicalactivity/strategy/eb11344/strategy_english_web.pd.
391 http://www.ncbi.nlm.nih.gov/pubmed/16990507.
392 http://www.ncbi.nlm.nih.gov/pmc/articles/PMC1540458/.
393 http://www.cbass.com/ResistanceMitochondrial.htm.

day-to-day living. The consequence is 2.3 billion overweight and 0.7 billion obese people in the world.[394,395] Obesity is a serious problem today given the high incidence among children. In the United States, obesity and fear of obesity have created a $26 billion business of weight management, including health foods, exercise clubs, and other programmed services.

Almost 70 percent of the total calories we consume are used for critical body functions such as breathing, manufacturing cells, and maintaining body temperature. This corresponds to BMR, the rate at which our body uses calories for these critical and routine functions. BMR is determined by our genes, age, gender, and body composition. It does decrease with age and with loss of lean body mass. More muscle simply means high BMR. Low BMR among the overweight and obese is often associated with problems of high blood pressure, type 2 diabetes, high-density lipoprotein below 30 mg/dl, cholesterol above 150 mg/dl, and chronic insomnia.

The calories beyond BMR do not have to be all the remaining 30 percent, or six hundred calories per day for a two-thousand-calorie diet. If we use only four hundred calories (not 600) beyond the BMR, we can keep off sixteen pounds in weight per year. Even a fifty-calorie daily reduction translates to a potential weight reduction of four pounds per year. Consumption of extra calories certainly requires more exercise to create a balance by burning them off.

A high calorie consumption requires more oxygen to burn it off. The heart needs to beat more, but it should do so at a reasonable rate. Even with a fifteen-beats-per-minute reduction, the heart needs to beat 7.884 million times less per year, a great news for a hundred-year life span because of a savings of 788.4 million beats for the tired old heart.

17.01 Effect of Exercise on Genes

The human genome evolved about 100,000 years ago for our Paleolithic ancestors in view of obligatory physical activity. The modern-day lifestyle, however, is much too sedentary, and our genes are maladapted for this physical inactivity. Exercise restores the homeostatic mechanisms by DNA demethylation of promoter genes PGC-1a, PDK 4, and PPAR δ; it affects at least 7,000 genes out of 25,000 on the genome. Demethylation, it should be

---

394 http://www.who.int/mediacentre/factsheets/fs311/en/.

395 http://www.who.int/dietphysicalactivity/media/en/gsfs_obesity.pdf.

pointed out, depends on the intensity and regularity of exercise, though. Chronic loneliness and stress also affect expression of many genes.[396] Exercise in a collective club is much better for our nuclear families today.

Also, exercise is known to boost dopamine and serotonin levels for good mood and low stress. It boosts endomorphins and norepinephrine modulation and control. Exercise can indirectly help build new neurons.[397] Also, exercise affects how we metabolize glucose and store fat in white fat cells by epigenetic effects. The genes of fat cells, in particular a mutated gene of CIDEC protein, can be changed in favor of small fat droplets by exercise. Let us keep in mind that only a fraction of genes are active at a given time. Exercise, even for five minutes twice a day, can help activate our good genes.[398]

17.02 Exercise and Hormonal Balance

Poor physical activity and lack of exercise cause stress, which increases under conditions of metabolic demand, cravings, high blood sugar, fat storage, emotional eating, and routine fast-food intake. Control of cortisol, insulin, and thyroid are critical to good health under these conditions. Exercise reduces anger-born stress, subdues tense moments and scariness (conditions of the fight-or-Flight situation), and helps create hormonal balance. Stress is genetic, and different individuals react to the same stressors differently.[399] Hormones that are balanced by exercise are cortisol, adrenalin, insulin, estrogen, dehydroepiandrosterone (DHEA), and thyroid hormone. If not mitigated by stress reduction, they can be responsible for the following:

1. High cortisol and high stress can cause: (1) increased heart and respiratory rate; (2) increased fat, protein, and carbohydrate metabolism for energy production; (3) increased use of glucose by the brain; (4) increased insulin secretion and release; (5) increased appetite; (6) increased storage of fat under the belly; (7) a weak

---

396 http://www.medicalnewstoday.com/articles/82496.php.

397 http://commonhealth.wbur.org/2013/06/brain-hundreds-new-neurons.

398 http://healthland.time.com/2012/03/07/how-exercise-can-change-your-dna/.

399 http://www.ncbi.nlm.nih.gov/pubmed/19750552.

immune system; (8) an upset digestive system; (9) disturbed mood, motivation, and fear; and (10) tense muscles.

2. Adrenalin means that one just had a fight, making him or her extra alert and focused, with the ultimate consequence of exhaustion and a need for more energy.
3. Insulin and cortisol together increase appetite.
4. Estrogen loss is common in menopausal women.
5. Underactive thyroid causes weight gain.
6. Abnormal under stress, low dehydroepiandrosterone (DHEA) leads to low sex hormones. The DHEA level is normal when one feels good.

Stress can cause a suppressed immune system, cancer, constricted arteries, heart attack, delayed wound healing, inflammation, weight gain and type 2diabetes, obesity, and loss of cognition. It hastens cellular and genetic aging processes via telomere shortening due to the deficiency of the enzyme telomerase. *Exercise, yoga, and meditation can reduce stress and improve longevity by a decade.* Yoga is a stretch exercise that allows for physical and mental flexibility and better memory. Meditation quiets down the mind, improves awareness, brings feeling of joy and peace, brings out the power of the present, and helps build confidence. In simple terms, diet nourishes and exercise nurtures.

17.03 Modes of Exercise for Breathing and Relaxation

Exercise is breathing, and breathing is relaxation with oxygen control. Slow breathing links mind and body, regulates heart rate, and helps a person to relax. Stress stimulates the sympathetic nervous system and activates the parasympathetic system simultaneously. Deep inhalation and exhalation is a bridge between the sympathetic and parasympathetic nervous systems.[400,401]

Breathing should be steady and aligned evenly. Abdominal or diaphragmatic exercise can be done in any posture. Short, fast, and rhythmic bellows breathing for energy, breathing in a sitting posture for relaxation, and breathing through both nostrils should be part of any daily exercise program. Each nostril is connected to different nerve centers in the brain.

---

400 http://www.yogajournal.com/practice/1523.
401 http://www.dbtsandiego.com/DBT2.pdf.

As such, alternate breathing through each nostril creates a balance if one stays mindful with full presence and punctuation of purpose or the goal of good health.

17.04 Physiology of Exercise

Every cell in our body breathes, and breathing, underlying oxygen and carbon dioxide exchange, is monitored in the cerebrospinal fluid by chemical receptors. It is under somatic and autonomic nervous systems' control. The air we breathe has only 21 percent oxygen. It is the oxygen that binds with hemoglobin in the lungs, gets to the heart, and finally gets to all cells in the body. We exhale carbon dioxide as a waste product. Quiet breathing influences the nervous system, slows the heart rate, and makes one feel relaxed.

Close to 0.5 liter of gas volume moves in and out during one breath. This translates to seven to eight liters of air consumption per minute, or 11,000 liters per day. Average respiratory reserve is 3.3 liters, and average lung capacity is six liters. Breathing and the manner of breathing, therefore, need to be controlled for maximum lung volume. Our lungs are 50 percent full during exhalation and 80 percent full during inhalation. Since inhaling depends on muscular activity, posture during breathing matters a lot. We should breathe such that exhalation takes twice as long as inhalation. Thoracic or chest breathing is accompanied by rib-cage expansion, where hands are locked behind while breathing in a standing posture. Most common and easy is diaphragmatic breathing. One should breathe through each nostril for two seconds for inhalation and and four seconds for exhalation. Sitting upright in a chair or lying flat on the back is a good natural posture for breathing. Breathing exercises and calorie-restriction diets are good for longevity.

Given an average of 0.5 liter of air per breath, arterial oxygen can increase to 120 mm Hg from 100, and carbon dioxide pressure can decrease to 35 mm Hg from 46 after six breaths, permitting around 20 to 25 percent exchange.

I know from a five-year-long practice that even three-times-a-day alternate nostril breathing calms me down and improves concentration. A two- to five-minute diaphragmatic (abdominal) breathing using each nostril alternatively can make a huge difference in oxygen supply to the brain and heart by increasing blood flow and metabolic reactions, and optimizing the heart and lung workload. Exercise and physical activity—say, accounting for 20 percent extra blood flow—automatically help manage the workload for the heart and lung. *Exercise increases blood flow to the brain, spurring the release of brain-derived neurotrophic factor (BDNF), responsible for formation of new neurons in the hippocampus, repair of cell damage, and strengthening of synapses that control brain cells.*[402]

17.04.01 Body Mass Index (BMI): Body Mass Index (BMI), as a measure of body fat, is proportional to body shape. It is large for overweight and obese people and small for muscular people. Normal numbers range from 18.5 to 24.9. Expressed as Kg/meter², it is weight in kilograms divided by the square of height in meters. Given my weight of 70 kg and height of 1.676 meters, my BMI is 24.91, and I am at the high end of the range. Obviously I need to exercise and breathe more.

17.04.02 Percent Body Fat: A body fat of 2 to 5 percent for men and 10 to 13 percent for women is essential. Normal fat storage in white fat cells should not raise it above 18 to 24 percent for men and 25 to 30 percent for women.[403] These numbers represent actual fat measures as opposed to shape-dependent BMI, which is less reliable.

17.04.03 Exercise and Nitric Oxide: Blood nitric oxide, the body's own vasodilator, at around 150 micromoles/liter is a good antiatherogen.[404] Exercise induces production of nitric oxide synthase, an enzyme necessary to produce nitric oxide from the arginine present in proteins in our diet. Endurance athletes have more nitric oxide derivatives (nitrites and nitrates) stored in their hearts. A diet with high-arginine protein from peas, lentils, nuts, red meat, and eggs is advisable for routine endogenous nitric oxide production.

17.04.04 Body Weight and Energy Intake: Carbohydrates, protein, and fat are part of our daily meals. All of them can be sources of energy, although carbohydrates are preferred. Fat is our main source of energy reserve. *The higher our weight, the higher is the energy expenditure for upkeep of the body.* Weight is often stable when energy intake equals (is in balance with) energy expenditure. In other words, there is a normal weight for every individual for a given energy intake.

---

402 http://www.ncbi.nlm.nih.gov/pubmed/15896913.
403 http://emedicine.medscape.com/article/123702-overview.
404 http://www.ncbi.nlm.nih.gov/pubmed/14599231.

Usually weight loss corresponds to 66 percent due to burning fat and 34 percent due to burning of lean body mass. Whereas energy expenditure for building one pound of fat tissue is 4,545 calories, it is only 909 calories for building lean body mass. A loss of one pound of weight, then, equates to 3,308.7 calories (0.66 X 4545 + 0.34 X 909), and this corresponds to food consumed over 1.65 days if one were on a 2000 calories per day diet. In theory then, one can lose maybe one pound of weight, say, in two days. Given too many factors that determine it, a value of 3,500 calories removed from fat for a one-pound weight reduction is a good guide. An increase in energy intake of 100 calories/day (say one cappuccino per day) may account for an increase of 6 Kg over 420 days. We should keep in mind that the physiological situation is much more complex and weight gain or loss is not linear and straightforward. Low energy intake being the same, heavier people lose more weight faster than lighter people, and people with more lean body mass find it difficult to lose weight. Losing one pound of weight per week or fifty-two pounds per year is possible.

As we lose weight, the fatty tissue-to-lean mass ratio changes, which decreases energy expenditure per unit of body weight. As a matter of fact, fatty tissues, skeletal muscle, and smooth muscles affect weight loss differently. The following account is a good approximation:

> **Total Energy Expenditure (TEE) = [70% of Resting Energy Expenditure (REE) +20% of Active Energy Expenditure (AEE) + 10% of Diet-Induced Energy (DTE)] per unit of fat-free mass.**

Variations in REE, AEE, and DTE limit weight change. Progressive weight gain is a matter of weight recycling, except that weight gained often has a higher ratio of fatty adipose tissue to lean mass, opposing weight loss. Since basal or resting metabolism accounts for most of the energy expenditure, what varies with body weight is the resting energy expenditure. AEE is related to physical exercise, and DTE is related to digestion of the foods we eat. Both can be managed in order to reduce weight. *It stands to good reason then that exercise and selection of foods with less net energy (proteins and dietary fiber that need more energy for digestion) can help reduce weight.*

17.04.05 Exercise, Weight Loss, and Weight Management: Our body is an energy conversion machine, and the requirements for energy at rest, for different bodily functions, and for different kinds of physical exercises and works are different.[405] Managing weight simply means managing energy conversions by way of controlled food intake and mental and physical exercises.

Pancreatic lipase makes glycerol and fatty acids from fat we consume. Fatty acids get coated with chylomicron for solubility, and then they go to the lymphatic system, then to the veins, and finally to the blood. Big-size fat is always broken into fatty acids and glycerol before it can get across cell membranes. Intermediary chylomicron lasts only for eight minutes. They too are under genetic control.[406,407] Weight gain is simply what happens to fatty acids and fat—the eventual fat storage in the white fat cell. Fat is also the logical choice for reserve energy because all free glucose amounts to only forty calories, which is good for only a few minutes of the body's operation; all glycogen, amounting to six hundred calories, is good for a day; all muscle protein, amounting to 25,000 calories, is good for four days; but twenty-four pounds of fat in a 154-pound male accounts for 24 pound X 453 gram X 9 Calories per gram fat = 98,935 calories, good for 49.46 days. Clearly then we are designed to store fat and live off it during a short supply of food.

Our 30 billion white fat cells are originally only *0.1 millimeter* in size. We gain weight as they grow in size, not in number. They can grow to be 80 percent pure fat. They act as a hormonal unit integral to our endocrine system and choose to store fat because fat costs the least energy to store. To convert 25 grams of glucose to fat, fat cells would have to spend only twenty-three calories. Storing fat costs only 5.5 calories.

Some fat cells are made during the third trimester of pregnancy. All the rest are made past puberty at age sixteen years or so to a maximum of around 30 billion. Original fat cells weigh around thirty pounds at the most, but as we gain weight, fat cells grow by filling up to 80 percent of their volume. Only 10 percent of them die, and 10 percent of newborn cells keep their number constant. They just grow in size with more fat, with a tiny nucleus on the periphery.

The genetics of exercise are different for different people.[408,409] America's gene pool did not change during the last six decades and cause rampant obesity. The problem of being overweight and obese, therefore, comes largely

405 http://cnx.org/content/m42153/latest/?collection=col11406/latest.

406 http://www.ncbi.nlm.nih.gov/pubmed/17292734.

407 http://www.ncbi.nlm.nih.gov/pubmed/22505585.

408 http://www.plosgenetics.org/article/info%3Adoi%2F10.1371%2Fjournal.pgen.1003607.

409 http://www.hindawi.com/journals/ijpedi/2010/138345/.

from overeating of refined foods with high-glycemic index carbohydrates and high fat, the doubly dangerous trans fat in particular. Energy conversion, the power consumed by different organs, and the power consumption in various exercise activities can vary with individuals.[410]

Fat cells train the brain to produce the appetite hormone leptin, but the training fails when there is too much glucose and insulin in the blood due to overeating. Fat cells make other peptides, such as resistin, apelin, and adidonectin also. Obese people have defective Ob(Lep), the leptin gene.[411] Heavy mothers produce heavy offspring.[412]

The answer to weight gain is a regular program of exercise that can optimize physiology for homeostasis. Exercise can help manage the hypothalamic response involved in methylation of 7,663 genes, nearly a third of all genes on the human genome. It can prevent insulin and leptin resistance when feast dominates famine. Feast is what dominates when we succumb to overeating. This shift to a "no famine" situation for Paleolithic hunter-gatherer human beings is changing genes that make fat cells, genes that help metabolize and store fat, and genes that help neurotransmitters mediate in all that has to do with fat.

A small molecular-weight leptin peptide (only 167 amino acids long) works with receptors on the hypothalamus for energy balance and controls hunger, metabolism, and even behavior. We know that circulating leptin is proportional to fat in the body.[413] It counteracts peptide ANANDAMIDE (N-archidonoethanolamine), which is a neurotransmitter for feeding.

The much smaller peptide ghrelin, of twenty-eight amino acids, is a circulating hunger hormone. Its receptors are everywhere—stomach, pancreas, intestine, and pituitary. Ghrelin increases before meals and goes down after meals. There is no doubt that a high-carbohydrate diet increases fat reserve because of increased insulin supply in response to fat breakdown and ketone production.

Can we stop the growth of fat cells? Can we reduce the number of fat cells? Can we change genes of the antistress neuropeptide Y (NPY) for a calming effect? Can we kill fat cells altogether? These are the major questions today in regard to weight reduction. While there are no easy answers, exercise can definitely make white fat cells shrink and help reduce weight.

410 http://cnx.org/content/m42153/latest/?collection=col11406/latest.

411 http://www.ncbi.nlm.nih.gov/pubmed/12421342.

412 http://www.sciencedaily.com/releases/2013/08/130804080952.htm.

413 http://www.sciencedaily.com/releases/2013/08/130804080952.htm.

It is well-known that exercise influences mitochondrial activity and energy production.[414]

We should recognize that the concept of dieting is contrary to evolutionary dictates because we are designed to store fat and gain weight in the process. Since one pound of weight change involves approximately 3,500 calories, we should plan to get rid of 3,500 calories from our diet in order to effect one pound of weight reduction. To lose one pound per week, one should either burn at least an extra five hundred calories per day or not consume five hundred calories that day. A combination of the two is much better, though. Resistance to change in weight is largely due to metabolic rate, which decreases in response to less energy intake, and so does our body's need for energy. Daily exercise helps solve this resistance problem.

Calories stored in reserve are used when we go on a restricted-calorie diet—a diet of fewer calories than required for critical body functions. The first reserve used is glycogen in the liver. Since one pound of glycogen loss corresponds to three pounds of water loss, weight loss during the first week is largely due to water loss. The second reserve is fat. Fat reserve begins to deplete when all glycogen is depleted. Depletion of fat involves production of ketones for energy; ketones decrease appetite and produce a feeling of well-being. Too much ketone in the blood should be counteracted by copious water intake in order to avoid nausea.

The resting metabolic rate decreases with weight loss on any kind of calorie-restricted diet as we focus on reducing weight. Thyroid hormone is produced for conserving energy, and one feels cold when on a calorie-restricted diet. That is why people tend to overeat and gain weight again. Also, poor thyroxin production by and of itself has an effect of weight gain.

Our body is 58 to 60 percent water, and we need to keep this level undisturbed. The good news is that most of the food we eat is largely water, and we can select foods for more water. As pointed out earlier, there is loss of water during depletion of the glycogen reserve when we go on a diet initially, but once it is all depleted and we begin to deplete fat, causing loss of ketone in urine, our water intake needs to increase. Failure to drink extra water as tea, coffee, or bottled water causes a drop in blood pressure, constipation, and a feeling of faintness. Success in weight reduction by a good diet plan is possible only by including extra water intake and by a sustained and simple exercise regimen.

---

414 http://www.ncbi.nlm.nih.gov/pmc/articles/PMC1540458/.

High-intensity interval training. (Professor Izumi Tabata of Retsumeiken University; Professor Martin Gabala of McMaster University; and Professor Jamie Timmons of University of Birmingham are proponents of short bursts of high-intensity training. The approach has been very effective in maintaining my weight now at age seventy-one. I do it twice a day. A routine HIIT for just five minutes a day helps me with cardio, fat burning, and muscle toning. The beauty is that we can design our own range of motion using dumbbells and yoga exercises.

A great way to burn belly fat is probiotic foods based on lactobacillus gasseri[415] by way of unpasteurized sauerkraut. Actually, probiotics that include bifidobacteria infantis and lactobacillus rhamnosus are very useful in toning the body and fighting digestive diseases.[416,417]

17.05 Exercise for Diabetics

Physical activity is a necessity for the young, adults, and senior citizens alike. We can't afford to remain inactive. A fun-filled thirty-minute-a-day exercise period or short bursts of high-intensity exercises help promote cell growth, lose weight, sleep well, lower cholesterol, lower blood pressure, lower blood sugar, lower body fat, prevent osteoporosis, prevent cancer, reduce risk of heart attack and stroke, and prevent diabetes. Exercise gives more energy, better joint health, increased bone density, improved muscle efficiency, better heart function, and improved balance; it helps build muscle for type 2 diabetics.

Exercise manages power-producing mitochondria in our cells. A routine and continuous exercise of thirty minutes per day for a diabetic person at least three times per week and up to one's 80 percent capacity engages his or her cells' power plant, mitochondria, in action for efficient energy production. Well-managed mitochondria, by the way, can give us 50 percent more power for living. To be able to do so, exercise must manage our genes. It can reverse the genetic imprint back to a younger age.[418] As a matter of fact, you are only as old as your genes. Exercise helps us prevent our genes

---

415 http://publix.aisle7.net/publix/us/assets/feature/probiotic-fortified-yogurt-may-trim-belly-fat_15985_4/~default.

416 http://www.medicalnewstoday.com/releases/46998.php.

417 http://www.ncbi.nlm.nih.gov/pubmed/11799281.

418 http://www.webmd.com/fitness-exercise/news/20091201/molecular-proof-exercise-keeps-you-young.

from growing old. Thus, exercise can help reverse the aging process and enhance cognition and neurogenesis (making of neural systems). Over the long term, exercise provides stamina, energy, and flexibility. An exercise schedule inclusive of ten times slow inhaling through the right nostril and exhaling through the left nostril is a great tool for toning the body for right gene expression.[419]

Resistance exercise, working with weights and using your own weight in exercise, involving repetition of the chest press, leg extension, shoulder press, lateral pull down, calf raising, abdominal crunch, back extension, and ten repetitions of arm flexion and extension is easy to do and to repeat. Strength exercises of push-ups and squats reduce body fat and weight, help reduce pain, and help reduce blood pressure and cholesterol. Yoga, an exercise in flexibility and muscle movements around the naval, is known to prevent inflammation, create more neurotransmitters (dopamine, serotonin, norepinephrine), and reduce depression.[420] It allows for neural growth, brain-volume growth, and cardiovascular fitness, and promotes youthfulness by age reversal also. We should constantly listen to our body when we exercise.

17.06 Foods for Weight Management

Our daily diet must provide for 55 to 60 grams of balanced protein for building muscles and other protein-based hormones and neurotransmitters. This requirement is a must to meet as we get older. The next item is a minimum of 15 grams of *essential fatty acids* (omega-3 and omega-6), necessary for cell replication and functioning of the nervous system. Reduction in daily calories consumed should be by way of eliminating fat and excess carbohydrates. Specific nutrients we must have in our diet for weight reduction are folate, B_{12}, selenium, and zinc—the gene-expression nutrients. We should eat these foods with all our five senses of touch, feel, smell, look, and taste engaged. We should opt for shape, color, flavor, and texture as criteria of food choice. Food nutrients listed in table 17.01, constituting the essence of the methylation diet, are good for gene expression.

---

419 http://smghealthadvantage.com/issues/2012/december-2012/the-benefits-of-mindful-breathing.aspx.

420 http://www.arthritistoday.org/news/yoga-lowers-inflammation035.php.

Table 17.01 Foods for Good Metabolism and Weight Loss

Common Foods	Active Ingredients	Ailments and Diseases Helped
Acetic acid	Vinegar, apple cider, pickled cucumbers	Lowers body weight, BMI, and visceral fat
Spinach and collards	Alpha-lipoic acid	Promotes weight loss
Apples	Pectin-soluble fiber	Lowers cholesterol and stabilizes blood sugar
Black pepper	Piperine, vitamin K, manganese, iron	Promotes energy metabolism
Butter	Conjugated linoleic acid	Reduces obesity
Cardamom, capsaicin, curry leaf	Capsaicin	Increases metabolic rate, reduces weight, inhibits carbohydrate conversion to fat
Celery seed and the trinity of celery-onion-bell pepper	Flavonoid, apiol	Promotes weight loss, helps reduce blood pressure
Chili powder	Capsaicin, vitamins A and C	Increases metabolic rate, suppresses appetite, helps lose weight
Cream cheese, meat, dairy, legume foods	L-carnitine	Helps lose weight, maintains heart health
Meat, mushrooms, green vegetables	Coenzyme Q_{10}, B_2 (riboflavin)	Helps energy metabolism and ATP production, very good for skin, eye, and hair health
Proteins and amino acids	L-acetyl carnitine	Powerful antioxidant across blood-brain barrier, helps burn fat and manage energy balance
Nonfat dry milk	Calcium, B_{12}, protein	Accelerates satiety and weight loss
Peanuts	B_3, manganese, tryptophan	Maintenance of cell function, protection from free radical damage, and stress control

Common Foods	Active Ingredients	Ailments and Diseases Helped
Peppercorn, black pepper	Selenium, manganese, beta-carotene, piperine, pinene, terpene, lemonine	Helps produce blood cells, promotes respiration; heart-healthy foods that promote nervous-system activities
Protein foods, beans, cheese, fish	Satiety molecules	Help lose weight
White kidney beans	Phaseolamin, an inhibiter of α-amylase	Help manage weight loss

17.06.01 Soy Protein in Weight Management: Soy and other proteins account for low-energy intake of high-satiety foods. They induce ghrelin suppression and elevate GLP-1 and cholecytokinin for increased satiety. Also, soy protein hydrolysate induces higher diet-induced thermogenesis, energy expenditure above basal metabolism. Daily consumption of 20 grams of soy protein containing 96 mg isoflavone can reduce BMI, decrease fat mass, lower cholesterol, reduce risk of coronary heart disease, and lower triglycerides. Black soybean in particular is a great food, with dietary fiber, protein, vitamin A, magnesium, zinc, and almost 40 mg isoflavone per serving. With other common foods, a good diet should be designed around

1. Least possible saturated fat, trans fat, salt, and refined sugar.
2. Conjugated linoleic acid (CLA) from dairy products as antioxidants and anticarcinogens.
3. Hot pepper dips with dietary fiber.
4. Catechin from tea.
5. Pickles, sauerkraut, and yogurt for probiotics.
6. Wine and grapes for stress reduction.
7. Whole grains, soy foods, seeds, lentils, and beans.
8. Fruit and vegetable smoothies for water and fiber.
9. Flavonoids from green leafy vegetables, tea, apples, berries, and onions.
10. Polyunsaturated fatty acids: omega-3s like alpha-linoleic and omega-6 like gamma-linoleic acid.
11. Sprouted beans for extra enzymes helpful to digestion.
12. Chocolate, blackberries, and tea for memory improvement.

13. Good emotional support, an active life, and good sleep.
14. Portion control and calorie restriction, with sufficient protein intake.

17.06.02 Commercial Products: Common foods with high fiber, fruits, and vegetables can do wonders for weight control. We should stay away from commercial products like *Insea,* which are supposed to prevent alpha-amylase and the glucosidase action of hydrolyzing starch. Instead, we should eliminate refined carbohydrates from our diet and use beans and lentil sprouts, which supply partially hydrolyzed proteins and other vitamin and mineral nutrients. The success depends on how well we avoid high-calorie diets, high-glycemic index food ingredients, and high-saturated and trans fat foods.

17.07 Exercise for Stress Reduction

Stress promotes weight gain in perceptible and measurable ways of increasing inflammation and levels of corticosteroids, insulin, and cytokinase production. As a matter of fact, stress puts our pituitary and thyroid glands in overdrive.[421] High consumption of fat and highly refined carbohydrates further leads to weight gain. Obesity itself causes inflammation. Much of this happens under control of our sympathetic nervous system. Exercise helps manage stress. Family members, friends, and faith in planned social interaction can effectively make stress and type 2 diabetes go away.

Current AARP polls show that boomers who are today's senior citizens are not as committed to physical fitness and exercise as they ought to be[422]. Although 43 percent of them feel fit, only 16 to 18 percent find it necessary to belong to a health club or commit to daily exercise. Longevity by and of itself is a drag if you are not physically and mentally fit.

17.08 Major Reference Institutions

Institutions listed below are at the fore front of research on exercise as it relates to health and longevity.

Stanford Center for Longevity, Copper Institute in Dallas, Texas, Mayo Clinic, University of Pittsburgh, and Harvard School of Public Health in the United States; Karolinska Institute in Stockholm and Lund University of Sweden.

---

421 http://www.freshlife.com/content/complex-connections.
422 file:///C:/Users/Owner/Downloads/FitAfter50Facts.pdf

CHAPTER 18

Health Values of Meditation and Yoga

"It is through your body that you realize you are a spark of divinity."
—B. K. S. Iyengar

Meditation and yoga bring about therapeutic values to gene expression for the mind and the body-mind connection by way of reducing stress, hormonal equilibrium, and connectivity of body systems. They are very closely related practices. Meditation deals with the mind, and yoga deals with both body and mind.[423]

Meditation is a five-thousand-year-old practice for health maintenance via building a better immune system and means of mental receptivity. I called it *akaagrataa* (Sanskrit word for "mindfulness") during my childhood in India.

It involves relaxation, focus, breathing, examining thoughts, and witnessing what comes and goes; it enhances creativity, integrates brain function, helps navigate through distractions, squashes anxiety, and improves concentration and contemplation.

Yoga is a mind-body exercise inclusive of oxygen intake and programmed mobility of the human body.[424] Yoga helps connect our mind and body for base physiology and homeostasis.[425] Yoga helps integrate the physical, mental, and spiritual also. It disciplines us for managing circulation, flexibility, mind

---

423 http://www.psychologytoday.com/blog/use-your-mind-change-your-brain/201305/is-your-brain-meditation.

424 http://healthandwellness.kaplan.edu/articles/yoga/Yoga%20-%20Connecting%20Mind,%20Body%20and%20Spirit.html.

425 http://www.amazon.com/Yoga-Meditation-holistic-approach-homeostasis/dp/1780883064.

control, breathing, and awareness.[426] Rooted in very old practices in India, close to nature, nutrition is part of yoga practice for health, well-being, and quality of life. Yoga helps us to relax by reducing stress. C. E. Patanjali's 195 aphorisms have now become rules for healthy living the world over. They teach us how to live well with our body and mind; they teach us meditative focus, breathing and posture control, and even adhering to rules of conduct for a better society.

Yoga emphasizes that the mind transcends the body by a creative construct and we can connect mind and body by managing our perceptions.[427] Neurobiology and the somatic psychology of meditation, concentration, and mindfulness (akaagrataa) is what Patanjali of India professed some five thousand years ago. He taught that mental focus, breath control, and embodied awareness can give a practitioner psychological insight and emotional balance for day-to-day living. Also, he professed connectivity with the community. He taught us to keep destructive behavior away from our being as we think and act. We now have come to confirm that his teachings are rooted in gene expressions.[428] Yoga and meditation thus can help free us from disease by control of neurotransmission via the somatic nervous system, homeostasis, and cell growth. Meditation and yoga boost mitochondrial performance[429] because they can modify the functional characteristics of mitochondria.[430] Meditation helps mitochondrial resilience[431] and cellular stability.

We are human because of 100 billion neurons in our brain. Our brain is unique, but much of it is often redundant. Our daily job is to remove this redundancy, enhance the brain's adaptability, and manage its sensitivity to blood-borne materials. Our brain functions only with good blood flow carrying sufficient glucose and oxygen. As a communications network, it regulates our memory, perceptions, learning, concentration, thoughts, and consciousness. With neuropeptides—the molecules of emotion—the brain

---

426 http://thesportdigest.com/archive/article/functions-yoga-exercise.

427 http://therapists.psychologytoday.com/rms/name/Transcend%3A+Mind+Body+Training+and+Beyond_Fairfield_Connecticut_106286.

428 http://www.huffingtonpost.com/2013/04/24/yoga-immune-system-genetic-_n_3141008.html.

429 http://www.newscientist.com/article/dn23480-meditation-boosts-genes-that-promote-good-health.html.

430 http://www.hellawella.com/meditation-genetics-and-dangers-misinformation-misinterpretation.

431 http://jonlieffmd.com/tag/meditation-helps-mitochondrial-resilience.

communicates with all 10 trillion cells in our body organs: heart, lung, kidney, liver, immune system, and the rest. Patanjali, some five thousand years ago, was right in saying, *"Yoga is to still the patterning of consciousness."*

We need to master our senses by breathing, focus, and relaxation because only then can we interconnect our senses and perceive fully. We can cognize, we can have thoughts, and we can perceive through our organs of sense. The power of perception leads to concentration, which in turn leads to meditative absorption. We can experience meditative absorption only in a state of maximized consciousness. We can thus reflect and realize friendliness, compassion, equanimity, and delight—the bases of good health without stress.

18.01 Gene Expression by Meditation and Yoga

Yoga has definite effects on brain chemistry. It can boost mood much more effectively than walking says Chris Streeter of Boston University School of Medicine[432]. Yoga increases gamma amino butyric acid (GABA), a neurotransmitter that is found to be deficient in conditions of depression.[433] We don't know enough about it yet, but genes have got to be involved in these antidepression processes.

Yoga and meditation are linked to low oxygen consumption and added nitric oxide production. The main mechanism behind the benefits of yoga rests on change in gene response to stress. We know that genes respond to stress. Meditation and yoga thus alter gene expression.[434] They alter both genes and gene expression[435] according to a Harvard study.

We lose memory with age, but it can be managed by diet, exercise, and yoga. Dendrites in our brain can be built by proper food. Much of our health is preserved by food and the environment. Alzheimer's and dementia are only 10 percent genetic; the remaining 90 percent can be managed by exercise, meditation, yoga, and belief.[436]

A single gene, monoamineoxidase A, causes accumulation of

---

432 http://www.ncbi.nlm.nih.gov/pmc/articles/PMC3111147/

433 http://www.ncbi.nlm.nih.gov/pubmed/20722471.

434 http://www.theatlantic.com/health/archive/2013/05/study-how-yoga-alters-genes/275488/.

435 http://www.mdmbac.com/meditation-yoga/harvard-study-shows-meditation-immediately-alters-gene-expression/.

436 http://www.sciencedaily.com/releases/2013/11/131118141817.htm.

neurotransmitters of violence.[437] We know that the cortex can be stimulated by electrical stimuli, which awaken neuropeptides and cell receptors. Meditation and yoga can induce such awakening.

18.02 Practice of Meditation and Yoga

Yoga facilitates conscious control of breathing, oxygenation, and gas exchange. It involves posturing, posing, breathing, muscle toning, and concentrating away from stressors in daily life. Yoga often includes exercises for musculoskeletal, nervous, respiratory, and cardiovascular systems. Since all our actions and responses are expressed in the musculoskeletal system, its daily conditioning is a physiological necessity. The yoga practice should include:

1. Exercises for back, spine, and legs
2. Exercises for knees, ankles, and feet
3. Exercises for elbows, shoulder, neck, and eyes
4. Exercises for circulation by breathing, nostril breathing, shoulder stand, and head stand
5. Exercises for muscle tone: side raise, back push-up, locust posture, and sit-up
6. Internal exercises: abdominal lift
7. Exercises for constipation: squatting abdominal lift and sitting abdominal lift.

Yoga thus is an act of practicing synchronicity with respect to events and people to create a sympathetic and parasympathetic balance. The sympathetic should never over-dominate, and stress should not be allowed to imbalance neurotransmitters. Abdominal breathing, an involuntary and unconscious activity, is very helpful. We should let the abdomen rise and fall and breathe up and down the body. Each of us needs to develop a yoga program for ourselves in order to review each day, prioritize each activity of the day, and manage time. Life should be made simple by conserving energy whenever possible. Thus, yoga can help enforce discipline and mindfulness in daily life, which is a critical part of health and wellness.

A little bit of stress from yogic exercise is good. Scanning the body

---

437 https://members.mhn.com/web/public/default/Mayo/MH00072.

every day, staying socially active, contemplating ideas of love and passion, recreating for fun to the maximum, and interacting with people and the world around is a sure way to good health. We have to remove social and psychological stress from our life by managing the intensity and frequency of practical yoga.

18.03 Meditation and Yoga for Stress Control

Yoga integrates all that is meditation by muscle relaxation, calorie burning, and enhanced oxygenation. Like meditation, yoga is good for stress relief and better health. In the practice of yoga, we stretch arms, extend legs, and breathe by both nostrils. We secure new postures and come up with new ways of exercising. Once made integral to daily living, yoga can secure good health and emotional balance; as a matter of fact, it can extend life.[438] Yoga, involving deep breathing exercises, muscle relaxation by movement and stretching, mindfulness and meditation, and imagery of things and places, makes us feel good; it conditions our body and mind. Yoga is for activating cellular receptors for new neural connections, for calming the mind, and for connecting the left and right hemispheres of the brain. Yoga thus integrates cellular functions and creates s mind-body connection. A skillful practice of yoga depends on perception of taste and smell, which is intercellular communication by chemicals and chemical reactions.

Amid growth and decay, our bones renew every alternate year. We get brand-new muscle cells three times a year. Red blood cells renew in 120 days. Platelets, thrombocytes without DNA responsible for routine growth, do so every ten days. Exercise and yoga boost renewal of body cells for growth, and therein lie the health values of meditation and yoga. Like exercise, yoga prevents decay and reduces stress. Yoga awakens receptors on our cells. Yoga practices can and do balance our physiology and help maintain homeostasis.

"To meditate is to master our psyche" is an axiom of mine. Meditation, I have believed from childhood, during my morning and evening single-minded sandhya (a practice of self-control) brings about conscious control of the somatic and autonomic nervous system and its interactions with skeletal muscle, involuntary nonstriated smooth muscles of the heart and blood vessels, and our emotional and mental activities. Neurotransmitters are the molecules of emotion. Yoga sets them in motion, directed to higher

---

438 https://members.mhn.com/web/public/default/Mayo/MH00072.

intelligence, by maintaining elasticity and flexibility of our spine, the communication cable between our mind and body. That yoga helps get rid of stress and thus helps us become active and young is now supported by a great deal of clinical research on mindfulness-based stress reduction (MBSR), led by Dr. Jon Kabat-Zinn, a program that is now practiced by the average public, company executives, and even by the Department of Energy.[439,440] There are one thousand centers in thirty countries on pursuit of mindfulness as a means to reduce stress, anxiety, and depression. Companies like General Mills Inc., J. P Morgan-Chase Bank, and Google and individuals like Apple's former CEO Steve Jobs and Congressman Tim Ryan from Ohio[441] are great testimony as to the value of meditation for mindfulness. The National Institute of Health has funded projects with more than $5 million, and there are 477 scientific journal publications on the topic[442], reports *Time* magazine. To me, the Sanskrit word *akagrataa* (single-mindedness) says it all. It is focus. It is being in the "present and now." It is mindfulness for brain health, brain's neuronal plasticity, balanced cortisol level, and an enhanced immune system. Mindfulness is integral to meditation, and it mitigates stress. For mastery and maximum benefit, the practice of commanding our breath, mind, thoughts, and time, I believe, should begin at an early age. Mindfulness should pervade our life as we select food, cook food, eat food, and even discuss foods.

18.04 Foods for Meditation and Yoga

Our diet affects our body, mind, and emotions, and thus our meditation. What we eat must serve our cells well because meditation and yoga deal with cells and cellular activity. I have emphasized in previous chapters that *fresh and plant-based foods that can deliver all nutrients for DNA methylation are good for meditation.* This then would include a variety of green vegetables, fruits, beans and legumes, whole grains, nuts, and seeds that provide the best balance of major ingredients of protein, fat, and carbohydrates plus necessary vitamins, minerals, and antioxidants. *Phytonutrients*, I should reiterate, is

---

439 https://www.google.com/#q=the+art+of+being+mindful+time&tbm=nws.

440 http://www.psychologytoday.com/basics/mindfulness.

441 http://www.innerresilience-tidescenter.org/documents/Mindful%20Nation_IRP_Excerpt.pdf.

442 http://www.huffingtonpost.com/joanna-piacenza/time-mindfulness-revolution_b_4687696.html

another name for plant-based foods—fruits, vegetables, seeds, and nuts. Methylation diet components like nuts, seeds, fruits, and vegetables have been the main diet for yoga practitioners in India for centuries.

18.05 Major Reference Institutions

Meditation and mindfulness are areas of very active research today. The following institutions perform cutting-edge research in this area.

Benson-Henry Institute of Mind-Body Medicine (Massachusetts General Hospital), Beacon Israel Deaconess Medical Center, Center for Healing and Spirituality at the University of Minnesota, Mayo Clinic, National Jewish Medical Research Center, University of California at San Diego, University of Colorado, University of Rochester, and Vanderbilt University.

CHAPTER 19

Stress Control by Foods

"Pressure and stress is the common cold of the psyche."
—Andrew Denton

Hormonal imbalance can cause anaerobic processes to dominate, the acidity in our cells to increase, mitochondria (the energy power plant) to slow down, and calcium to increase in the cytoplasm. All body systems are involved in stress, and imbalance of adrenalin, serotonin, and cortisol underlie all that is stressful. Stress transfers health problems to all body systems—muscular, cardiovascular, renal, central and sympathetic nervous system, immune system, and the second brain of the digestive system.[443,444]

We need to control stress to a level that is good for our physiology because stress and the stress hormone corticosteroid (cortisol) can be dangerous to our being. Mindfulness and meditation can reduce stress. Adaptogenic stress-control foods that have been time-tested in various old civilizations are now part of routine alternative medicine even here in the United States. Well-known antistress methylation and gene-expression foods are orange juice and camu camu powder for vitamin C, salmon and flaxseed for omega-3 fatty acids, spinach and soy foods for magnesium, and other foods like black tea, nuts including pistachios, walnuts, and Brazil nuts, avocado, raw vegetables, herbs and spices that deliver antioxidants, antimicrobial molecules, and molecules of vitamins. Mitochondria control stress by fusion and fission. They fuse to mitigate stress by mixing and diluting damaged mitochondria and they subdivide in order to create new mitochondria, removing damaged

---

443 http://www.mayoclinic.org/stress/art-20046037.
444 http://www.stress.org/stress-effects/.

ones.[445] What a clever construct of quality control when a complement is born in nanoseconds!

In order to avoid stress, we need to stay happy, to connect our body systems by exercise, to move in rhythm, to meditate, to organize, to socialize, and to sleep well. We need to be creative and social, and we need to keep moving. Most important of all, we need to be regular in building strength and keeping our body and mind connected. To be stress-free is to introspect daily, a sure way for freedom from high-cortisol stress. If not, stress can cause anxiety and depression by compromising the blood-brain barrier and disrupting the 60,000-mile-long blood vessels carrying oxygen and nutrients to our cells. Stress can kill our brain cells.

To be stress-free, we need to exercise, breathe, concentrate and meditate, sleep well, learn to control our mind, and introspect. Introspection needs to become integral to our lifestyle for the mind-body connection and for maximum consciousness.[446] Stress and its progenitor, anxiety, can thus be rooted out for better health by willful stress control.

Brain cells create good ideas, but stress is not one of those good ideas. We have evolved as human beings with stress as a defense. So the idea of stress is hardwired in our DNA, and the fight-or-flight response is part of our being. There come argument and anger when we choose to fight, and we feel depressed when we withdraw in defeat. This is where akaagrataa or single-mindedness, variously described as mindfulness, helped me, a strict Hindu child when I was hardly twelve years old. I was not a Buddhist, and akaadrata to me was rooted in Vedic scriptures.[447,448] I was instructed and told to digest the ideas of objects I could touch and hold and also the transitory objects of thought. The latter, I was taught to control and willfully transition to sustained single-pointedness for getting rid of whatever depressed me. I used to do so twice a day, and later only once a day up to mid-1963.

I came to the United States in September 1964 and, under the pressures of daily life, discontinued committing to akaagrataa for almost three decades. But I stayed who I was, and the "inner me" compelled me to introspect daily for fifteen minutes. For me introspection, single-mindedness, purity of mind, and mindfulness are all the same. Doing what needs to be done with

445 https://www.sciencemag.org/content/337/6098/1062.
446 http://www.sciencedaily.com/releases/2007/10/071008193437.htm.
447 http://www.swamij.com/yoga-sutras-30916.htm.
448 https://webspace.utexas.edu/shp9/www/pages/yoga/ekagrataaurobindo.pdf.

purity of mind and not getting bogged down in anticipation of rewards, says Shrimad Bhagawatgeeta,[449] keeps one stress-free.

Stress, excessive eating, and eventual obesity are the unpleasant consequences when the energy balance begins to favor fat storage. Upregulation of mitochondria involving genetic mutation is a prerequisite for a negative energy balance.[450] There are many stressors in our day-to-day life, such as added responsibility for higher-ups in an organization, workload, deadlines, sleep deprivation, family relationship problems, job loss, financial problems, death in the family, and one or another daily emergencies. Workload, routine problems of making ends meet, family friction, high credit card debt, and morning rush hour are reoccurring dominant stressors in our lives. Unfortunately, actually the bottom forty-four percent of low income America lives under constant stress, which is a chronic disease by itself.[451] Close to 60 percent of our trips to physicians are related to physical or psychological stress. Repeated stressful conditions are now well-documented as causes of loss of appetite, alcoholism, drug dependence, overweight conditions, obesity, indecision, poor concentration, helplessness, inflammation, heart disease, cancer, liver diseases, and a host of other illnesses.[452,453] Even a newborn baby can have abnormal myocardial contractions and irregular heart rhythm due to stress.[454] Medical science has established that even a fetus is vulnerable to traumatic stress.[455]

19.01 Genetics and Epigenetics

There are genes for stress, and stress can change gene expression.[456] A stressful life in one's early years gives rise to genetic variants of the serotonin transporter gene, and in most cases, depression can now be traced to gene-environment interaction. Stress genes cause coronary atherosclerosis. There

---

449 http://advaita-academy.org/Articles/shrImad-bhagavad-gItA---Part-20.ashx.

450 http://www.ncbi.nlm.nih.gov/pmc/articles/PMC2824926/.

451 http://www.apa.org/monitor/2011/01/stressed-america.aspx.

452 http://www.sciencedaily.com/releases/2012/04/120402162546.htm.

453 http://www.medicinenet.com/stress/related-conditions/index.htm.

454 http://pediatrics.med.nyu.edu/conditions-we-treat/conditions/persistent-pulmonary-hypertension-newborn.

455 http://www.psychiatry.emory.edu/PROGRAMS/GADrug/Feature%20Articles/Mothers/The%20effects%20of%20maternal%20stress%20and%20anxiety%20during%20pregnancy%20(mot07).pdf.

456 http://www.medicalnewstoday.com/articles/270418.php.

are genes that make proteins that can prevent and help recover from stress. There are genes for manic depression and hypertension. *It is now known that child abuse modifies stress genes permanently* [457]. A mother's problems can become the problems of the newborn baby because maternal stress during pregnancy transfers itself to the hypothalamic, pituitary, and adrenal activities of the fetus. Although mostly nongenetic, both genetic and epigenetic factors are at play in stress development.[458] This can be remedied, and there can occur counter-stress changes by relaxation. Food, energy balance, exercise and yoga, and simple meditation are tools of stress remediation.

19.02 Mechanism of Stress

Our mental, psychological, and emotional states are governed by molecules in action. Neural (hypothalamus) and biochemical reactions always attend to situations when we win a lottery or when we quarrel with someone. Stimulated by the neuropeptide Y (NPY), the hypothalamus secretes corticotrophin releasing hormone or *CRH* in response to a stressful event. This can have two consequences. One is that brain-stem and spinal-cord cells tell the adrenal gland sitting atop our kidneys to produce epinephrine and glucocorticoid hormones, which cause increased heart rate, breathing, and muscle response. The second action involves the pituitary gland, which releases *cortisol*, a hormone that increases metabolism. This is a molecular response to overactivity or its anticipation, and it has numerous metabolic manifestations, all controlled by the limbic or the emotional brain. Molecules of emotion are relevant to the mind-body connection because emotions link mind and body via the information network.[459] They are intertwined, and they relate to stress directly.

19.03 Personality Attributes and Stress

People can be classified with respect to propensity for stress in four ways.[460] One can easily see his or her stress propensity and take remedial measures by food, exercise, yoga, meditation, and day-to-day social interaction.

457 http://www.genome.gov/27554258
458 http://www.hopkinsmedicine.org/news/media/releases/chronic_stress_may_cause_long_lasting_epigenetic_changes.
459 http://www.sciencedaily.com/releases/2008/07/080715152325.htm.
460 http://www.midus.wisc.edu/findings/pdfs/221.pdf.

1. The high-serotonin explorer-type people are spontaneous. They sleep less, and they are impulsive. Such people are prone to heart attack, diabetes, kidney disease, accidents, and cardiovascular problems.
2. The type-A director is a workaholic, with the least social interaction. He or she is bound to have a compromised immune system and propensity to having ulcers and dizziness. His or her neurons have a tendency to shrivel.
3. Hardworking emotional negotiators are verbal, compassionate, and kind, but they have a propensity to lupus, multiple sclerosis, chronic stress, and obesity.
4. Frugal and laid-back types are interested in the status quo. They are builders and followers.

For most of us, the activities of mitochondria in neurons with respect to personality types have yet to be deciphered.[461].

19.04 Unhealthy Consequences of Stress

Hormonal imbalance can cause anaerobic processes to dominate, the acidity in our cells to increase, mitochondria (the energy power plant) to slow down, and calcium to increase in the cytoplasm. Stress can have damaging effects on reproductive performance and cardiovascular functions. It can even kill brain cells.[462] It can cause atherosclerosis of the carotid artery and thus stroke. Chronic stress causes inflammation and interrelated diseases, including compromising effects on the blood-brain barrier as to:

1. Imbalance of calcium, potassium, and sodium ions.
2. Enlarged mitochondria, affected Golgi apparatus and cytoskeleton around cells, and slowing down of atopsis.
3. Damaged cells.
4. Accelerated shortening of telomere DNA.
5. DNA damage and a compromised immune system.

---

461 http://www.ncbi.nlm.nih.gov/pubmed/22388959.
462 http://www.fi.edu/learn/brain/stress.html.

High concentration of cortisol beyond the optimum can kill brain cells, those in the hypothalamus in particular.[463] The hypothalamus has been found to shrink by 14 percent under excessive stress. Men are at higher risk than women. The optimum level of cortisol is listed in table 19.01.

Table 19.01 Stress Hormone Cortisol in the Human Body

Adult	Morning	5–23 mcg per deciliter
	Afternoon	3–13 mcg per deciliter
Child	Morning	3–21 mcg per deciliter
	Afternoon	3–10 mcg per deciliter
Newborn		1–24 mcg per deciliter

The stress hormone cortisol rises with age. It affects memory retrieval and function, and too much of it over a long period of time is really bad.

19.05 Stress Reduction

Stressed folks have a high level of cortisol. Back pain, chest pain, stiff neck, increased appetite, constipation, upset stomach, dry mouth, light-headedness, tiredness, sweating, high blood pressure, palpitation and accelerated heart rate, short breath, weight gain, and sex drive problems can result from excessive stress in daily life. Stress can cause heart disease, sleep problems, digestive problems, depression, obesity, memory impairment and Alzheimer's disease, and unhealthy skin. Stress affects the brain, and the brain affects the body. How we react to stress is itself a killer because it is an admission of defeat, a kind of mental illness. In consequence, we feel and behave differently, and our body is sensitized to stressors that engender one or another type of emergency.

Today our society promotes stress by default. Prolonged stress, in physiological and evolutionary terms, is an equivalent of famine. In a way, modern society is designed for stress, a case of false alarm without famine. Some people with mirror neurons thrive on stress because they can turn it off and on at will. A mirror neuron is a neuron that fires irrespective of whether one acts or one observes the same action performed by another. Thus, the neuron "mirrors" the behavior of the other, as though the observer

---

463 http://veteranssuicides.weebly.com/ptsd---cortisol-kills-brain-cells.html.

were itself acting. Such people manage stress more successfully. I am of Indian descent, and my own Eastern belief is that people who have mirror-touch *synesthesia* have very active mirror or empathy neurons.[464] We have to fill our daily lives with cheer and joy, or else stress can alter our ability to cope with stress. We need to be bold, busy, and valiant, and we need to relax. Relaxation in our life must dominate tension and stress.

The relaxation response is used to treat high blood pressure, anxiety, depression, insomnia, menopausal hot flashes, backache, headache, and phobias.[465] This simple idea of relaxation is integral to practices in tai chi, qigong, and Hinduism's chanting of mantras during my akaagrataa sessions for meditation and prayers. It is about progressive muscular relaxation, biofeedback, and simple deep breathing. Since our body responds to how we feel, think, and act, stress can be routinely avoided by doing things that oppose built-in fight-or-flight reactions. We can do this by matching activities of the mind and the body and by creating a physical state of deep rest. Examples of such defensive activities are

1. Meditating and doing yoga.
2. Accessing the psychosomatic network or brain-body connections.
3. Staying emotionally wholesome and being with nature.
4. Committing to daily introspection.
5. Acknowledging family and friends, living unselfishly, being true to oneself, practicing emotional self-care, relating to the body, exercising, and eating wisely.
6. Seeking challenge, having a commitment, and being passionately curious.
7. Not allowing stress to affect personal biology.

Relaxation by visualization and imagery, yoga, and tai chi help lower heart rate, blood pressure, and muscular tension; slow down breathing; increase blood flow; improve concentration and focus; reduce anger; and build confidence.[466,467]

---

464 http://www.daysyn.com/Banissy_Wardpublished.pdf.

465 http://www.psychologytoday.com/blog/heart-and-soul-healing/201303/dr-herbert-benson-s-relaxation-response.

466 http://www.cliving.org/teensadults.htm.

467 http://my.clevelandclinic.org/heart/prevention/emotional-health/stress-relaxation/stress-management-your-heart.aspx.

Good relaxation can easily come about by healthy snacks, keeping faith in daily work, music, preparation for the future, recalling past successes, resting, swimming, and using polite and courteous language. Homeostasis and the mind-body connection can be maintained by simple breathing and resistance exercises. We must inhale for tense muscles and exhale to relax them.[468]

19.06 Food for Stress Control

Diet can change behavior. We know that endomorphins are the body's natural painkiller. Proteins make DNA, carry information, digest food, and protect the immune system. Beyond comprehension, the molecularity of life is amazing and enigmatic. As a matter of fact, all cellular communications underlie health and well-being. An example is in order here. Our retina reacts within a quadrillionth of a second when rhodopsin changes to bathorhodopsin. Retinol and vitamin A modulate this process, and what intervenes is the vibration of a single carbon to a carbon double bond.[469]

No doubt stress has its own molecular dimension,[470] and our daily food is very critical to stress chemistry. Amino acid arginine in our protein component of diet helps produce nitrous oxide(NO), a multifunctional, lipid-soluble molecule. Nitric oxide has a half-life of only thirty seconds. It is an autocrine or paracrine mediator of homeostasis and, therefore, of health and well-being. Gama amino butyric acid (GABA) regulates anxiety, fear, and stress. Proteins from lentils and legumes, almonds, and tree nuts provide glutamic acid and glutamine for gamma amino butyric acid. Other complementary gene-expression nutrients can come from bananas, broccoli, spinach, halibut, and oats and other whole-grain products.

A daily dose of an orange for vitamin C, spinach for magnesium and B vitamins, omega-3 fatty acids from salmon and flaxseed, and complex carbohydrates from whole grains constitutes good stress-control foods. Adaptogen aswagandha, according to the ayurvedic medicine of India, stabilizes cortisol.[471] These are foods with antioxidants for prevention of oxidative damage and stress. These are foods for gene expression.

468 http://www.health.harvard.edu/newsletters/Harvard_Womens_Health_Watch/2008/July/relaxation_techniques_breath_focus.
469 http://www.cell.com/biophysj/abstract/S0006-3495(79)85205-4.
470 http://www.wholehealthchicago.com/4590/the-chemistry-of-stress/.
471 http://naturalmedicinejournal.net/pdf/NMJ_JUNE10_TC.pdf.

19.07 Major Reference Institutions

Stress is a huge public health problem. The following institutions are at the forefront of chronic stress research.

The University of Pittsburg is a great center devoted to stress control. Other prestigious institutions include the Heart and Stroke Foundation of Ontario, Canada, the Canadian Institute of Heart Research, the National Institute of Health in the United States, Yale Stress Center, Stockholm Stress Center at the University of Stockholm, and the Center for Anxiety, Depression, and Stress Research at McLean Hospital, an affiliate of Harvard Medical School.

CHAPTER 20

Sleep Therapy

"Early to bed and early to rise makes a man healthy, wealthy, and wise."
—Benjamin Franklin

Eight hours of sleep daily is essential to the health of our mind, and sleep can be managed by foods. Foods can help maintain the circadian rhythm and our chronobiology, our puberty, and our essential task of efficient and faultless species perpetuation.

Sleep is a fundamental for building and integrating, a set of processes common to all life, including plants and animals. Whereas plants are always in an anabolic state, mammals among animals are only minimally conscious in sleep, In terms of anabolism though, they are at maximum. Research at the University of Helsinki shows that *sleep is the brain of our immune system,* and the cytokines produced by the immune system promote sleep.[472] A good night's sleep is necessary for both mind and body. The hypothalamus in our brain has got our biological clock and control of our circadian rhythm. The neurons in the brain stem connect with the spinal cord and produce serotonin- and norepinephrine-type neurotransmitters that keep us awake. As a matter of fact, temporal organization—the sleep and awake states—is just as important as is cellular organization in day-to-day living. Solar and lunar rhythms matter, and our neurons oscillate in seconds, minutes, hours, days, months, and years as they time our life's master clock.

Unfortunately an average American is sleep-deprived, and sleep loss causes disastrous consequences of poor mind-body control, reduced consciousness, weight gain, type 2 diabetes, obesity, a compromised immune

---

472 http://www.sciencedaily.com/releases/2013/10/131023183908.htm.

system, stress, loss of granulocyte-type white blood cells, and other chronic diseases. Sleep deprivation causes mitochondrial malfunction as well.[473,][474] Sleep deprivation can even cause death.[475] We know that sleeplessness is the cause of at least 10,000 car accidents in the United States per year.[476]

Sleep is responsible for brain function and repair, daily metabolism, and functioning of our immune system. Neurons and brain proteins are made and memory is consolidated and optimized during sleep. Brain, our super computer, needs to be defragged every day. The mind and consciousness may be a consequence of quantum computing.[477] Eight hours' sleep is responsible for an efficient mind and body. REM sleep allows complete charge of the brain battery. Adenosine, a homeostatic sleep factor, puts us to sleep at night for coping with trauma, solving problems, learning and organization, and dreaming.[478] The state of sleep is a dynamic event as it relates to manufacture of immune cells and interaction with vagus nerve cells, which when stimulated, prevent the spleen from making inflammatory tumor necrotic factor.[479] The spleen and brain are in constant communication, and sleep is the brain of our immune system.[480]

Sleep, nutrition, and exercise are key to our physical and mental health and to our longevity. Hormone functioning, homeostasis, metabolism, and brain activity depend on good sleep, and it is the least expensive medicine for us to use. Dependence on Ambien and Lunesta for daily sleep is a bad idea.

Our brain requires sleep for consolidation of long-term memory. Yawning, which happens to 40 million Americans every day, is a sign of not being awake and of the brain trying to check out. Food can train our biological clock and affect health issues related to the daily circadian rhythm and sleep, a necessity for good health.[481] Nutrition in our daily diet and exercise are about electrical and chemical signals that mediate the dance of our neurons. Diets do dictate this dance, and sleep is part of the dance.

---

473 http://www.ncbi.nlm.nih.gov/pubmed/20176368.

474 http://www.ncbi.nlm.nih.gov/pubmed/20176368

475 http://www.scientificamerican.com/article/how-long-can-humans-stay/.

476 http://www.medscape.com/viewarticle/503105_2.

477 http://www.quantumconsciousness.org/penrose-hameroff/quantumcomputation. html.

478 http://www.jneurosci.org/content/26/31/8092.full.

479 http://www.ncbi.nlm.nih.gov/pmc/articles/PMC2696570/.

480 http://www.ncbi.nlm.nih.gov/pubmed/12498102.

481 http://www.mylocalhealth.com/stress_less/train_your_brain_to_sleep_recognizing_ the_circadian_rhythm_448.

All mammals with the exception of dolphins and whales sleep. A seven- to eight-hour-a-day sleep is ideal for human beings. Our body and mind crave for sleep at night and around midday. Our neurons use conductance-based signals. Electrons and electricity prevail even when we are asleep. Low-amplitude high-frequency oscillations become large-amplitude and low-frequency during what is called low-wave sleep. Perceptual awareness is gone during such a state of sleep. However, the rapid eye movement (REM) sleep is, in a way, a wakeful state.

Histamines and other chemicals stop flowing, and the *night hormone melatonin* secretion begins in the pineal gland when we go to sleep. Our mind's biological clock asks for the *day hormone serotonin* when there is light. A balance of melatonin and serotonin is essential for good sleep, and the cycle of the two is part of the circadian rhythm and our chronobiology.[482,483]

Sleep is a very dynamic state, characterized by changing electrical activity and flow of messenger molecules for control by two structures in the hypothalamus of our brain. The *hypothalamus* controls the circadian rhythm and chemicals promoting sleep and arousal. The *thalamus* blocks inputs from the senses, and the pineal gland produces melatonin when it is dark. The *hippocampus* helps form memory during REM sleep, *pons* are involved in arousal and dreams, the *cerebral cortex* works with pons, and the *retina* sends signals when light falls on it.

A neural dance in the hypothalamus puts us to sleep. The dance represents a dynamic state, with shifting levels of electrical activity and flow of chemicals. The hippocampus goes to work for memory formation. The brain is awake, but other voluntary senses are paralyzed even during rapid eye movement (REM) sleep.

Low melatonin is a problem, and so is a low level of human growth hormone. Both cause sleep disorders that lead to diabetes, hypertension, obesity, memory loss, depression, anxiety, and accelerated aging.[484]

Time (solar and lunar rhythms) affects our activities in a day and during our lifetime because temporal organization is just as important for our daily life as is cellular organization.[485] There is time to puberty, to menopause, and to signs of aging. Oscillations in neurons can be of the order of milliseconds

---

482 http://sustainablebalance.blogspot.com/2012/09/sleep-series-part-3-serotonin-melatonin.html.

483 http://delicatebalance.com/?dictionaryterm=melatonin.

484 http://umm.edu/health/medical/altmed/supplement/melatonin.

485 https://www.cell.com/current-biology/abstract/S0960-9822(13)00754-9.

or minutes, up to hours or even two days. Certain genes are expressed at certain times of the day, and humans can be chronotypes of morning or evening people.[486]

Sleep balances leptin (appetite) and ghrelin (hunger) hormones, lowers stress hormones, lowers other inflammatory chemicals, lowers cholesterol, lowers hypertension, lowers depression, and lowers obesity.

20.01 Genetics of Sleep

We should be thankful to our maker that no more than forty families on the planet have the fatal insomnia gene, a gene that produces malformed *prion* proteins, proteins that affect the structure of the brain and other neural tissues. The protein attaches to the thalamus, causing fatal sleep insomnia. Gene DBQ-0602 is variant in 25 percent of the population, but very few with the gene have sleep disorders. It keeps the potassium ion channel open during the period we are awake and closed when we are asleep. A sleep gene is known to affect homeostasis, and although the mechanism is not known, a "clock" gene controls circadian rhythm. Mitochondrial genes may be involved in the sleep-wake state transition.[487]

Sleep deteriorates 27 percent by each decade as we grow older, and its patterns are determined by the circadian rhythm[488]. Sleep regulates emotions, enhances memory, processes experiences, tunes up the nervous system, and organizes information. Sleep is necessary for emotions, decision making, concentration, and social interactions. Seniors need to have a more punctual sleep habit.[489]

20.02 Why Do We Sleep?

Sleep is vital both for our body and our mind. Sleep is necessary for synaptic plasticity—time to rest and to dream. Since the brain runs on glucose, a drop in glucose after more than twenty-four hours without sleep triggers a number of deleterious events: the brain cortex doesn't process emotions and reason, the ability to fight free radicals decreases, the brain remains undetoxified, the leptin level falls and increases appetite, and hormonal

486 http://www.chronobiology.ch/wp-content/uploads/publications/2003_12.pdf.
487 http://www.ncbi.nlm.nih.gov/pubmed/9602159.
488 http://www.webmd.com/sleep-disorders/features/adult-sleep-needs-and-habits
489 http://www.sciencedaily.com/releases/2008/08/080825203918.htm.

imbalance causes weight gain. A twenty-four-hour sleep deprivation causes homeostasis problems also. Reaction time increases to 0.5 seconds as opposed to 0.25 seconds when we go for eighteen hours without sleep. This effect is equivalent to 0.08 percent alcohol in our blood. Melatonin lowers temperature and blood pressure during sleep. Serotonin is produced during REM sleep. Serotonin converts to melatonin, the sleep hormone. Sleep, therefore, is not an inactive period. Too many critical things happen when we sleep.

1. Sleep is not a default state. It is an active biological process necessary for staying alert. Circadian rhythm and sleep homeostasis are designed into our routine physiology, and our mind is active during the sleep, still using energy.
2. Muscles are repaired, hormones are released, the immune system is boosted, and our 10 trillion cells are energized.
3. We perceive, reflect, recall, and organize our brain cells. More learning needs more synapses, which in turn require high glucose for running the brain. Memory is replayed, and information is consolidated and formed. Ineffective neurons are repaired and pruned off during sleep, and memory is consolidated during the REM phase of sleep. Depleted circuits are repaired.
4. Biogenic amines and neurotransmitters regulate sleep, appetite, and sex. Sleep is necessary to maintain the balance.
5. Metabolic waste is flushed out during sleep.
6. Critical molecules depleted and used during wake time are synthesized, and cells are repaired for their critical functions.

20.03 What Makes Us Sleep?

Although the molecular mechanism of sleep is not known yet, research centers at UC San Diego, the Siegel Lab at UCLA, the University of Wisconsin in Madison, Washington State University in St. Louis, Missouri, Harvard Medical School, Stanford University, University of Pennsylvania, NIH National Center for Sleep Disorder, American Sleep Research Institute, and the famous CFSR in Australia are busy trying to decipher the mysteries of sleep. We have yet to fully understand what goes on in our jellylike brain floating in 150 ml of cerebrospinal fluid secreted to the extent of 500 ml per day. The limitation today is that out of 125 million neurons, we can study only five hundred of them in a day by implanting mini electrodes.

Sleep and metabolism are closely connected. There are circadian and homeostatic mechanisms that control sleep, where circadian determines timing and homeostasis determines need.[490] Sleep is truly the brain of our immune system.[491]

Sleep is controlled by our hippocampus center through mediation of a neuronal group. There is a delta sleep-inducing peptide as part of the neuronal group. Even cytokines are involved in sleep. The serotogenic system controls sleep, emotion, and appetite. Human beings enjoy more REM sleep approaching the dawn period, whereby procedural, visual, perceptual, and performance aspects of memory are restored. Non-REM sleep is good for declarative memories (remembering names and dates). We need an average of eight hours of sleep per day for our health and well-being.

We do know that (1) circadian rhythm is controlled by the hippocampus region of our brain, (2) neurons that fire together are wired together, (3) neurons do not regenerate, and (4) daily pruning and strengthening of neurons during sleep is a physiological need.[492]

20.04 What Happens in Our Brain When We Sleep?

We have five phases of sleep: phase 1 is 10 percent of sleep duration, phase 2 is 50 percent, phases 3 and 4 are 10 percent each, and REM (rapid eye movement) is 20 percent. Since infants are busy programming, they spend 50 percent of sleep time in REM, responsible for procedural memory, pattern recognition, and motor skills (Weisman Institute, Rehovot[493]). Non-REM sleep is devoted to declarative memory. Both REM and non-REM sleep are required for visual and perceptual task performance. It all happens with the brain. Whereas REM is a state of high-frequency brain waves, stages 3 and 4 of non-REM sleep are slow-wave patterns that increase with age. During sleep, 50,000 neurons control the body's master clock, sense light, coordinate cellular clocks, rhythm, and autonomic functions, and spur endocrine functions.

---

490 http://www.acnp.org/g4/GN401000075/CH075.html.

491 http://www.ninds.nih.gov/disorders/brain_basics/understanding_sleep.htm.

492 http://www.nature.com/scientificamerican/journal/v309/n2/full/scientificamerican 0813-34.html.

493 http://www.sciencedaily.com/releases/2012/08/120826143531.htm

20.05 Effects of Sleep Deprivation

Sleep deprivation leads to poor wound healing, a poor immune system, hypertension and heart disease, and obesity-type serious public health problems. We know that regions with high sleep deprivation also suffer from obesity. We need sleep, exercise, and nutrition for staying alert and active. Sleep-deprived folks overeat.

Sleep deprivation and long-term loss of sleep cause poor reaction time and other problems, including low cortisol secretion, poor growth, weight loss or gain, risk of obesity, poor functioning, onset of ADD/ADHD, an impaired immune system, variation in heart rate and risk of cardiovascular disease, heart attack, fatigue, clumsiness, irritability, type 2 diabetes, Alzheimer's disease, and depression. They cause changes in body temperature and changes in cortisol level. Sleep deprivation leads to:

1.   A poor immune system.
2.   Imbalance in metabolism, high blood glucose, and obesity[494] (Columbia University study).
3.   Hormonal imbalance that leads to weight gain and obesity[495] (Stanford University Study).
4.   Pituitary function causing low cortisol levels and reduced thyroid-stimulating hormones.
5.   More insulin production and increased risk of type 2 diabetes.
6.   Metabolic syndrome, glucose intolerance, and snoring. Obesity and obstructive sleep apnea go hand in hand.
7.   Increased appetite due to changes in hormones ghrelin and leptin (Stanford University[496], University of Brussels[497]).
8.   Poor wound healing.

---

494 http://www.nhlbi.nih.gov/guidelines/obesity/ob_gdlns.pdf
495 http://www.huffingtonpost.com/2012/10/26/sleep-deprivation-obesity-leptin-ghrelin-insulin_n_2007043.html
496 http://med.stanford.edu/news/all-news/2004/stanford-study-links-obesity-to-hormonal-changes-from-lack-of-sleep.html
497 http://www.thenakedscientists.com/forum/index.php?topic=3633.25

20.06 Ideal Foods for Good Sleep

No doubt, diet can tweak the body's master clock and sleep patterns, wake neurons up, and manage our ninety-minute REM and non-REM sleep cycles. *Lycopene from watermelon and tomatoes, vitamin C from oranges and lemons, and powerful antioxidant-like selenium from Brazil nuts can turn short sleepers into ideal seven-to-eight-hour sleepers. Dark chocolate with caffeine can help obese and overweight people who oversleep. Tart cherries, a rich source of natural melatonin at 0.8 percent, are therapeutic sleep aids. Melatonin fights insomnia, cancer, and depression and is a good antiaging nutrient. It is also a powerful natural antioxidant.* Sleep-inducing foods include almonds, bananas, sunflower seeds, chickpeas, cottage cheese, cheese, warm milk, yogurt, whole eggs and egg whites, soy milk, tofu, soybean, nuts, seafood like salmon and trout, lean meats, poultry, peanut butter, beans, lentils, sesame seeds, flaxseed, and sunflower seed. This list can be used to design a high-protein breakfast, a high-fiber and high-complex carbohydrate lunch, and a low- to medium-protein dinner. Prostaglandin, melatonin, L-theanine, and GABA (gamma amino butyric acid), all of which reduce stress, are good sleep aids. Melatonin is also an antioxidant.

On the other hand, high-fat diets alter the circadian rhythm. Also, serotonin imbalance is known to be a cause of anxiety, migraine headache, depression, addiction, and schizophrenia. Low stress enhances sleep by reducing neuropeptide Y(YPN). However, normal sleep can be secured by selecting the daily breakfast, lunch, and dinner properly.

20.06.01 Breakfast: A high-protein breakfast with amino acid tryptophan is necessary for synthesis of serotonin and melatonin. These are the neurotransmitters that slow down brain traffic and help bring about sleep. One boiled egg can be a great source of 6 grams of protein. Other high-protein breakfast options could be French toast, melon with cottage cheese, pita bread with hummus, Greek yogurt smoothie, and Gouda cheese bites.

20.06.02 Lunch: A good lunch must offer complex carbohydrates, dietary fiber, and well-balanced protein. Fish, egg salads, bean soups, and chicken are great examples. We should get a nap between lunch and dinner even if it is a short one in the park. Caffeine after lunch should be avoided, limiting it to breakfast only. We do not need more than 25 mg caffeine for daytime alertness. A 40 mg dose in a large coffee is plenty.

20.06.03 Dinner: Meals that are high in carbohydrates and low-to-medium in protein help us relax in the evening and set us up for a good night's sleep. Included in dinner can be warm milk with honey, pasta with parmesan cheese, scrambled eggs and cheese, tofu stir-fry, hummus with whole-wheat pita bread, whole-wheat bread, seafood, pasta, cottage cheese, meats and poultry with veggies, a tuna salad sandwich, chili with beans, high-tryptophan sesame seeds sprinkled on salad with tuna chunks and whole-wheat crackers, bananas, potatoes, oatmeal, almond, flaxseed, mushrooms, lemon, and basil as herb. One can use a good combination of the above-mentioned foods for dinner each day of the week.

Foods that are high in carbohydrates and calcium and medium-to-low in protein make ideal sleep-inducing bedtime snacks, such as whole-grain crackers with hummus made with chickpeas and sesame. Hummus-like dips can also be prepared with a blend of cottage cheese and pureed blend of pumpkin seed, sunflower seed, and nuts. Consumption of cherries for melatonin, bananas and almonds for muscle mineral magnesium, warm milk and cottage cheese, and low-glycemic index foods are variously described to bring about good sleep.

20.07 Sleep Management

Keep a punctual sleep schedule and avoid sleeping pills. One should exercise and relax before bed, keep a cool and dark environment for sleep, and avoid heavy eating, smoking, and alcohol. Sleep should commence within twenty minutes of going to bed, and then sleeping until sunlight is a great idea.[498] We need to slow the mitochondria down for falling asleep,[499] Tufts' health plan, I believe, offers a great program for sleep management[500].

Diet pills, pain pills, pesticides, herbicides, and heavy metals deplete neurotransmitters, create hormonal imbalance, and cause abnormal sleep patterns. We should avoid them at all costs.

20.8 Major Reference Institutions

SRI International, Menlo Park, California, is an authority on sleep and circadian rhythm; and the University of California, San Diego, School of Medicine has a Center for Functional MRI that monitors health values of sleep. Any one of us can sign up for their newsletters and continuous updates.

---

498 http://www.tuftshealthplan.com/providers/pdf/mng/CPAP_BiPAP.pdf.
499 http://drmyhill.co.uk/wiki/CFS_-_The_Central_Cause:_Mitochondrial_Failure.
500 http://www.tuftshealthplan.com/providers/pdf/mng/Multiple_Sleep_Latency_Test_Sleep_Studies.pdf

CHAPTER 21

Belief and Faith Therapy

"Health is a state of complete harmony of body, mind,
and spirit. When one is free from physical disabilities and
mental distractions, the gates of the soul open."
—B. K. S. Iyenger

Every religion has rules about food, rules about timing of food
consumption, and even foods for essential nutrition.[501] In fact, religious
laws like Lent in Christianity, Vratas in Hinduism, and Ramadan in Islam are
about programmed calorie- and carbohydrate-restriction for bringing about
basal metabolism. This practice promotes growth of good bacteria in the
colon, and it is good for reducing risks of chronic diseases.[502] Furthermore,
the practice directly manages mitochondrial activity. Often such practices
are in line with a "feast and famine" schedule embedded in nature and
human evolution. Actually fasting is a way for restricting calories and thus
promoting longevity. Fasting can influence markers for temperature and
insulin level.[503] There is evidence now that supports the concept of spiritual
healing by intermittent fasting for reduced-calorie intake, energy balance,
and optimal homeostasis.[504] The result is reduction in stress.

Health is not simply an absence of disease; it is, rather, a state of complete
physical, mental, social, and spiritual well-being. Anything that affects the
body affects the mind, and vice versa. Faith, it is widely believed, affects

---

501 http://library.thinkquest.org/11960/facts/religion.htm.
502 http://www.butlersguild.com/index.php?subject=103.
503 http://www.nature.com/nature/journal/v457/n7230/edsumm/e090205-11.html.
504 http://www.med.umich.edu/umim/food-pyramid/.

the mind, body, and health.[505,506] Lifestyle and dietary practices have a lot to do with health because functioning of organs per se has diet dependencies. Furthermore, we know that the mind has the power and mechanism to (1) connect to the body, (2) harmonize emotions, and (3) improve physical well-being. Mind, noted scientists like Candice Pert believe, is distributed throughout the body and not just the brain.[507] An obvious example is the enteric nervous system. Maybe that is why organ recipients develop the tastes of the donor.

To a certain group of believers, prayer therapy works, and it should work because if prayer in faith can make us positive and cheerful, socially interactive, and emotionally balanced, it can definitely help us control anxiety and stress; it can confer powers of commitment and mind control. Faith can thus modify undesirable emotions that cause stress-dependent ailments and diseases. Faith is part of emotional epigenetics.[508] Prayer has a direct effect on our mind, and therefore, it affects our body.

From a practical point of view, prayer and meditation are almost synonymous. Like meditation, prayer can align mind and body, and it can define and create personality; with practice, it can become part of a set of beliefs. Conscience, matter, and miracles have power because *emotional epigenetics* can link mind and matter, affecting personal health.[509,510] If the subconscious mind processes 20 million stimuli per second and faith is preprogrammed in the subconscious mind, emotional states of spiritualism must affect cell functions in our body. Since religion impacts our social, dietary, and physical practices, it must impact health by making genes and the environment act in tandem.[511] The experience of faith per se can affect gene expression, personality, behavior, and even cognition.[512] Actually, the experience of faith can shape inheritance via signals outside the cells in our body.[513]

Belief triggers mental states, emotions, and attitudes. The latter affect

505 http://www.jstor.org/stable/1412143.

506 http://umm.edu/health/medical/altmed/treatment/mindbody-medicine.

507 http://www.psychologytoday.com/blog/the-science-willpower/201208/is-your-mind-separate-your-body.

508 http://www.thegod720.com/GenieInYourGenes.pdf.

509 http://www.ncbi.nlm.nih.gov/pubmed/21835681.

510 http://www.nature.com/ejhg/journal/v14/n6/full/5201569a.html.

511 http://www.ncbi.nlm.nih.gov/pmc/articles/PMC2654994/.

512 http://www.sciencedaily.com/releases/2012/03/120312114119.htm.

513 http://www.popsci.com/science/article/2013-08/what-twins-reveal-about-god-gene.

our body's biochemistry and its messengers and messaging systems. The parietal lobe of the human brain, which is very active during meditation and deep prayer, helps modulate the body's biochemistry.[514],[515] Along with the thalamus and frontal lobe, it deals with matters of faith. All three combined help connect mind and body for positive gene expression. Routine physiology is controlled by the cell membrane and receptor proteins on it. The membrane around our cells is a programmable liquid crystal, kind of a silicon chip. The proteins and the membrane interact with the environment and help gene expression. They are part of our epigenome. Faith-based individuality thus can be an epigenetic manifestation.[516]

Emotions are related to heart disease, diabetes, depression, and obesity. Faith and spirituality lead to better cognition and memory by yet unknown mechanism(s) involving cellular-energy production and communication. Vibrations on the cells' surface matter in this process of communication, just as do signals from outside the cell. Faith as an element of consciousness has dimensions of time-space-energy, and consciousness is pure energy. Faith can help create signals, it can calm us, and it can reduce stress. We can change our mind and health by spiritual engagement. A cell is alive because it has energy, and so are we as a vibrating composite system. Faith, friends, and family members, I believe, can help manage our body's energy. Faith can heal because it is critical to our energetics, physiology, and neurobiology; it is our body's power, underlying health.

We behave because we eat, think, and believe. It appears reasonable to believe, therefore, that we can heal ourselves with thoughts and feelings. We can unstress ourselves by mind control, and we can heal ourselves by compassion, expectation, hope, imagination, joy, commitment, and optimism. Faith and spirituality—call it a special domain of compassion— can train our mind and body; they can energize the body as a system of molecules. Faith can thus help us practice brain chemistry for improved health and well-being.

It is in human nature to depend on and to believe in a higher power. The belief of spiritual healing is common to all major religions. People travel to Lourdes in France, to Mecca in Saudi Arabia, to Varanasi in India, and to synagogues all over the world and pray for health, the sine qua

---

514 http://www.npr.org/templates/story/story.php?storyId=104310443.

515 http://www.createvibranthealth.com/services/neurotransmitter-assessment/.

516 http://www.vitalityconcepts.com/index2.php?option=com_content&do_pdf=1&id=23.

non of our very existence. This belief, it appears, influences consciousness, emotions, and our daily behavior; it influences our mind, body, health, and daily physical energy.

21.01 Religion, Diet, and Health

Jewish Yom Kippur, Muslim Ramadan, Hindu Vratas (exercise of conviction by fasting), and Catholic Lent have prescriptions of raw and cooked foods, fasting, and foods for pre- and postnatal periods. All such prescriptions attempt to bring the body to basal metabolism, provide for nutrients, and enhance our immune system. Religious practices of aura awareness, breath works, pranic healing, reiki, grounding and ecology, vaastu (place and directions), and yuen (Chinese energy medicine) teach us to manipulate and master "biochemical energy" or ATP for balance and harmony. Diet control, exercise, and oxygen balance for stress control are common denominators to all such practices and are summarized in table 21.01.

Table 21.01 Multireligious Practices for Health and Healing

Religion and Faith	Prescriptions and Practice	Effect
Christianity	Pilgrimage to Lourdes, France	Low stress and indirect cure
Hinduism	Yoga and meditation	Stress reduction by stilled consciousness, self-reflection, oxygenation, muscular movements
Islam	Pilgrimage to Mecca and Ramadan	Mental focus, stress reduction, metabolic control
Qigong, China	Physical and mental training	Breathing (qi) for relating matter and energy and for using our intrinsic energy for better health by emotional control
Reiki, Japan	Mind and emotional control	Use of intrinsic energy for spiritual reflection
Shamanism	Mending the soul by balance	Treatment of ailments and diseases

Religion and Faith	Prescriptions and Practice	Effect
Sangoma	Herbal medicine, divination, and counseling	Sangoma is the healer in Swazi, Zulu, Southern Africa, and Zhosa. This is analogous to practice of alternative medicine.
Nganga	Herbal medicine and social counseling	Nganga is the healing counselor in Brazil, Haiti, and Cuba.

21.02 Energy Healing

Human beings are actually electric and magnetic beings. Their bodies, running in reality on electrons and protons, are endowed with electric and magnetic fields. If not energy, what else could be the logical modality for healing and maintaining homeostasis? It is widely speculated that our biological clock has quantum attributes.[517] The steady state of homeostasis could be simulated only by a quantum computer.[518] Keeping balance and maintaining physical stability, practicing force and speed in body movements, doing meditation, and practicing tai chi are directed to oxygenation by breath control and motion of body parts. This is why modern exercise programs call balancing human body's energy for healing effects[519]

21.03 Bridge between Belief and Science

To me prayer is an act of mind-body connection.[520,521] Prayer in effect is meditation. Prayer can stabilize emotions and connect with the brain's hardwired center for spiritualism; it may even heal. Let us reflect on a few specific cases of effect of prayer.

---

517 http://www.glimmerveen.nl/le/biological_clock.html.
518 http://en.citizendium.org/wiki/Homeostasis_(biology).
519 http://www.thewildrose.net/alternative_healing.html
520 http://www.csh.umn.edu/mindbody/prayer/theo/th01.htm.
521 http://ww2.odu.edu/ao/instadv/quest/StirringTheSoul.html.

1. Prayer raises the level of the stress *hormone cortisol* on demand for empowering one to fight diseases better.[522] Religious-service attendees have better stress control and better health.[523]

2. Faith and perceptions can bridge belief and biology. Faith influences gene expression and epigenetics.[524]

3. Faith-based fasting can change gene expression and help manage immunity. *The gene for ubiquitin is a case in point.* Ubiquitin is only seventy-six amino acids long, a small polypeptide responsible for antigen processing and destruction of other foreign proteins. Its four encoding genes are affected by prayer.[525,526]

4. The brain is engaged with both emotion and belief. Belief is an element of thought, and the brain is an engine of thought. Both are designed for our survival, including fighting a disease by our immune system.

The nucleus in our cells is a fine-tuned memory disc. The cell membrane, as a three-layer skin around it, is its brain. Integrated membrane protein (IMP) is our cells' sense organ or the nano-antenna. The effector proteins of the membrane are a kind of central processor, and the cells can program themselves. Although they do respond to the environment, the cells link gene activity and biological behavior.[527] It appears that fasting and prayer can affect the workings of the nucleus and cell membrane.[528]

Fasting and prayer can affect the behavior of cells, and therefore, help us reduce the risk of diseases. DNA has no instructions for building a tumor cell, for being confident, or for being a racist. It does not code for cellular activity. It does not code for plan, morphology, or the form of the body. The cell membrane does it, and faith can modulate cells' behavior and change a lot in our being, including health.

Our DNA has instructions for 100,000 electromagnetic proteins, molecules that keep us on the go. But the human genome has only 25,000

522 http://www.ncbi.nlm.nih.gov/pubmed/20391859.

523 http://www.concordiaplans.org/graphics/assets/documents/Church_Attendance.pdf.

524 http://www.popsci.com/science/article/2013-08/what-twins-reveal-about-god-gene.

525 https://wellspentjourney.wordpress.com/tag/ubiquitin/.

526 http://news.techniont3.com/2012/01/10/nobel-prize-winning-ubiquitin-in-action/.

527 http://www.sciencedaily.com/releases/2013/11/131127170135.htm.

528 http://www.heartmath.org/templates/ihm/e-newsletter/publication/2012/winter/emotions-can-change-your-dna.php.

genes, not enough to code for 100,000 proteins. Obviously something else is involved. How can a minor difference of only 1,500 genes account for the biological existence of a highly ordered human being compared to a poorly ordered roundworm? The mystery lies in the protective sleeve of the DNA that can be changed by environmental factors. Fasting and prayer, I believe, are dominant environmental factors that affect the DNA sleeve.[529]

Disease is the malfunctioning of the body, its organs, and organ systems. The body's immune system is designed to fight diseases. Faith, by controlling emotions and stress, can influence and enhance our immune system. Adrenalin, morphine, serotonin, and melatonin have been around in living organisms throughout their evolutionary span in time. Our limbic brain is the survival brain. It is superior to the reptilian brain, which deals largely with fear and anger. Also, the limbic brain has a lot to do with our biology of emotion and reason. The mind, thinking, and feeling are inseparable. By sustained practice, we can enshrine love, joy, pleasure, pain, emotions, and beliefs in our DNA . Fasting and prayer can lead to purposeful connections with consciousness.

21.04 Case Histories of Divine Interventions

By faith, we can train our mind, change our genes, change our physiology, and therefore, change our health.[530,531] Faith has kept inoperable liver-cancer patients alive in happiness and made a lump half the size of a baseball disappear. Ole Nielsen Schou, a Danish pharmaceutical company production employee, had his melanoma spread to liver, abdomen, lungs, bones, and ten spots in his brain (2002)[532]. The tumors regressed spontaneously with vitamins and other food supplements because of his faith. Geneticists are proposing that there is a gene for the membrane protein VMAT2, which transports dopamine-like neurotransmitters from body cells to brain cells, and that this gene is responsible for inheritable spiritualism.[533] Current controversy notwithstanding, it is common knowledge that people of faith have less stress, the cause of many chronic diseases.

No doubt, faith, friends, and family can help us use our energy to keep

---

529 http://www.sapphyr.net/peace/peace-thinkcreate.htm.

530 http://content.time.com/time/magazine/article/0,9171,1879202,00.html.

531 http://www.centerforinquiry.net/forums/viewthread/5451/.

532 http://www.forbes.com/fdc/welcome_mjx.shtml

533 http://www.ncbi.nlm.nih.gov/pmc/articles/PMC2262126/.

the immune system at its best and fend off diseases. But in case of an endemic attack that has exceeded the power of the body's defense system, modern medicine must be allowed to come to our rescue. We don't understand the healing power of faith as much and as clearly as we have to. What is clear, though, is the fact that faith can make our mind work better and tame it positively. However, it can't cure epidemics due to bacteria and viruses because they destroy our mitochondria and DNA and disrupt homeostasis.

21.05 Major Reference Institutions

Healing by faith is a bit controversial but a few major institutions listed below are doing research on human brain as it may affect healing.

The Center for Spirituality and Mind, University of Pennsylvania, devoted to brain-scan research; the Center for Spirituality and Healing, University of Minnesota (shaman outreach with Hmong immigrants); the Center for Spirituality, Theology, and Health, Duke University; the National Center for Complementary and Alternative Medicine; the Templeton Foundation (private); the National Institute of Aging in Baltimore, Maryland; Manchester, United Kingdom study on two-day diet; and Shanghai Jiao Tong University.

The practice of fasting recommended to her by French doctor Pierre Dukan helped reduce prewedding weight for Kate Middleton[534], the duchess of Cambridge. The diet plans and fasting are also recommended by Reverend George Handzo, Healthcare Chaplaincy, city of New York; Dr. Andrew Newberg, University of Pennsylvania; and Dr. Richard Sloan, Columbia University

---

534 http://www.stylecaster.com/dukan-diet/

CHAPTER 22

The Methylation Diet for DNA Stability and Gene Expression (Recipes for DNA Methylation and Gene Expression)

"To make your genes happy is to learn to willfully regulate your stomach in terms of what, how much, and when to eat."
—Triveni P. Shukla

Much of the cure for chronic diseases is right in our kitchen. The methylation diet is for stable DNA and for a difficult-to-corrupt epigenome, and complementary to the microbiome of gut bacteria. All food is matter, and all cooking is about transformation of matter. The range of transformation is wide, including change of states of matter (solid to liquid to gas); joining and self-assembly of molecules, something we do not know as much about as we should; rearranging of ionic and hydrogen bonds; making of gels, emulsions, and foams; production of new flavors and textures; production of caramel-like color; and creation of a fancy food plate for appeal, aesthetics, taste, flavor, and presentation. The kitchen thus can do a lot for our health. What has been known as culinary art should be a matter of more practical science for all who wish to control appetite, satiety, food safety, and nutrition for proper gene expression.

Our kitchen, I believe, can help modulate digestive homeostasis. The final dishes should have no more than 40 percent carbohydrate calories; they must not activate genes for tumor-cell angiogenesis, carcinogenesis, and inflammation. Unlike high-carbohydrate foods, they should prevent genes from working overtime. Most of the food that we consume is for energy and building and repair of our body, but there are nutrients in it that keep our genes stable and expressible. We must incorporate such nutrients in our daily diet because we need to allow our mitochondria to function optimally for proper supply of electrons and protons and for quenching of any electrons gone amok by food-based antioxidants.

Our daily dishes should be capable of delivering dietary fiber, probiotics, antioxidants, omega-3 fatty acids, conjugated linoleic acid, proteins and peptides, lycopene, lutein, amino acids arginine, tyrosine, and tryptophan, and phytosterols, along with vitamins A, B_1, B_2, B_3, B_5, B_6, folic acid (B_9), C, D, E, and K, and minerals calcium, chromium, copper, iron, magnesium, selenium, and zinc. This then makes a methylation diet. Senior citizens, in particular, should not suffer from any nutrient deficiency. Nutrients most widely researched in relation to gene expression to date are vitamins A and D, folic acid, choline, vitamin B_{12}, betaine, and amino acids serine and threonine.[535,536,537,538] The current US law requires that wheat flour be fortified with folic acid, but food selection for proper magnesium, vitamin D_3, vitamin K_2, copper, chromium, selenium, and zinc rests in our hands. So does food selection for antioxidants like kaempferol, pyrroquinilin, ubiquinone, apigenin, leutin, and lycopene.

A good dish should avoid excess of cholesterol, glucose, and saturated and trans dietary fats because they have adverse effects on gene expressions. High-antioxidant diets prevent DNA or gene damage, and calorie-restricted diets affect gene expression via even better DNA stability. Genes are expressed by first making a transcript (messenger RNA) from the gene (little DNA), which then helps produce proteins essential for life. As a matter of fact, DNA makes life, and proteins help it keep going. Clearly metabolism and physiology are connected to diet. A dietary nutrient can directly act as a ligand for transcription factor (help in decoding DNA for making protein), or it can get converted to an intermediate that then alters gene transduction factors and finally the expression of the gene.

A combination of artichokes, asparagus, avocados, beans, blueberries, broccoli, cauliflower, edamame, figs, flaxseed, garlic, Greek yogurt, kiwi fruit, milk, olive oil, melon, nuts, oatmeal, quinoa, salmon, sauerkraut and coleslaw as probiotics, red beets, red bell pepper, sweet potato, strawberries, tomato, and tofu in our breakfast, lunch, or dinner dishes can provide all the

---

535 http://www.nap.edu/openbook.php?record_id=10299&page=32.

536 http://www.ncbi.nlm.nih.gov/pubmed/10089110.

537 http://learn.genetics.utah.edu/content/epigenetics/nutrition/

538 http://umm.edu/health/medical/altmed/supplement/betaine.

gene-expression nutrients listed throughout this book. Antioxidants coming from fruits and vegetables can effectively maintain reduction-oxidation homeostasis, a state that affects gene expression. We need to select nutrients for our genes because they help our cells decide which parts of DNA—call it genes—need to be expressed; they thus decide the state of our health. Diets affect the expression of genes for mitochondrial superoxide dismutase,[539] which quenches free radicals and prevents damage to mitochondrial DNA. Healthy mitochondria simply mean good health and long life. The methylation diet is, in a sense, merely a diet for healthy mitochondria.

The unnecessary cost of $200 million on vitamins included in $11.5 billion spent on food supplements[540,541] and $1.5 billion in antacids can be mitigated by simple selection of foods for our daily dishes. Antacids make us sick, declares the FDA, by causing diarrhea due to magnesium and constipation due to aluminum[542]. Nutrient data tabulated in various chapters earlier are categorized in this chapter in order to consolidate them all and to clarify the case for the methylation diet. The categorized data can be used in daily food selection and preparation so that we can exercise our power of thought and imagination in creating a daily diet that keeps us healthy and helps avoid costs of fighting ailments and diseases. We need to plan our healthy eating and good living, and this chapter is about how to do it by choosing what and how much to eat. The question of when to eat is often embedded in societal norms. We do need, however, to normalize it.

Information on what to eat is reorganized, classified, and presented here as a package on methylation foods. This includes the information on how much really should come from our sensory organs of shape and color from eyes, smell from nose, taste and texture from tongue, and even touch from fingers. Just imagine the pleasure from the color and crunch of an ordinary pickled cucumber. Our brain can help us determine how much to eat, but only when we exercise some control. We must restrict total calorie consumption and yet enjoy our daily food. The details on foods and nutrients are presented here even at the risk of duplication and repetition, along with

---

539 http://www.ncbi.nlm.nih.gov/pubmed/18684339.

540 http://www.nutraingredients-usa.com/Markets/Supplement-sales-hit-11.5-billion-in-U.S.-report-says.

541 http://newhope360.com/site-files/newhope360.com/files/uploads/2013/04/TOCEXECSUMM110930.supp%20report%20FINAL-2.pdf.

542 http://www.healthguidance.org/entry/9854/1/Indigestion-and-Antacids.html

some model recipes as a ready guideline and reference for creating a dish for the best of nourishment and health benefit.

The nutrition information on every package sold in the US retail stores contains information on how much to eat. This is given as total calories per serving, total fat calories, total carbohydrates, dietary fiber, sugars, and protein. At the end of the day, it is up to us to tally total calories from good fats, carbohydrates, and protein that we consume, avoiding an excess of refined sugar, sodium, cholesterol, trans fat, and saturated fat. Very few of us do this tallying with any punctuality and rigor. We should note in this regard that the USDA and FDA use Calories with a capital C to denote kilocalories. Although a recommendation to use the measure of kilojoules was made thirty years ago and Codex Alimantarious uses both calories and kilocalories side by side, the nutrition information panel in the United States lists per serving energy content as calories. An entry of 2,000 calories in fact then means 2,000 kilocalories or 2,000,000 calories (7,946 BTU per day). This is a lot of energy, which ranges up and down depending on how strenuous a work we do. So varies the range of calories that we should consume and our body has means to optimize and control, but only when our daily intake is within reasonable limits.

When total fat consumption doubles on a given day, we quickly get the signal of heartburn; we need antacids when the stomach produces too much acid in expectation of digesting too much protein. To keep a record of what and how much we eat is good for monitoring body response in terms of allergies, discomfort, and poor digestion.

22.01 A Diary of Medical Records

Health is a matter of serious planning, and planning requires a good account of health data and information on the foods we eat. We must know the details of the state of our health in order to select our daily foods. Personal data in a well-kept diary should include personal targets for

- blood pressure of 125/82;
- total cholesterol up to 240 mg/dl;
- total cholesterol-to-HDL ratio below 4;
- blood sugar around 100 mg/dl;
- triglyceride level of 150 mg/dl;
- annual physical examination (including BMI, weight, and height);

- diabetes screening and checks on colorectal cancer and bone density;
- comprehensive eye examination; and
- pneumococcal vaccine and vaccines for flu.
- Pap smear, pelvic examination and mamogram for women.

REM sleep disorder, loss of sense of smell, and persistent constipation may be early signs of Parkinson's disease. For senior citizens, inability to recall faces and names of celebrities is an indicator of dementia.

If pregnant, a woman should secure four hundred more calories from carbohydrates and protein and keep weight gain to no more than 25 percent of her prepregnancy period. Folic acid and omega-3 foods are a must, and they can come from salmon, swordfish, and cod only once a week. Flaxseed in soup, cereals, yogurt, and even a daily omelet can be a very viable source of omega-3 fatty acids also.

We need to test our body's response at home by testing urine, stool, breath, blood, hair, and skin regularly. They are indicators of many disease problems, as shown in the block below. We need to keep track of mitochondrial diseases, Alzheimer's, and Parkinson's, especially during our senior years.

Urine: Should be standard yellow; no pink or red color due to blood, a sign of kidney stone; dark color means you need water; Coca-Cola color could be due to a tumor; brown urine could be due to liver damage; sweet-smelling urine indicates diabetes.

Stool: Various shades of brown are normal; black suggests bleeding; iron supplements can darken the color; bright red means problems with the colon; narrow and thin may mean colon cancer; if it sticks to the side of the bowl, it indicates high fat consumption; constipation could be due to diet or medication.

Breath: Don't allow bacteria to grow; use hot drinks; use Listerine regularly; oral hygiene is a must.

Hair and Skin: Skin health is the greatest indicator of good health. Avoid the sun, and wear protective dress when in the sun; don't smoke.

A great health-care program can be instituted at home in terms of managing mitochondrial gene expression, and maintaining health of breath, hair, skin, stool, and urine by consumption of antioxidants, omega-3 fatty acids, vitamin D$_3$ foods, and minerals copper, chromium, magnesium, selenium, and zinc. Probiotics as routine foods help us much more beyond digestive health because they are responsible for the maintenance of our second genome. We can use our perceptions of smell, taste, sight, texture, and touch for appetite and portion control as we design our daily diet right in our kitchen.

Aided with oxygen by exercise and copious daily water intake, the methylation diet is a great cure for mitochondrial diseases. Antioxidants and nutraceuticals present in fruits, vegetables, seeds, and nuts add values to methylation diets, and sprouts have offered enzyme therapy for ages. Gene therapy with the exception of prostate cancer[543] is still in the making. It is not achievable by diet alone for a specific disease. Yet our daily diet does keep our DNA and genes stable.

22.02 Food Selection and Preparation

Nutrients for genome stability and gene expression can be incorporated in daily breakfasts, lunches, and dinners by proper choice of foods that are local, traditional, and ethnically relevant. We need to practice a brightly colorful and green gastronomy based on nutrient information in appendix I for fiber, appendix II for nutrients in fruits, appendix III for nutrients in vegetables, appendix IV for nutrients in grains and nuts, and appendix V for omega-3 fatty acids and selenium in fish. The appendixes are constructed to include nutrient information on items highlighted in the third paragraph of this chapter.

Flavor, taste, and texture should come from healthful phytochemicals, herbs, spices, and a blend of plant-food materials. The human sense of smell has evolved over years as a sophisticated molecular detector. We can detect three drops of n-butyl mercaptan added to two Olympic-size swimming pools. Each nerve in the nose builds its receptors from a single gene for olfactory neurons directly exposed to the air.[544] But learning of smell happens in the brain, not in the nose, and there must be other genes involved. There are genes that control cognition, sleep, metabolism, satiety, digestion, and

---

543 http://www.cancer.org/cancer/prostatecancer/detailedguide/prostate-cancer-new-research.

544 http://www.ncbi.nlm.nih.gov/pmc/articles/PMC2258179/.

digestive health, and all are expressed under the influence of what we eat. It is not clear if there are genes for food choice, but we do know that variants of FTO (gene on chromosome 16 position 12.2 for fat, mass, and obesity-associated mRNA demethylation) and BDNF (brain-derived neurotrophic factor) linked to obesity may influence food choice. We should have the will to make relevant food choices. The rule of eating red fruits like tomatoes, apples, and cranberries for antioxidants that promote heart health; yellow fruits for vitamin C; dark green vegetables for vitamin B complex and anticarcinogens; and seeds and nuts for good fat, protein, and trace minerals is a good one to follow.

A great example is President Bill Clinton's selection of roasted cauliflower, cherry tomatoes, spiced and herbed quinoa with green onions, shredded red beets in vinaigrette, Asian snow-pea salad, and an assortment of fresh-roasted nuts, plates of sliced melon and strawberries, and beans tossed with onions and olive oil. A green gastronomy not known to him while in Arkansas is now designed into his breakfast, lunch, snacks, and delightful dinner. He is for sure using his willpower. The idea is to consume fewer total calories by restricting total intake of foods, to consume snacks of seeds and nuts for stress control, and to consume breakfast, lunch, snack, and dinner foods on time. This includes consumption of copious amounts of water, a very critical food ingredient.

22.03 Selection from Common Foods

For due emphasis I should define the methylation diet again as a diet containing a good and balanced variety of low-calorie adaptogenic, anti-inflammatory, antioxidant, and antiangiogenic foods, including prebiotic dietary fiber and probiotic foods like yogurt and sauerkraut, fruits, vegetables, whole grains, seeds, nuts, and soy foods that meet RDAs (Recommended Daily Allowance) for minerals and vitamins and deliver all essential nutrients.

Chapter 6 on food functions has extensive details on foods for omega-3, water-soluble B vitamins and vitamin C, fat-soluble vitamins A, D, E, and K, and various minerals, along with their recommended daily allowance.

In terms of reemphasis and categorization, table 22.01 presents ailment- and disease-specific health values of nutrients in our common foods that I have compiled from a variety of credible sources throughout the last decade. In addition, table 22.02 presents the information in terms of specific foods containing nutrients and phytochemicals known to prevent and cure specific diseases and health problems. Table 22.03 goes a step further to list bioactive food nutrients and phytochemicals, along with their availability in common foods. We must aim for the following in designing our daily dishes for nuclear and mitochondrial gene functions:

1. Homeostasis of digestion, although seemingly simple, is very complex. Saliva is lightly acidic (pH 5–8 depending on diet), the stomach is highly acidic (at pH 1.9 or even below), pancreatic secretions and the intestine are alkaline (pH 7 to 8, and the colon (pH 5.5–7.0), where bacteria function to secure added homeostasis, is often cose to neutral. Probiotics flourish only under slightly acidic conditions. Wrong bacteria grow under alkaline conditions. Diet, including dietary electrolytes, helps maintain digestive homeostasis.

2. Dark green vegetables (brussels sprouts, broccoli, collards, chard, grasses, kale, sprouts) are alkaline. So are sweet potato and carrot. The cooking tradition calls for balancing alkalinity and even preserving color. The tradition of cooking has science embedded into it for our good health.

3. Imagine what drinking a large glass of Coca-Cola, which has a pH of around 4.00, and topping it with an acidity-prone high-sugar meal high in meats does to our acid-alkaline homeostasis. It can create havoc. We know for a fact that if the blood pH gets to 6.9, there can be diabetic coma; and if gets to 7.9 on the high side, death can occur due to lockjaw or tetany. The point here is moderation, and the rest can be left to the delicately balancing mechanism of homeostasis. Digestive physiology and BMR vary from individual to individual.[545] Research has shown that the basic metabolic rate (BMR) can vary hugely, from 1,027 to 2,499 calories per day, with a mean of 1,500 calories per day. Most of this variation, about 62 percent, is due to lean body mass, with some contribution due to fat mass. Furthermore, there are variations among people with the same lean body mass. It is a good idea, therefore, to target a 1,500 +/- 250-calorie daily diet. It should be even lower for a restricted-calorie diet. A stress-free life accounts for a balanced sympathetic nervous system and balanced BMR.[546] The sympathetic nervous

---

545 http://www-users.med.cornell.edu/~spon/picu/calc/beecalc.htm.

546 http://www.wisebrain.org/ParasympatheticNS.pdf.

system control by relaxation and heartbeat control are necessary for reliable BMR.[547] Our diet plans, therefore, must depend on how stress-free a life we live.

4. There is now evidence that two cups of coffee without sugar can be a great preventive against Alzheimer's disease[548].

5. The heartburn problem that a majority of Americans experience is often due to orange juice and morning coffee, hamburger and french fries for lunch, a lot of beer in the evening, followed by a huge dinner with appetizers of chips and salsa. We need to moderate this unhealthy eating behavior.

6. Design restricted-calorie diets of below 1,800 calories/day, a bit lower than the often-talked-about 2,000 calories. This is almost necessary for diabetics and senior citizens.

7. Include foods that remedy mineral deficiencies of chromium, cobalt, copper, iron, magnesium, manganese, molybdenum, selenium, and zinc.

8. Include foods that provide for daily RDA of omega-3 fatty acids and vitamins D_3, vitamin B complex including folate, and vitamin K.

9. Consume snacks of seeds, nuts, and Gouda cheese for cancer prevention and stress control. Use smoothies as a complement. A good example is a blend of pumpkin and sunflower seeds with two Brazil nuts every day.

10. Consume green vegetables for antioxidants and phytochemicals. Vegetables should be central to our daily eating pleasure and need to occupy a majority space in our dishes.

11. Be creative and substitute varieties by design in foods for breakfasts, lunches, snacks, dinners, and dessert recipes included in this chapter. The recipes in this chapter are only guiding samples.

12. Prepare fruit-based snacks including a variety of berries for routine consumption, or just use them as fruit servings. Consume fruit crumbled with oats as a fiber source at least twice a week.

13. Examine data on the health value of common foods consolidated in various tables in this chapter and select foods for variety but equivalent nutrient values.

---

547 http://www.ncbi.nlm.nih.gov/pubmed/18070759.

548 http://www.ncbi.nlm.nih.gov/pubmed/20182037

14. Use homemade vegetable soup in order to meet the requirement of six servings of fruits and vegetables per day. Soups in a way are vegetable concentrate. Use vegetable stock whenever possible.

15. Manage daily refined sugar consumption to less than 20 grams per day. The rest of carbohydrate calories should come from low-glycemic complex carbohydrates. Watch daily intake of processed foods loaded with sugar. For instance, one can of tomato soup delivers 30 grams of sugar, one blueberry muffin delivers 22 grams, ten thin mints deliver 26 grams, and one cup of cole slaw delivers 23 grams of sugar. Guard yourself from these hidden refined sugars in processed foods.

16. Periodically select the following at least twice a week:

    a. Artichoke, asparagus, avocado, broccoli, cabbage, cauliflower, flaxseed, ginger, mushroom, spinach, variety of green vegetables, and tomato in various food preparations.

    b. Sweet potato, soy products including green edamame, seeds and nut dips.

    c. Flavorful dishes by incorporating spices and condiments.

    d. Probiotics as yogurt, sauerkraut, kimchi, or fermented foods in general.

    e. Tables in this chapter summarize food values and thus selection criteria. Table 22.01 is about health values, table 22.02 is about active ingredients that mitigate health problems, table 22.03 is about sources of nutrients, Table 22.04 is about antioxidant RDA and its functions, Table 22.05 is about antioxidant classification, Table 22.06 is about free-radical scavenging power of antioxidants, and table 22.07 is about potential therapeutic value of our common foods. Use these tables as a guide for selecting foods when designing a methylation diet.

    f. Double antioxidant consumption for a better mitochondrial function.

    g. Meet RDA of nutrients and mitigate deficiencies by well-designed side dishes that go with lunch and dinner.

Table 22.01 Health Values of Everyday Foods

Health Value	Common Foods
Against Allergies	Onion, rich in antioxidants
Anti-osteoarthritic foods[549,550]	Antioxidants, vitamin C, lycopene, B vitamins, coenzyme Q_{10}, fiber and protein; betacryptoxanthine from apricots, bell peppers, nectarines, oranges, papaya, peaches, and watermelon. Flavonoids from apples, grapes, green tea, soy, and onion. Antioxidants from berries and licorice. Resveratrol from grape skin, wine, blueberries, mulberries, and peanuts. Gama linoleic acid from black currant and primrose oil. Boswellic acid from boswellia serrata. Vitamin C, phthalides, and coumarins from celery and celery seeds. Capsaicin from chili peppers. Caffeine and theobromine from coffee. Isoflavones from edamame. S-adenosylmethionine, calcium, and magnesium from nuts, seeds, fruits, and vegetables. Gingerol from ginger. Omega-3 fatty acids from flaxseed, salmon, trout, and walnuts. Oleocanthal from olive oil. Glycirrhizin from probiotics. Cucurmin from turmeric. Natural endomorphins— serotonin and norepinephrine. Glucosamine from crustacean skin. Escin from horse chestnuts and chestnuts. Pycnogenol from pine bark extract.
Against cancer and antiangiogenic foods[551,552]	Artichoke with folic acid and zinc has been used against skin cancer for a long time. Fruits and vegetables such as avocado, asparagus, apple, barley, broccoli, bok choy, brussels sprouts, cabbage, cauliflower, collard greens (leutin, beta-carotene, and vitamin C), carrot, celery, dark chocolate, kale, kiwifruit, flaxseed, Gouda cheese, grapefruit, pineapple, licorice, mushroom, nuts, orange (vitamin C and beta-carotene), lavender, parsley, pumpkin seed, red grapes, seaweed and kale, soy foods, blackberries, blueberries, cranberries, raspberries, strawberries, tomato, turmeric, and wine can offer daily remedies against cancer.

---

549 http://preventarthritis.org/10-super-foods-to-fight-osteoarthritis/.

550 http://preventarthritis.org/10-super-foods-to-fight-osteoarthritis/.

551 http://www.oprah.com/health/Prevent-Cancer-with-the-Right-Diet-of-Antiangiogenic-Foods/2.

552 http://helenpapas.wordpress.com/2011/02/08/190/.

Health Value	Common Foods
Anticoagulant	Cinnamon, clove, melon, tea
Anti-constipative	Apples, beets
Antidiabetic	Cinnamon promotes insulin activity; fenugreek seed and white potato peel with protease inhibitor are other suggested cures for insulin resistance
Anti-inflammatory[553]	Apples, clove, flaxseed oil, garlic, ginger, onion, pineapple
Antiviral foods	Kimchi, turmeric, garlic, chocolate, green tea, omega-3 fatty acids, vitamin E
Against urinary infection[554]	Blueberries, cranberries
Astaxanthin	A very powerful carotenoid antioxidant from from deep red foods. it is 100 times more powerful than vitamin E[555]. Good sources are sockwye salmon and microalgea.
Bone health	Celery (delivers silicates)
Coenzyme Q_{10}	Sesame oil, nuts, seeds, beans, egg, chicken, broccoli, and milk
Digestive health	Ginger
Energy balance	Pumpkin seed
Brain health	Lecithin, beans and legumes, white beans in particular
Heart health	Omega-3 fatty acids to prevent inflammation, improve heart health, better cognition; fish and fish oil, garlic, mushroom, olive oil, tomato, walnuts, other nuts, flaxseed, beans, lentils, eggplant, grapes with resveratrol and quircetin, grapefruit juice, oats, pumpkin with antioxidants, rice bran, sweet potato (beta-carotene), turmeric
Muscle relaxants[556]	High boron and magnesium foods: almonds, cashew nuts; reduce pain

553 http://health.usnews.com/best-diet/anti-inflammatory-diet.

554 http://www.med.nyu.edu/content?ChunkIID=21411.

555 http://www.nutrex-hawaii.com/natural-sources-of-astaxanthin

556 http://www.betternutrition.com/magnesium/columns/askthenaturopath/947.

Health Value	Common Foods
Nervous system regulation	Wheat gluten, lecithin
Immune response	Rice bran
Osteoporosis[557]	Turmeric
Painkiller[558]	Chili pepper (capsaicin), cloves, date with natural aspirin, raspberry, turmeric
Satiety	Protein foods for satiety and weight control; chickpeas against bacterial infection (soy foods)
Stimulants	Coffee
Fiber & Phytonutrients	Barley, beets, dates, fruits, vegetables, whole wheat
Probiotics	Yogurt, sauerkraut, kimchi, tempeh

Table 22.02 Common Foods against Specific Ailments

Common Foods	Active Ingredients	Ailments and Diseases Helped
Antidepressant Foods		
Common Foods	Active Ingredient	Function
Broccoli and eggs	Calcium, vitamin D	Prevent depression
Guarrana	Caffeine-like stimulant	
Red clover	Isoflavone	
St. John's wort	Antidepressant hypericin	Reduces depression
Indian ginseng ashvagandha	Adaptogen	Promotes sleep
Tart cherry	Anthocyanins and caffeic acid like antioxidants	Anti-inflammatory function of pain relief

---

557 http://opa.ahsc.arizona.edu/newsroom/news/2010/ua-study-shows-benefits-turmeric-preventing-osteoporosis.

558 http://www.nbcnews.com/id/26136767/ns/health-alternative_medicine/t/youre-pain-you-want-relief-naturally/.

Common Foods	Active Ingredients	Ailments and Diseases Helped
Foods that Fight Allergies		
Dark chocolate	Polyphenol, theobromine	Cures cough, prevents DNA damage
Marjoram	α-terpinine and flavonoids	Cures sinus congestion, indigestion, stomach pain, headache, dizziness, colds, coughs, and nervous disorders
Nettle with onion	Antiallergenic	Effective against allergies with quircetin
Red and yellow onion; consume baked onions	Quircetin, vitamin C, chromium, flavonoids; antioxidative and anticarcinogenic	Cures eye, nose, lung, and intestine allergies, reduces risk of Alzheimer's and Parkinson's disease
Eye Health and Vision		
Bilberry	Pycnogenol	Improved eye blood flow, improved oxygen level, reduced pressure
Sprouts	Enzymes	Soft tissue recovery, heart health, and joint health.
Spinach	Iron, folate, other phytochemicals	Prevent macular degeneration
Green beans	Leutin and zeaxanthin	Improves eye health
Kale	Beta-carotene, vitamins A, C, E, and K, folate, leutin, manganese	Powerful antioxidant, protects against macular degeneration
Swiss chord	Fiber, leutin, carotenoids, vitamins A, B_6, C, K	Lutein protects from macular degeneration and cataracts
Gall Bladder and Liver Health		
Milk thistle	Sylamarin	Prevents gall bladder disease

Table 22.03 Nutrients in Healthful Common Foods

Nutrients	Common Foods
Omega-3 fatty acids, omega-6 balance	Flaxseed, kiwifruit, purslane, and salmon are good sources of alpha-linoleic acid, EPA, and DHA. A newborn infant's brain is 50% DHA. A ratio of 1 omega-3 to 2 omega-6 is ideal.
Beta-carotene	Carrot, spinach, sweet potato, kale
Glucosamine	This supplement is antiwrinkle and wound-healing ingredient.
Hesperidins	Citrus fruits
Hyaluronic acid	Negatively charged amino-carbohydrate promotes longevity. Chicken bone and cartilage via soup are great sources of hyaluronic acid and glucosamine.
Peptides	Calpris and Evolus Sour milks are great sources of isoleucin-proline-proline and valine-proline-proline peptides. Tiens (albumin polypeptide) is sold in the Chinese market. Such peptides enhance immune health.
Polyphenols	Anthocyanidins, procyanidins like pycnogenol, catechins, flavonoids, and even tannins
Resveratrol	Prevents cancer. One serving of red wine a day is a good prescription.
Quircetin	A flavonoid antioxidant found in onion, red wine, and tea
Tocotrienol	A phytosterol from rice bran oil
Vitamin A	Sweet potato; very useful against acne. Beta-carotene, lycopene, and leutin are precursors. Consume mangos and frozen berries.
Vitamin B complex	Green vegetables; prevents scaly skin; biotin and niacin are essential for hair health. B_2 is a must for homocysteine maintenance.
Folic acid	Beans, legumes, lentils, green vegetables
Vitamin B_6	Trout, salmon
Vitamin B_{12}	Trout, salmon, supplements
Vitamin C (400 mg)	Broccoli, red bell pepper, brussels sprouts, papaya, cherries

Nutrients	Common Foods
Vitamin D (500 IU)	Take supplements.
Vitamin E (400 IU)	Almonds, sunflower seeds, hazelnuts, peanut butter; prevents psoriasis, boosts immunity
Vitamin K	Leafy green vegetables with K maintain skin health.
Calcium (1000 mg)	Milk, cheese, yogurt, potato, spinach, collard green, leavening agents
Minerals and vitamins	Copper, chromium, magnesium, selenium, zinc from . Soy, soluble-fiber artichoke, dried tomato, dried fruits, limnoids, mixed seeds of pumpkin and sunflower; mixed nuts, wasabi peas, or peanuts
Minerals and vitamins	
Copper	Baked potato is a good common source.
Selenium	Brazil nut and cottage cheese are great sources of this antioxidant.
Zeaxanthin	Antioxidants lutein and zeaxanthin from broccoli, corn, and spinach are good for eye health.
Zinc	Oysters, seeds, and nuts

Table 22.04, a very important compilation, lists antioxidants, their daily requirement, and their health value. Table 22.05 presents a chemical classification of antioxidants.

Table 22.04 Antioxidants in our Common Foods

Antioxidant	Daily Intake	Function
Alpha-carotene/ lutein	2 mg	Found in carrots, tomatoes, winter squash; great for eye health
Beta-carotene	6 mg	Carrots, winter squash, bell pepper, and pumpkin are good sources. It is pre-vitamin A carotenoids.

Antioxidant	Daily Intake	Function
Apigenin flavonoid	No recommended RDA	Thyme, tomato sauce, peppermint, and red wine are good sources. It is an antioxidant, anti-inflammatory, and anticarcinogenic agent; prevents gout by blocking uric acid.
Alpha-lipoic acid	0.5 gram	An antioxidant from broccoli, spinach, collard greens, chord
Anticarcinogens	Spice herb	Rosemary prevents anticarcinogen production during grilling and cooking.
Anthocyanins	12.5 mg	Cyanidin, delphinidin, malvidin, pelargonidin, peonidin, petunidin, pterstibene (resveratrol-like), stilbenoids, and resveratrol. Use bright-colored berries, red onion, apple, cocoa, grapes, and eggplant.
Proanthocyanidins	100 mg	Tea, black currant, cranberry, grape seed and skin
Arginine	No RDA but it is an essential amino acid	An amino acid from proteins is responsible for increased lean mass and bone density
Astaxanthin	4 mg	An approved food color as terpene carotenoids. Good sources of this eye-health antioxidant are salmon, trout, and shrimp.
Canthaxanthine	150 mg	Antioxidant
Capsaicin	0.5 mg	A capsaicinoid from chili pepper against pain, cancer, & psoriasis. It increases energy metabolism.

Antioxidant	Daily Intake	Function
Crocin (saffron)	No RDA for Crocin	Antioxidant, anticarcinogen, lowers triglyceride and cholesterol, improves memory
Cucurmin (turmeric)	75 mg	As a natural phenol, it is a great antioxidant, anti-inflammatory. Sources are yellow onions and turmeric.
C3G glucoside	No RDA	Anthocyanin flavonoid antioxidant from purple corn popular in Mexico, blackberry, and black currant; is good for eye health
Chalcones/quircetin	NO RDA	Great antioxidant in beer (hops)
Chicoric acid	NO RDA	Basil leaf and chicory root antioxidant
Coenzyme Q_{10}	150 mg	Natural electron transport chain antioxidant found in meat, oils, fish, nuts, beans, seeds, eggs, and chicken.
L-glutamine	NO RDA	Increased hydration by suatamine in Japan. Glutamine peptone has similar effect.
Glycosides	NO RDA	Perennial herb used to treat bronchitis and liver problems in India. Example is picroside 1 antioxidant.
Lactoferrin, glycopeptide	NO RDA	Eye health, prevents tumor growth, improves immune system, helps iron absorption. These are peptides from whey protein.
Lycopene	6-12 mg suggested RDA	Prevents lung and bladder cancer as an anticarcinogen.

Antioxidant	Daily Intake	Function
Omega-3 fatty acids	1 gram	Prevents leukemia and kidney cancer
Phenylethyl isocyanate	NO RDA	Prevents leukemia and kidney cancer
Phospholipids	300 g	Increased cognition
Selenium, a mineral	100 mcg	Works with vitamin E; an antioxidant that protects cells
Superoxide dismutase	NO RDA	An iron and manganese enzyme that quenches free radicals; found in broccoli and cabbage
Ubiquinol	60 mg	Reduced electron-rich form of powerful antioxidant; donates electrons and quenches free radicals. Like coenzyme Q_{10}. Sources include chicken, beef liver, and broccoli.
Vitamin C	120 mg	Donates electrons and quenches free radicals
Vitamin D_3	1000 IU	Acts like cancer drug transferrin; inhibits iron-catalyzed oxidation
Vitamin E, tocopherol	30 mg	Breaks free-radical chain reaction; present in most virgin oils
Zinc	10 mg	An electron-donating antioxidant to OH (hydroxyl) radicals; helps 100 enzymes, immune system, taste & smell, and growth

Note: Beta-carotene, lycopene, beta-cryptoxanthin, zeaxanthin, astaxanthin, and lutein are in the carotenoids family of antioxidant. Flavonoid antioxidants include quircetin, rutin, hesperidins, apigenin, and luteolin, and then there is the famous epigallocatechin-3-gallate from green tea. Isothiocyanates, resveratrol, and tannins are other antioxidants.

Table 22.05 Classification of Antioxidants in Common Fruits and Vegetables

Antioxidant	Average Daily Intake	Function
Fruit/Vegetable flavonoids	190–200 mg	Cancer prevention and protection from inflammation
Flavanones	15 mg	Antioxidant function
Hesperidins, naringinin	No RDA but it used widely as lemon zest	Anti-inflammatory and immune-boosting citrus antioxidant; grapefruit antioxidant
Flavonols	160 mg	Glycosilated flavonoids from black tea, buckwheat
Catechins, epicatechins, gallocatechins, kaempferol, quircetin, theaflavin	Varoius antioxidants from common fruits and vegetables.	Antioxidants in cocoa products and chocolate, peaches, and even vinegar. Kaempferol is a glucoside antioxidant against cancer and cardiovascular disease (in apples, broccoli, kale, cabbage, beans, tea, capers); quircetin is a flavonoid antioxidant found in capers, kale, dill, sweet potatoes; and theaflavins is a polyphenol antioxidant from tea against HIV and cancer. Daily onions, fruits, and vegetable can provide cancer-fighting flavonoids in sufficient amounts.
Flavones apigenin (4,5,7 trihydroxy flavone), luteolin, & tangeretin	20–50 mg	Beneficial effects against atherosclerosis, osteoporosis, diabetes mellitus, and certain cancers; cereals are a good source.
Eugenol	Up to 150 mg	Very potent antioxidant from clove
Pycnogenol	20 mg	A proanthocyanindin polyphenol antioxidant
Glutathione	500 mg supplement is not necessary.	Our body provides most of our need for this powerful antioxidant with our methylation diet.

Antioxidant	Average Daily Intake	Function
Isoflavones daidzein, genistein and glycetin	25 mg	Anticarcinogenic isoflavones natural to soy foods provide up 30 mg/day in Asiatic countries.
Lutein & zeaxanthin	2.5 mg	Yellow xanthophyl carotenoids of green vegetables, broccoli, kale, turnip, and Swiss chord are necessary antioxidants.
Lycopene	6-12 mg	Antioxidants from tTomatoes and watermelon for eye health.
Melatonin	5 mg	A terminal antioxidant and sleep-wake rhythm hormone from olive oil, beer, wine, grape skin, and walnuts
Pyrroloquinoline quinone	2 mg	Fermented soy foods, spinach, green pepper, green edamame. Pyrroquinoline quinone is a powerful antioxidant.
Phenolic acid esters	800–1000 mg	Antioxidants from daily fruits and vegetables
Chicoic, chlorogenic	2 cup coffee per day	Coffee, cinnamon, cranberry
Ellagic, ellagitannins, galic, and gallotannins		Common fruits and vegetablessupply enough of these antioxidants.
Rosemarinic acid	NO RDA	This is an antioxidant from rosemary
Salicylic acid	NO RDA	blackberries and blueberries supply this antiinflammatory ingredient.

Table 22.06 lists the power of antioxidants in terms of oxygen radical absorbance capacity. Herbs and spices used in preparation of our daily dishes have high antioxidant values beyond simple preservation of food flavor, quality, and shelf life.

Table 22.06 Oxygen Radical Absorbance Capacity (ORAC) Values of Antioxidants Present in Our Daily Foods

Common Foods	Micromole in ORAC =Trolox Equivalent, TE per 100 grams
Spices, cloves, ground	314,446
Cinnamon, ground	267,536
Oregano, dried	200,129
Turmeric, ground	159,277
Cocoa, dry powder, unsweetened	80,993
Unprocessed cocoa beans	28,000
Black pepper	27,168
Cinnamon, 1 tablespoonful	21,402
Oregano, 1 tablespoonful	16,010
Elderberries, 1 cup	14,697
Lentils, 1 cup	13,981
Turmeric, 1 tablespoonful	12,742
Chocolate, 30 grams	12,060
Pinto beans	11,864
Artichoke, cooked, 1 cup	7,904
Pomegranate juice 100%	5,923
Prune, ½ cup	7,291
Plum, dried	5,700
Black plum, 1	4,844
Red wine	5,693
Red kidney beans	13,727
Cranberry, raw, 1 cup	9,584
Blueberries, raw, and anthocyanins, 1 cup	9,019
Cherries	4,620
Sweet cherries, 1 cup	4,873
Apples, red, quircetin, and other polyphenols	5,900

Common Foods	Micromole in ORAC =Trolox Equivalent, TE per 100 grams
Granny Smith, 1 apple	5,381
Watermelon, a great source of lycopene	3,800
Raisins	2,830
Blackberries, 1 cup	7,701
Raspberries, 1 cup	16,058
Strawberries, 1 cup	5,938
Pears, 1 medium	5,255
Pecans, 1 oz	5,095
Kale	1,700
Spinach	1,260
Brussels sprouts	980
Broccoli	890
USDA and USDA define ORAC values as Trolox Equivalent (micromole TE per 100 gram). Trolox is a vitamin E analogue. Freeze-dried fruits and vegetables have high ORAC values because of concentration of antioxidants..	

Note: Vitamin D function requires magnesium, magnesium and calcium control heartbeat. Magnesium relaxes and calcium helps contract muscles. Magnesium is critical to type 2 diabetes in relation to conversion of sugar to energy. It is critical to the energy process as a whole. Magnesium is next to potassium in concentration as a positively charged ion in our cells. Antioxidant enzymes glutathione peroxidase, superoxide dismutase, catalase, and coenzyme Q_{10} from sprouts can be routinely used to reduce oxidative stress.

Table 22.07 is a master compilation to be used as an instant reference for selecting foods for designing our daily dishes. Common foods are listed along with key nutrients, known mechanism of action, and their therapeutic values. This table is a product of my five-year-long research. Major sources include (1) Andrew Chevallier, *Herbal Remedies* (New York: Metro Books, 2007), http://www.yogic-slim.com/phyto_profile.htm; and (2) Thompson Cynthia Thomson, Cheryl Ritenbaugh, James P Kerwin, and Robyn DeBell(1996), *Preventive and Therapeutic Nutrition Handbook* (New York: Chapman & Hall, 1996).

Table 22.07 Food Nutrients and the Mechanism of Their Curative Action

Food	Key Nutrient	Mechanism	Cure
Adaptogenic foods			**Promote epigenome stability and homeostasis**
Asafoetida	Ferulic acid, asaresinotanol	Antioxidants	Anti-flatulent, antibacterial, anti-constipative
Artichoke	Silimarin, dietary fiber	Antioxidants	Prevents skin cancer
Asparagus	Vitamins C, E, folic acid, B_{12}, and rutin [559], [560] (Swedish study).	Anti-inflammatory; effective against pancreatic cancer, histamine production	Telomerase activator, enhances immune system, improves complexion
Asparagus roots	Cytoastrogenol	Folic acid	Boosts telomerase activity
Asvagandha	Alkaloids, steroidal lactones, saponins, withanolides	Adaptogenic	Prevents DNA damage
Astragulas	Cycloastrogenol, fiber, telomerase activator	Antioxidants	DNA damage control, antiaging food, anti-HIV, anti-inflammatory, liver-protective food
Chinese berries	Achisandra berries	Antioxidants	Liver-protective medicine in China
Garlic		Anti-inflammatory	Promotes heart health
Ginger		Anti-inflammatory	Prevents pain and promotes heart health

---

559 http://www.healthsupplementsnutritionalguide.com/vitamin-deficiency-symptoms.html

560 http://juicerecipes.com/recipes/asparagus-delight-2

Food	Key Nutrient	Mechanism	Cure
Licorice		Antioxidants	Prevents DNA damage
Tomato	Lycopene, lutein	Antioxidants	Eye health
Shitake, maitake mushrooms	Vitamin D and B vitamins	Free radical quenching	Prevents DNA damage
Antiangiogenic Foods			
Almond oil	Vitamins E, A, B$_1$, B$_2$, B$_6$, omega-3 and omega-9	Antioxidants and essential vitamins	Very good for hair care [561].
Dark chocolate	Polyphenols	Antioxidants	Cures cough, prevents DNA damage
Kale	Vitamins K, A, and E and Minerals	Antioxidants	Helps prevent cancer
Red grapes	Resveratrol	Antioxidants	Heart-health value
Tomato	Lycopene, lutein	Carotenoid antioxidant	Improves eye health
Turmeric	Curcumin	Anti-inflammatory, anticarcinogenic	Prevents cancer, cures joint pain
Anticarcinogenic and Antioxidant Foods			
Fruits	Vitamins, minerals		Heart disease and cancer
Vegetables	Vitamins, minerals, phytonutrients		Heart disease and cancer
Anti-Inflammatory Foods			
Blackberries, blueberries	Anthocyanins (Tufts University[562]	Communications between neurons	Brain health

561 http://pubs.acs.org/doi/abs/10.1021/jf051692j
562 http://www.lef.org/magazine/mag2006/feb2006_report_blueberries_01.htm

Food	Key Nutrient	Mechanism	Cure
Antioxidants	Carotenoids, vitamins A, C, E, and K, and minerals selenium and zinc, eugenol, limonine, orientin, vicerin	Free-radical scavenging	Control DNA damage, manage telomerase length
Dahi (India), cheese, kimchi, mother's milk, sauerkraut, yogurt	Probiotic food	Builds immune system	Digestive health and enhanced immunity
Oysters	Zinc and iron	Enzyme function	Prevents DNA damage
Ash, poplar, and weed goldenrod extract	Salicin (aspirin-like) chemical	Anti-inflammatory	Reduces arthritic and joint pains
Almonds and peanut butter	Magnesium	Muscle relaxant	Prevents colon cancer
Almonds	Copper, magnesium, manganese, vitamin E, protein, dietary fiber	Anti-inflammatory	Muscle health due to magnesium, lowers cholesterol for heart health
Apples, berries, red grapes, tomato, pea pods	Quircetin, pectin-soluble fiber (Cornell University[563])	Regulates blood flow	Promotes weight loss, fights dementia and Alzheimer's, prevents H. pylori stomach ulcer
Avocado	Olive oil, fat-soluble beta-carotene and lycopene, fiber, folic acid, vitamins K, C, B_5, and B_6	Excellent methylation food[564, 565]	Heart-healthy food

563 http://www.nutritionj.com/content/3/1/5

564 http://www.whfoods.com/genpage.php?tname=foodspice&dbid=5

565 http://www.revaclinic.com/blog/detoxification-continued-the-power-of-the-methylation-cycle_101225.html

Food	Key Nutrient	Mechanism	Cure
Avocado	Vitamin B$_3$	Various mechanisms	Heart-healthy food, reduces inflammation
Banana	Potassium, vitamins B$_6$, C, dietary fiber, lutein, carotenoids, zeaxanthin, magnesium, manganese, molybdenum	Good source of trace minerals and monounsaturated fat	Reduces risk of breast cancer
Basal	Vitamin K, eugenol, limonene, flavonoid, orientin, vicerin	Antioxidants	Prevent DNA damage
Baker's yeast	Folic acid	Methionine synthesis	Fights allergies
Barley	Beta-glucan soluble dietary fiber, selenium	Antioxidants	Reduces cholesterol
Beans	Antioxidants, dietary fiber, copper, magnesium, manganese, molybdenum, folic acid	Dopamine, serotonin, melatonin synthesis	Blood sugar control, nerve function
Beet roots	Betaine, folic acid, manganese, potassium, dietary fiber, multiple studies [566], [567]	Antioxidants, increase blood flow	Prevents DNA damage, prevents lung and stomach cancer, increases blood flow, promotes brain and cardiac health
Bell pepper	High vitamin C and carotenoids	Antioxidants, anti-inflammatory	Build immune system, prevent cardiovascular diseases

566 http://www.whfoods.com/genpage.php?tname=foodspice&dbid=49
567 http://www.sciencedirect.com/science/article/pii/S1756464611000673

Food	Key Nutrient	Mechanism	Cure
Black currant, borage oil	Gama linoleic acid and omega-6	Anti-inflammatory	Builds bone, eye, skin, and hair health; improves bone health
Black pepper, peppercorn	Piperine, bioperine, chromium	Antioxidants	Interferes with shelf-renewal of cancer stem cells, antidiabetic, induces weight loss, promotes energy metabolism, reduces pain
Bosswella serrata	Oil and terpenoids	Antiinflammatory	Cures osteoarthritis
Brans (rice, cereals)	Phytosterols	Lowers cholesterol	Promotes heart health
Blueberries	Pterosilbene, elagic acid antioxidants	Resveratrol-like action	Improves cognition and memory, reduces stress, boosts immune system, protects capillaries from oxidative damage
Brazil nuts	Protein, dietary fiber, selenium, folic acid, vitamin E, magnesium, manganese	Antioxidant	Prevent DNA damage; involved in cognitive function, cell signaling; turns on gene Nrf2, which regulates hundreds of other genes
Broccoli, brussels sprouts, cauliflower, kale	Sulphoraphane, indol-3-carbinol Multiple studies[568], [569]	Histone acylation, fights free radicals, produces cancer-killing enzymes, antioxidants	Cancer prevention, genomic stability, indol-3-carbinol inhibits growth of estrogen-responsive cancer

568 http://www.cancer.gov/cancertopics/factsheet/diet/cruciferous-vegetables

569 http://www.diagnosticpathology.org/content/9/1/7

Food	Key Nutrient	Mechanism	Cure
Brown rice	Polyphenols, phytosterols	antioxidants	Prevention of obesity and diabetes type II
Bok choy	Bracissin, anthocyanins	Anticarcinogens	Reduced risk of cancer, reduced estrogen
Buckwheat	Polyphenol antioxidants	Quenches free radicals	Reduces stress, prevents DNA damage, prevents macular degeneration, prevents cancer, improves cognition
Butter	Conjugated linoleic acid	Helps reduce body fat and improves blood lipids and insulin metabolism	Reduces obesity, improves weight loss
Buttermilk	Probiotic lactic bacteria	Our second genome	Digestive health
Cabbage (Swedish Study[570], [571])	Isocyanate, vitamin K	anticarcinogenic	Protects against prostate and pancreatic cancer
Cacao	Theobromine	Liver detoxification	Boosts mental capacity
Canola oil	Monounsaturated fat and oil	Anti-inflammatory	Promotes heart health
Cardamom, very common in Indian diet	Pine, linatol	Antioxidant, antibacterial, lowers cholesterol	Promotes heart health, improves weight loss
Carrots, red and yellow	Provitamin A, beta-carotene, vitamins A, K, and C	Antioxidant	Reduces risk of cancer, reduces risk of heart disease, DNA repair

---

570 http://www.whfoods.com/genpage.php?tname=foodspice&dbid=19

571 http://www.slu.se/en/collaborative-centres-and-projects/ekoforsk/projects-2008-2010/cabbage-and-onions-pests/

Food	Key Nutrient	Mechanism	Cure
Cashew nuts	Monounsaturated good fat, copper, B_5, B_6, riboflavin, thiamine, manganese, zinc, selenium, tryptophan	Antioxidants and neurotransmitter precusors.	
Celery-onion-bell pepper trinity	Apiol, flavonoid	Antioxidants prevent blood lipid damage	Promotes weight loss, reduces blood pressure
Cereals	Dietary fiber, copper, manganese, magnesium, selenium, vitamin B_1	Improves immune system and digestive health	
Cheese, egg, beef	Conjugated linoleic acid	Anticarcinogenic	Prevents atherosclerosis, builds immune system
Cheese, yogurt	Vitamin B_{12}	Methionine synthesis, antistress vitamin	Fights allergies, helps mental health and memory
Chickpeas and red pepper hummus	Protein, iron, vitamin C	Protein nutrition	Builds red blood cells
Chicken soup, beef, trout	Amino acid cysteine, coenzyme Q_{10}	Anti-inflammatory	Promotes muscle health
Chia seeds	Omega-3 fatty acids	Protein and fiber for appetite control	Helps burn fat, improves metabolism
Chili pepper, cayenne pepper	Capsaicin, carotenoids, vitamins A and C, B vitamins, magnesium, potassium, iron	Mediates and boosts metabolism, multiple effects in cardiovascular health	Blood sugar control, improves insulin function, and reduces cholesterol., improves cognition, protects from cancer, acts as anticlotting agent, fights pain, induces weight loss, increases metabolism

Food	Key Nutrient	Mechanism	Cure
Chocolate	Polyphenols	Antioxidants	Prevents DNA damage
Cinnamon	Proanthocyanins, eugenol, cinnamaldehyde UC Santa Barbara[572]	Analgesic, antidiabetic, hypocholesteremic	Prevents stress, reduces adipose tissue (Lund University[573])
Citrus peel	Flavonoid mobiletin, limonene, tangeretin; vitamin C	Antioxidant, inhibits key proteins that promote cancer of skin	Reduces cholesterol, LDL-associated apoloprotein B, and triglycerides
Cloves	Eugenol, omega-3 fatty acids, manganese	Antibacterial, antidiabetic	Promotes digestive health
Coconut, green, freeze-dried	Cococin, vitamins, minerals, high lauric acid	Emulsion stability, creaminess	Promotes skin health[574]
Cocoa powder	Antioxidants, arginine, fiber	Prevents DNA damage, promotes heart health as antioxidant	Reduces stress and promotes heart health.
Coconut oil	Low-calorie medium-chain fatty acids	Helps burn calories, doesn't go in adipose fat	Promotes fat metabolism
Compfrey ointment	Allantoin	Anti-inflammatory	Promotes growth of healthy tissue
Cottage cheese	Selenium, protein, vitamins, minerals	Antioxidants and protein nutrition	As a satiating food, helps reduce weight.
Cranberry	Anthocyanins	Antibacterial, anticarcinogenic	Cures urinary infection, prevents cancer

---

572 http://www.ia.ucsb.edu/pa/display.aspx?pkey=3022

573 http://www.lunduniversity.lu.se/o.o.i.s?id=24732&postid=649022

574 http://www.nsf.ac.lk/newsletter/VOL3NO3/conut.pdf

Food	Key Nutrient	Mechanism	Cure
Curry powder	Curcumin Multiple studies[575], [576]	Anticarcinogens, antioxidant, Anti-inflammatory	Prevents Alzheimer's disease and cancer.
Coffee, tea	Antioxidants Finnish Study[577]	Central nervous system stimulation and cognition	Promote memory, reduce dementia
Collard greens	Calcium, fiber, antioxidants	Antioxidant effect Fiber effect on digestive health	Promotes cardiovascular health
Coriander	Phthalides, coumarins, terpenol	Antibacterial, Anti-cholesteremic	Promotes digestive health, prevents cancer
Cucumbers	Lignans lariciresinol, pinorecinol, cucurbitacins, vitamins A, B_1, B_6, C, folate, calcium, magnesium	Anticarcinogenic, anti-inflammatory, antidiabetic, hypochlesteremin	Prevents cancer, lowers cholesterol and joint pain. Consume cucumber and carrot in combination.
Cumin	Luteolin, piperine, pipene, terpine, iron	Antioxidants	Prevents prostate cancer, improves digestive health
Dark chocolate	Polyphenols	Antioxidants, a natural opiate anandamide	Prevents DNA damage, promotes heart health, reduces stress, reduces pain
Dietary fiber	Butyric acid in colon	Histone deacylation	Longevity
Dill seed	Quircetin, kaempherol	Antioxidants	Prevents DNA damage
Edamame	Antioxidants, isoflavones	Antioxidants and protein nutrition	Promotes weight loss, enhances cardiovascular health

575 ww.ncbi.nlm.nih.gov/pubmed/24139527

576 http://www.crd.york.ac.uk/crdweb/ShowRecord.asp?LinkFrom=OAI&ID=
12013062549#.U6aFgPldUtw

577 http://www.ncbi.nlm.nih.gov/pubmed/19158424

Food	Key Nutrient	Mechanism	Cure
Egg yolk	source of choline for neurotransmission.	Neurotransmission, methyl group donor	DNA methylation
Egg white	Almost pure protein	Satiation	Low food intake and weight control
Egg, peanuts, green vegetables	B_5	Energy production	Antistress vitamin
Fruits and vegetables	B_6, COX-2 inhibitors, prebiotics	Anti-inflammatory, methionine synthesis, homocysteine in blood	Fights allergies
Crimin mushroom	B_2 or riboflavin	Homocysteine balance in blood	Improves heart health
Garbanzo beans	Molybdenum, protein, dietary fiber, choline, vitamins A and C	Maintenance of oxidation state and oxygen transfer.	Supplies protein and amino acids for endogenous endomorphins
Garlic	Allinase	Anticarcinogens	Boosts immunity, protects blood cell membranes, fights cancer
Ginseng	An adaptogen	Antiinflammatory	Prevents DNA damage
Ginger	Gingerol, gingerone	Antioxidant, anti-inflammatory	Digestive aid, fights arthritic pain, protects DNA
Gouda, Jarlsberg cheese	B_{12}, vitamin K	antiangiogenic	Protective of lung cancer
Grapes	Resveratrol	Anti-inflammatory, helps produce nitric oxide vasodilator	Promotes heart health, antiaging, anticarcinogenic
Green peppers	Vitamin C, beta-carotene, lycopene	Antioxidants	Very high vitamin C, fights allergies
Green tea	Epigallocatechin-3-gallate, vitamins A and C	Antioxidants	Promotes heart health and prevents cancer, DNA repair and gene stability, kills cancer cells

Food	Key Nutrient	Mechanism	Cure
Gouda, Emmentaler, and Adam cheese	Menaquinone, menatertrenone		Prevents cancer, promotes heart health, fights diabetes
Flaxseed	Lignan, omega-3 fatty acid, B_6, copper, manganese, lignan	Anti-inflammatory, antioxidants	Promotes cardiac and brain health, reduces cardiac arrest and cholesterol, lowers blood pressure, prevents prostate cancer, promotes kidney health, builds immune system
Fenugreek		Slows glucose absorption	Satiety and hunger control
Fermented soy products	Nattokinase, enzyme from natto (University of Chicago Work[578])	Breaks down fibrin, Isoflavone Daidzein prevents cancer	Blood-clotting enzyme
Fruits and vegetables: broccoli, yellow squash, red bell pepper, leafy greens, tomato, potato peels, citrus fruits	Prebiotic fiber, sulphoraphane, isothiocyanates, probiotic bacteria, vitamins A, C, coenzyme Q_{10}, selenium	Antioxidants, anticarcinogenic, antiangiogenic, homocysteine reduction	Prevent cancer, boost immune system, improve cellular communication, improve glucose and fat metabolism, good gene expression, cellular mobility by binding tubulin protein for digestive health, arterial elasticity

---

578 http://www.chicagoreader.com/Bleader/archives/2010/12/23/one-bite-natto

Food	Key Nutrient	Mechanism	Cure
Garlic	Diallyl sulfide, allicin, vitamin C, chromium	Histone acylation, antiplatelet enzyme, antidiabetic	Cancer prevention, reduction of blood pressure, protects arteries and veins, prevents clotting
Ginger	Gingerol	Blocks prostaglandins	Antiarthritic, anti-hypertensive
Grape juice	Polyphenols U of Cincinnati[579, 580]	Antioxidants: anthocyanin, phenolic acid, flavonoids, flavonols, myristin, resveratrol, and quircetin	Heart health, helps prevent depression and bipolar disorder
Honey	Low-glycemic sugar	Satiation	Antidiabetic
Hop extract	Perluxan	Inhibit pro-inflammatory chemicals	Fast pain relief
Horseradish	Ally isocyanate, glucosinolates, Vitamin C, calcium, magnesium, phosphorus	Anticarcinogen, glucosinolates in horseradish prevent cancer.	10X better than broccoli, helps liver do detoxification
Inulin (Jerusalem artichoke)	Oligosaccharides	Regulates blood sugar and lowers cholesterol	Promotes digestive and colon health, builds immune system
Isomaltulose	Low-glycemic sugar	Regulation of blood sugar by slow energy conversion	Weight and obesity control
Kohlrabi	High vitamin C, fiber, folic acid, minerals, sulphoraphane, indol-3-carbinol, vitamins, isothiocyanates, and B vitamins	Antioxidant vitamin C, other phytochemicals	Healthy connective tissue, teeth, and gums; prevents asthma and bronchitis

579 http://www.phytochemicals.info/plants/grape.php
580 http://www.psychiatry.uc.edu/FacultyStaff/FacultyProfile.aspx?epersonID=MTQ1OA%3D%3D

Food	Key Nutrient	Mechanism	Cure
Kokum	Luteolin, flavones, apigenin, antioxidants, hydroxycitric acid	Balm formulations, promotes cell oxygenation	Soft skin
Leafy green vegetables	Zeaxanthin	Powerful antioxidant	Prevent skin cancer
Legumes, peas, soy, beans	Folic acid and fiber, indol-3-carbinol	Homocysteine control	Reduced cholesterol and reduced risk of heart attack; protects against breast, ovarian, and colon cancer
Lentils	B vitamins and folic acid	DNA methylation	Weight and obesity control
Licorice	Anethol, glycirrhizin, isoflavone	A nutrient dense antiinflammatory food	Protects liver and reduces cholesterol
Lipowheat ceramide	A fatlike molecule	Rehydration, inhibition of enzyme elastase	Smooth and soft skin
Meat, mushrooms, green vegetables	Coenzyme Q_{10}, B_2	Critical to electron transport and reduction of oxygen to water	Energy metabolism, ATP production, great for skin, hair, and eye health
Maple syrup	Sugars	Antioxidants	Cures inflammation and cancer
Mango powder	Citric acid, malic acid, oxalic acid	Natural acidulant	Digestive aid
Millet	Fiber	Bile acid secretion	Removes cholesterol
Mint[581]	Menthone and menthol. Mint chutney is a great marinating agent.	Regulates gut bacteria. Quenches free radicals.	Antibacterial against sinusitis, bronchitis, and pneumonia; digestive aid

581 http://www.ajol.info/index.php/tjpr/article/viewFile/93279/82692.

Food	Key Nutrient	Mechanism	Cure
Mushrooms	Vitamins A, B, and C, protein, dietary fiber, copper, potassium, iron, selenium, zinc	Anticarcinogenic	Prevents cancer
Mustard, garden greens, kale	Selenium, protein, oil, good emulsifier, B_2, B_3, B_5, vitamin C and K	Antioxidants Antiinflammatory anticarcinogenic	Open tumor-cell receptors to chemotherapy drugs for effect and efficiency
Natural food colors	Carotenoids used for ages	Antioxidants, anti-inflammatory	Prevent DNA damage
Nonfat dry milk	Calcium, B_{12}, protein	Balanced amino acid from high quality protein	Satiety and weight loss
Nutmeg	Atropine-like bioactivity	Antifungal	Digestive aid
Nuts and seeds	Minerals and vitamins; widely approved	Antioxidants, good fat, and trace minerals	Mental health and improved HDL
Oats	Beta glucan soluble fiber	Satiety enhancer	Reduces blood pressure, reduces cholesterol, digestive health, boosts immunity
Olive oil	Oleocanthal and squaline (Monell Chemical Senses Center, Philadelphia PA[582])	Anticarcinogenic, Anti-inflammatory	Relieves pain, heart health, mental health
Onion	Quircetin	Antioxidant	Antidiabetic
Oranges	Vitamin C, hesperidins	Antioxidants	Prevents DNA damage, prevents cancer, prevents varicose veins, improves digestion
Oregano	Vitamin K, carnasol, quircetin, thymol	Antioxidants; antifungal	Digestive aid

---

582 http://www.monell.org/news/news_releases/trpa1_receptor

Food	Key Nutrient	Mechanism	Cure
Oysters	Zinc, protein, Vitamin A	Prevents zinc deficiency	Heals wounds Cures infant diarrhea
Papaya	Vitamin C, Vitamin A, and folic acid	Antioxidant and antiinflammatory	Promotes health of eye, digestive and cardiovascular systems
Parsley	Apigenin, apiin, cresoviol, luteolin, flavonoids, vitamins A, C, and K	Antioxidants	Prevents DNA damage, builds immune system, enhances white blood cells
Peanut butter, peanuts	Unsaturated plant fat, protein, fiber	Nutrient dense peanut butter delivers good fat,	Promotes heart health, induces weight loss
Pine bark extract	Pycnogenol	Antioxidant, nitric oxide vasodilator	Lowers blood pressure, reduces osteoarthritic pain
Pistachios	Protein, fiber, oleic acid, vitamins E, K, B_6, and thiamine, minerals	Antioxidants (Stanford University[583])	Reduces stomach fat Increases HDL
Pomegranate	Antioxidants, elagic acid to urolithin A & B conversion by gut bacteria	Antioxidants	Prevents DNA damage, prevents prostate cancer, improves heart health, speeds up skin-cell synthesis
Potato	Potassium, copper, B vitamins, starch	Vitamin C and potassium nutrition	Cures acne Promotes digestive health
Protein, high	Methionine, tyrosine, tryptophan, phenylalanine, arginine	Dopamine synthesis, satiation, produces nitric oxide vasodilator	S-adenosyl methionine for stable epigenome, antidiabetic, satiation, anti-hypertensive

583 http://www.musclemagfitness.com/nutrition/healthy-eating/lower-ldl-cholesterol-naturally-on-the-pistachio-diet.html

Food	Key Nutrient	Mechanism	Cure
Prunes/dried plums	Boron, vitamin K, antioxidants	A natural laxative with high antioxidant power	Bone-density improvement
Pumpkin seeds	Zinc, protein, vitamins A, C, E, K, and folic acid	Works with anabolic receptors	Protects bladder health, reduces pain
Red yeast	Statin-like monacolin K, trade name Cholestin	Lovastatin-like mechanism of action	Reduces cholesterol
Red grape skin	Resveratrol	Anti-inflammatory COX-2I inhibitor	Reduces pain
Red wine	Resveratrol	Histone deacylation	Longevity
Rice bran	A good source of phytosterol oryzanol, tocotrienol, ferulate, vitamins, minerals	Anticarcinogenic FDA and (Pennington Biomedical Research Center[584])	Heart health promoter, reduces LDL cholesterol and triglyceride
Rosemary	20 antioxidants	Anti-inflammatory, antibacterial	Improves cognition and memory
Salmon, sardines	Omega-3 fatty acids U of California Berkley[585]	Anti-inflammatory	Brain structure and function, improves memory and cognition
Sage oil extract	Rosemaric acid, cornosol, thymol, carvacol	Inhibits enzyme that destroys neurotransmitter acetylcholine	Improves cognition, neuroprotective, cures Alzheimer's
Salt, black	Typical table salt	Carminative, anti-flatulent	Improves digestive health
Seafood, milk, egg, meat, fish	L-taurine	Brain and blood cell function	Crosses blood-brain barrier, reduces cholesterol, prevents congestive heart failure, reduces blood pressure

---

584 http://www.scribd.com/doc/88478479/rice-bran

585 http://www.bri.ucla.edu/bri_weekly/news_081031.asp

Food	Key Nutrient	Mechanism	Cure
Sesame seed	Molybdenum, copper, calcium, iron, zinc, manganese, vitamins B_6 and C, protein, dietary fiber, choline	DNA methylation	Skin health, bone health, oral and digestive health, heart health by blood pressure reduction, prevention of DNA damage
Shitake mushroom	Lentinan, vitamin D	Anticarcinogens slow tumor growth	Boosts immune system, improves cognition
Soy, barley, wheat, rye	Lunacil, isoflavones (Dr. Benito, UC Berkeley[586])	Anticarcinogenic	Kills cancer cells
Sprouted grains	B_{12}, folic acid, other B vitamins, vitamins A and K, manganese	DNA methylation	Stable DNA
Soy products and fiber	Genistein, proteins, saponins, isoflavones to uquol conversion by gut bacteria, iron, molybdenum, manganese, potassium, zinc, niacin, folic acid	Methylation, regulates cell death, cell-cycle control, DNA repair	Cancer prevention, estrogen-like effect, prevents breast and prostate cancer, regulates tumor-suppressor gene, antiprotozoal (giardia) agent
Spinach, green pepper, soy foods	Pyrroloquinoline quinone	A very powerful antioxidant	Prevents DNA damage
Spinach, Swiss chord, kale	Folic acid, vitamin K, luteolin, magnesium	Blood clotting	Promotes weight loss, prevents calcification of arteries, promotes bone health, fights Hodgkin's lymphoma, prevents dementia, enhances immune system

---

586 file:///C:/Users/Owner/Downloads/541291.pdf

Food	Key Nutrient	Mechanism	Cure
Squash, yellow vegetables	Carotenoids	Antioxidants and anticarcinogens	Prevents cancer
Stevia sweetener	Steviol base sweet glycoside	Antioxidant, anti-inflammatory	Antidiabetic
Strawberries	Antioxidants, ellagic acid	Antioxidants Anticarcinogens	Antidiabetic, boosts immune system, protects capillaries from oxidative damage
Sugar beets	Betaine	Breakdown of toxic products from SAM synthesis	Protects from osmolytic stress
Sunflower seeds	Folic acid, selenium, copper, B_{12}	Methionine synthesis, homocysteine in blood	Fights allergies
Sweet potato	Copper, iron, magnesium, manganese, protein, good fat, dietary fiber, vitamin A, carotenoids	Antioxidants and anticarcinogens	Promotes eye health, promotes heart health Protects from cancer
Tarragon	Charvicol, polyphenol	Antioxidant	A great heart-healthy spice
Tea, black & green	Catechins, theanine, flavonoids	Phenolic antioxidants	Prevents DNA damage, improves cognition
Thyme	Geraniol, borneol, thymol, carvacol, vitamin K	Antimicrobial, antioxidant	Carminative, anti-spasmodic
Tomato	Lycopene, chromium	Antioxidant (Harvard Med School[587])	Promotes heart health, prevents breast and lung cancer

---

587 http://www.health.harvard.edu/newsweek/Tomatoes_and_Prostate_Cancer.htm

Food	Key Nutrient	Mechanism	Cure
Turmeric	Curcumin, folic acid, vitamins C, K, and B$_6$	Antioxidant, antibacterial, anti-inflammatory, anticarcinogenic	Prevents cancer, fights osteoarthritic pain, prevents enlargement of heart, protects liver, digestive health promoter, reduces osteoarthritic pain
Vinegar, apple cider, pickled cucumber	Acetic acid,	Antidiabetic and anti-atherosclerotic	Prevents lung and prostate cancer, promotes weight loss
Walnuts	Vitamin E, omega-3 fatty acids, manganese, copper, tryptophan	Anti-inflammatory, antioxidants	Cardiovascular health, brain structure and function, cognition, memory, immune system enhancement
Watermelon	Lycopene	Antioxidant	Promotes eye health
Watercress	Isothiocyanates	Anticarcinogen	Triggers enzyme that stops cancer-cell growth
Whey protein (hydrolyzed)	Ace inhibitor-like polypeptide	Immunoproteins	Blood pressure control
White kidney beans	Great source of molybdenum, folate, fiber, copper, and manganese	Prevents DNA damage	Helps weight loss Lowers cholesterol Maintains blood sugar and cures diabetes type II
Whole grains	Antioxidants, omega-3 fatty acids, polyphenol antioxidants, minerals	Antioxidants	Digestive, cardiac, and mental health
Yogurt	Vitamin B$_{12}$	Our second genome	Digestive aid

Food	Key Nutrient	Mechanism	Cure
Fruits, vegetables, and whole grains	Dietary fiber	Prevent free radical damage to lipids, DNA, and proteins	Enhanced immune system, enhanced humoral immunity, increased immunoglobulin concentration, short-chain fatty acid supply
Spinach and collard	Alpha-lipoic acid	Boosts ATP production	Promotes skin health, promotes weight loss
Fruits & Vegetables	Isocyanates	Affects genes that produce telomerase	Effective in preventing cancer-cell growth
Fruits and Vegetables	Bioflavonids	Block receptor sites	Prevent infection
Coconut and palm kernel oil with 8–12 carbon long fatty acids	Medium-chain triglycerides	Eases fat metabolism	Antiaging cream formulations for healthy skin
Bacterial exopolysaccharide	Abyssine	The polysaccharide postpones cellular aging	Antiaging ingredient in BB or Blemish Balm cream

Note: This master table should come in very handy as we proceed to cook and prepare our daily food. The methylation diet, it is clear from this master table, helps prevent major chronic diseases and ailments and mitigate deficiencies of B complex vitamins including B_{12}, vitamin D, magnesium, and zinc. Almost 40 percent of Americans have vitamin B_{12} deficiency, 54 percent suffer from pain, and 64 percent suffer from constipation. This table helps create a methylation diet by proper food selection. The methylation diet eliminates problems of nutrient deficiency, promotes proper gene expression, maintains a stable DNA and genome, and fights any epigenetic problems. The methylation diet keeps us healthy.

22.04 Time Line of Foods for Human Consumption

Consumption of an increasing variety of foods, albeit gradual, is a story of conquest in human evolution. Although fire was discovered one million years ago [588] and hearths appeared 250,000 years ago, the practice of roasting, which predates boiling and stewing, began 50,000 years ago. Cooking, it appears, made us a large-brain human. This happened via the enhanced bioavailability of nutrients expedited by sharing and social interactions. Cooking became an intentional and social event. This then lead to the domestication of foods.

On the basis of the rate of mutations of mitochondrial DNA [589] and the fact that we have had at least one thousand generations since our beginning, table 22.08 suggests that food-dictated mutation rates will accelerate in proportion to changes in the rate of new food introduction and food processing for consumption. Too much sugar, salt, trans fat, and omega-6 fatty acids, too much saturated fat via processed foods for the masses, and unnatural preservatives may have increased mutation rates even faster. Mitochondrial DNA sits right next to the free radical-producing mitochondrial electron and proton processing factory, and thus it is very prone to mutation. Mitochondrial DNA mutations, we are now finding out, seem to be the cause of chronic diseases [590] and cancer. [591] As somatic variations in mitochondrial DNA increase, there is also an increase in *age-related disorders such as heart disease, Alzheimer disease, and Parkinson's disease. Also, research suggests that too many of these mutations over our own lifetime may play a role in the aging process.* [592] The methylation diet can reverse such effects.

Foods of Vikings, foods of Robin Hood and his merry men, Anglo-Saxon and Norman British foods, medieval foods, and Shakespeare's food descriptions reveal foods of combinations, variety, multiple formats, and multiple cooking procedures. These are truly methylation diets in composition and nutrient characteristics. A great example of my personal knowledge

588 http://www.cbsnews.com/news/humans-used-fire-1-million-years-ago-says-study/.
589 http://www.cs.unc.edu/~plaisted/ce/humanity.html.
590 http://www.ncbi.nlm.nih.gov/pubmed/10939569.
591 http://www.academia.edu/232929/Mitochondrial_DNA_mutations_in_cancer--from_bench_to_bedside.
592 http://ghr.nlm.nih.gov/mitochondrial-dna.

are fruit and nut ball preparations for pre- and post-pregnancy women[593] in India. These are preparations based on almonds, pistachios, walnuts, cashews, cardamom, fenugreek, fennel seed, turmeric, brown sugar, edible gums, melon seeds, oats, turmeric, ginger, pulses and lentils, finger millet, and sesame seed, almost complete with respect to trace minerals, vitamins, protein, fiber, and antioxidants. What an ancient wisdom! We are slowly learning more about our ancient genome[594] and the methylation value[595] of ancient diets consumed for years[596]. Table 22.08 includes a chronology of human foods as determinants of our genetic makeup. DNA methylation for sure varies with various ethnic diets.[597]

Table 22.08 Chronology of Domesticated Foods

9000 BC	Middle East	Plant cultivation begins[598]. http://allthatcooking.com/history-of-cooking/
7000 BC	Central America and Mexico	Domestication of gourds, pepper, avocado, grains, and amaranth begins. Beer and wine making in Mesopotamia. http://www.eolss.net/sample-chapters/c06/e6-34-09-09.pdf
6500 BC	Turkey	Peas were already domesticated.
6000 BC		Cattle are domesticated.
5000 BC	India	Vedas talk about butter oil, dahi (yogurt), milk, sugar, and wheat. Panjeeri is a nut-based heat-processed food.
4000 BC	Egypt	Yeast-leavened bread evolves.

593 http://nopr.niscair.res.in/bitstream/123456789/11517/1/IJTK%2010(2)%20339-343.pdf.

594 http://geogenetics.ku.dk/latest-news/epigenomes/.

595 http://advances.nutrition.org/content/1/1/8.full.

596 http://suppversity.blogspot.com/2013/06/nutrigenomics-let-food-be-thy-medicine.html.

597 http://jn.nutrition.org/content/132/12/3814S.full.

598 http://allthatcooking.com/history-of-cooking/

3500 - 2300 BC	Mesopotamia	Apples, barley, broad bean, butter, cheese, corn, grape, fig, garlic, herbs, honey, leeks, lettuce, millets, onions, peas, rye, turnips, vegetable oil, and wheat. Corn crop is cultivated. Yale University has records of Akkadian recipes. Source: Stephen Bertman, *Handbook of Life in Ancient Mesopotamia* (New York: Facts on File, 2003) pp. 291–293.
3100 BC	Egyptian foods	Grains, legumes, beer, fish, vegetables, fruits, pickled salmon and catfish, Egyptian flat bread, and sesame rings are used. Irrigation begins, Sumerians use thyme, and dates are cultivated in the Middle East. Source: Joan P. Alcock, *Food in the Ancient World* (Westport CT: Greenwood Press, 2005), pp. 136–8.
2700 BC	China	There is already an herbal listing of 365 plants.
Pre-2500 BC	Sumerian	Beans, barley, beer, chickpea, cress, cucumbers, lentils, lettuce, and mustard are cultivated.
Pre-2500 BC	Babylonians	Babylonians improved upon Sumerian food habits. Source: Alan Davidson, *Oxford Companion to Food* (Oxford: Oxford University Press, 1999), p. 47.
2000 BC	India	Sanskrit literature describes methods of water purification by boiling. Sanitation is embedded into daily meditation in India even today.
2000 BC	Egypt	Bread production begins.
1500 BC	Egypt	Coriander is used.
1000 BC	Global practices	Preserved potatoes by freezing, Germans used geese, and Chinese made alcohol from rice.

1500–1700 AD	Global practices	Sugarcane, chocolate, cocoa, mustard, coffee, biscuit, and potato processing are well-established; British and Dutch East India Companies are in operation.
600 BC	Asia	Cheese making begins.
1955–2014	United States and post–World War II period	McDonald's franchise and fast food begins, and so do mass food production and retailing. The curse of the refined sugar, cholesterol, saturated trans fat, low fiber diet begins.

Sources: http://www.foodtimeline.org/foodfaq3.html.
Don Brothwell and Patricia Brothwell, *Food in Antiquity: A Survey of the Diet of Early Peoples*, (Baltimore: Johns Hopkins University Press, expanded edition 1998), pp. 109, 124.
Stephen Bertman, *Handbook of Life in Ancient Mesopotamia* (New York: Facts on File, 2003), pp. 291--293.
Andrew Dalby, *Food in the Ancient World from A-Z* (London: Routledge, 2003), p. 217.
http://allthatcooking.com/history-of-cooking/.

22.05 Select and Cook for Pleasure, Health, and Wellness

It is very clear from the previous section that we should use whole grains, fruits and vegetables, beans and legumes, nuts, seeds, fish, and poultry for the best of nutrient delivery, with emphasis on antioxidants, vitamins A, B complex, C, D, E, and K and minerals copper, chromium, selenium, and zinc as pointed out earlier. Colorful fruits, vegetables, herbs, and spices add to our dishes antioxidants and phytochemicals. The dishes can be further improved by adding protein and fiber ingredients and by minimizing use of fat and refined carbohydrates.

Beyond selection and design of a dish with food ingredients per se, we need to examine our method of cooking for preserving most of the nutrients, promoting best flavor development, and consuming a dish that controls appetite and satiety. Our primary objective in cooking is to maintain volume by reducing and dehydrating foods for optimum nutrient delivery per serving by concentration. For instance, reducing a soup dish can be used as a means to accommodate at least three servings per day of vegetables. We should therefore

1. Always build amino acids alanine and glutamic acid from good proteins for gut health and hydration, blanch vegetables to kill pathogenic bacteria, and roast or toast for enhancing taste and flavor.
2. Always avoid an excess of exotic items, saturated fat, and red meat at all costs. Always avoid additives and rare chemicals listed on the nutritional panels of retail foods.
3. Always create aesthetics, taste, flavor, and color with ingredients of a long history as food use because chronology of evolution, it appears, parallels chronology of food use by human beings.

22.06 Learn a Little Kitchen Chemistry

Baking, broiling, grilling, and poaching are time-tested methods in food cooking for flavor and texture development. Grilling, we now know, should not be excessive but optimized for "just-right" flavor and texture, or else grilled food becomes a source of acrylamide, a carcinogen. The Chinese and Japanese practice of high temperature-short time stir-frying preserves nutrients during food preparation, saves preparation time, and extracts flavors and fragrance. Mass production schemes not only fail to do so but end up creating unhealthy chemicals.

22.07 Use as Much Science as You Can

Food is not just for existence; it is for taste, texture, flavor, portion, and proportion. The new American plate, says the American Cancer Research Institute, should have two thirds fruits and vegetables, whole grains, and beans, and ⅓ or less animal proteins. *Vegetables have to get to the center of all eating pleasure.* Plant- and animal-based foods have to be balanced in our daily food choices. Fat, sodium, and sugar must not gain dominance in the name of flavor and texture. Good food selection, including fiber, protein, and plant fats, can replace the need for food supplements except for rare vitamins like D_3 and B_{12}.

We can manage our menu and daily diet with oats, oat bran, rice bran, brown rice, and wheat bulgur made into porridge and pilaf. To which can be added dried fruits and fresh fruits and vegetables for nutrients and also for color appeal. The color variety can come from roasted almonds, blueberries, grapes, peaches, and all kinds of readily available colored vegetables. A 1,300-calorie diet followed with regularity can help reduce weight while keeping the pleasure of eating intact. The dish should include around 65 grams of protein, 175 grams of carbohydrates, 28 to 30 grams of fiber, and less than 30 grams or 2 tablespoonfuls of fat. Daily sodium and cholesterol should be below 1,500 and 50 mg per day respectively. Such a restricted-calorie diet leaves enough room for once-in-a-while splurging on impulse food.

Just look at how President Clinton lost thirty pounds following a diet of smoothies for breakfast made from blueberries, protein powder, and almond milk, a lunch made of green salads and beans, a snack of nuts, hummus, and raw vegetables, and a dinner with quinoa and veggie burger without sacrificing his dessert of colorful fruits. But it did require a presidential drive and commitment.

22.08 Follow a Few Rules

I believe that our excessive weight and obesity problems have come about because of lack of physical activity, high consumption of refined sugars and high glycemic carbohydrates, consumption of too many calories, use of too many unnatural substitute foods, and a poor lifestyle deprived of natural light and optimum sleep. Following a few rules on consuming nutrient dense and calorie restricted methylation diets on day to day basis can enhance our healthful and disease free life.

1. Design your daily plate with commonly available non-exotic foods.
2. Get maximum nutrient values from side dishes of chutneys, dressings, dips, sauces, vegetable juices, vegetable soups, salads, and a variety of fruit-based snack items.
3. Prepare your main dishes containing as little refined sugar as possible, and restrict daily intake to less than 20 grams per day or less than 4 percent of a two-thousand-calorie-per-day diet.
4. Use herbs and spices for maximum food healing power in all preparations of breakfast, lunch, snack, dinner, dessert, and side dishes. Include black pepper, basil, cinnamon, clove, nutmeg, oregano, rosemary, and turmeric for sure.
5. Breakfast: We need a relatively high-protein breakfast of 50 to 60 percent complex carbohydrates, 30 percent protein, and 10 percent good fat. The proteins containing tyrosine allow for high dopamine for energy and alertness.

6. Lunch: We need a lunch that is comprised of complex carbohydrates and good-quality protein; the lunch dish can be a simple crisp salad with eggs (for brain-food choline). Gene-expression nutrient choline can also come from sesame seed tahini, wheat germ, and whole grains. We can add whey protein powder to our smoothies, if part of lunch, for the amino acid serine.

7. Dinner: The dinner plate needs to secure 500 calories coming from complex carbohydrates and 200 calories from protein in a 2:1 ratio. A simple dinner before 7:00 p.m. of baked sweet potato along with a salad can do wonders for us. A simple butternut squash (loaded with vitamin A, vitamin C, and 2.8 g fiber) can be a great alternative.

8. Always use seeds and nuts in your snacking program. Crumble made of a variety of fruits crusted with oat flakes can be used as an alternate snack during the week.

22.09 Allow Your Genes to Express

Design a high-fiber and high-protein diet of a total 1,300 calories including breakfast, lunch, and dinner. Use 300 to 400 calories for snacks and desserts. Leave room for 200 calories of impulse food. Use a variety of foods and target for RDAs of enough vitamins A, C, D, E, B_{12}, and folate, and minerals calcium, copper, chromium, magnesium, potassium, selenium, and zinc by a combination of the following:

1. Green foods (spinach, chard, bok choy, zucchini, cruciferous veggies, kale, cabbage): Green foods are rich in folate and good for brain health, DNA formation, cell repair, and growth.

2. Red foods for vitamin a, lycopene, and leutin: Eye health and vision from red peppers, tomatoes, red apples, cherries, strawberries, and raspberries.

3. Yellow and orange foods for vitamin C for healing and against cold, and collagen-building nutrients from cantaloupe, oranges, carrots, mango, papaya, squash, and apricots.

4. Blue and purple foods like berries, grapes, pomegranates, eggplant, and so on for immune-boosting antioxidants and other phytochemicals.

5. Seeds and nuts offer multiple nutrients: Flaxseed for omega-3 fatty acids and fiber lignan, pumpkin seed for protein, Brazil nuts

for selenium, walnuts for omega-3 fatty acids, and almonds for magnesium. Soy foods, edamame, and flaxseed can become part of our breakfast, lunch, and dinner dishes. Prepare as often as possible a dip from soaked and pureed pumpkin seed, sunflower seed, Brazil nut, and walnut and use as a dip with chips or use seeds and nuts as a snack. The blend offers (1) amino acids arginine, tryptophan, and tyrosine; (2) vitamin E; (3) minerals magnesium, copper, zinc, and selenium; (4) choline; and (5) homocysteine-optimizing betaine. This is a great pain-fighting combination of foods because of high magnesium and salicyclic acid.

6. Cinnamon added to oatmeal and fruit spreads over Greek yogurt and sesame paste added to hummus can be incorporated into our daily snacks.

7. A daily dose of carrots with high-protein dip made out of soaked nuts and seed puree can be prepared as an anti-inflammatory snack item. Carrots, bell pepper, celery, rosemary, and thyme contain luteolin, a flavonoid that reduces inflammation. To this can be added dishes containing betaine-rich beets and folate-rich asparagus to delay onset of dementia and cognitive impairment.

8. Vitamin B complex and B_{12} foods such as yogurt at breakfast and sauerkraut regularly at lunch can be regular components of our weekly dishes.

What follows is a list of representative and typical breakfast, lunch, snack, dinner, and dessert items. Many replacements are possible depending on what we have in the kitchen pantry on a given day.

22.10 Make and Follow A Weekly Dietary Plan

What follows is a plan for eating low calorie per day methylation diet each week Sunday through Saturday. The dessert is skipped on certain days of low physical activity. Any dessert item can be selected if there was sufficient physical activity from the list in section 22.12 that includes a number of recipes.

Sunday: A high-protein breakfast, high-protein vegetable-based lunch, snack for mineral nutrition, and high-omega dinner. The Greek yogurt

and fruit dessert two hours before retiring is a great complementary addition.

Breakfast: Egg white veggie scramble.
Lunch: Hummus veggie pita sandwich of one half whole-grain pita spread with one teaspoon mustard, stuffed with vegetables bell pepper, sprouts, lettuce, two slices of avocado; two tangerines; unsweetened herbal tea.
Snack: One sheet of graham cracker with a nut and seed dip.
Dinner: A three-ounce grilled salmon fillet with citrus glaze (orange juice and honey); one half cup cooked brown rice; one cup cooked winter squash or toasted artichoke (topped with lemon, olive oil, onion, salt, or balsamic vinegar).
Dessert: One half cup fat-free plain Greek yogurt with two teaspoons fruit spread.

As an alternative, we can use a breakfast of four ounces of orange juice, multigrain pancake, green tea; a lunch of salad with vinaigrette dressing, grilled chicken, fresh fruits, strawberries; and a dinner of roasted vegetables, mushrooms, meatloaf, mixed salad, and two cups of spiced pear. What a simple dinner!

Monday: Begin day two with a high-protein breakfast, high-protein lunch via Greek yogurt, and a snack of vegetables. A dinner of a small portion of chicken or equivalent tofu with herbal tea completes the day.

Breakfast: Low-fat cottage cheese with pineapple chunks.
Lunch: Pasta with eggplant-tomato relish with four ounces of tofu; one tablespoon low-fat Greek yogurt; water with mint and lemongrass.
Snack: One piece of fresh fruit or one cup mixed vegetables with two teaspoons of vinaigrette.
Dinner: Three ounces of grilled chicken or three ounces of tofu; one medium baked potato; two cups toasted greens topped with one tablespoon of vinaigrette, followed by unsweetened herbal tea.

A good alternative is a breakfast of bran flakes, soy milk, one cup banana, whole-wheat toast, apricot fruit spread; a lunch of corn spaghetti, olive roll, and fresh orange; a snack of a peach-pineapple-apricot crisp; and a dinner of grilled salmon, green snap beans, whole-wheat couscous, and a pumpernickel roll. The desset can be skipped.

Tuesday: Breakfast provides both protein from Greek yogurt and soluble dietary fiber from oats, lunch delivers vegetable proteins from lentils and vegetables, lunch provides protein and antioxidants, and the dinner of vegetable burger and vegetable salad is a combination of light food for the day.

Breakfast: One cup oatmeal with low-fat Greek yogurt, coffee or green tea or clove toddy.
Lunch: One cup lentil soup, two cups mixed salad greens, herbal tea.
Snack: One medium orange.
Dinner: Veggie burger on whole-wheat bun, spinach salad with onion and orange, dressed with one teaspoon olive oil.

An alternative is a breakfast of nut bread, two slices of fruit bits; lunch of grilled chicken, mixed green salad, vinaigrette dressing, large apple; a snack of whole-grain crackers with one half tablespoonful peanut butter; and dinner of pork tenderloin, chicken, or tofu with wild rice pilaf, steamed asparagus, and cinnamon applesauce. Use fruits for dessert or skip it.

Wednesday: A high-antioxidant breakfast, high-protein lunch, fruit-based snack, and fish protein and powerful broccoli dinner is just right for multiple nutrients. The snack may include Brazil nuts for selenium and other minerals from pumpkin and sunflower seeds.

Breakfast: Buckwheat pancake, sliced kiwi.
Lunch: Turkey sandwich (1 slice of bun, 2 oz turkey) with tomato and mustard.
Snack: One piece fresh fruit: banana or apple. May include Brazil nuts.
Dinner: Baked fish, Mediterranean mashed potato with broccoli.

An alternative is a breakfast of fried egg sandwich, grilled whole-wheat muffin, grapefruit, blackberry banana smoothie; a lunch of tofu steak sandwich, seven-vegetable slaw, one cup sliced kiwi/blueberries; a snack of fruit and spicy nut trail mix, with one half cup orange juice; and a dinner of curried winter squash soup, cracked wheat bread, and spinach and mushroom salad with vinaigrette. No dessert.

Thursday: Berries with antioxidants and yogurt with probiotics begin the day, lunch adds protein and fiber value of beans by midday, an artichoke and cheese snack adds protein and phytochemicals, and a light dinner is a great combination for the midweek.

Breakfast: Fresh berries and nonfat yogurt.
Lunch: One ounce low-fat cheese with five whole grain crackers.
Snack: Stuffed artichoke (bread crumbs, diced provolone or Gouda cheese, beans, parsley, olive oil, garlic).
Dinner: Sweet and sour vegetables with tofu tossed green salad and orange.

An alternative is a breakfast of two slices of whole-wheat toast with peanut butter and strawberry fruit spread, apple-cranberry juice; a lunch of chipotle chicken chili, baked tortilla chips, fruit juice, oatmeal raisin cookie; a snack of toasted nuts and seeds; and a dinner of restaurant crostini (Italian ciabatta bread), sea bass wild rice pilaf, steamed broccoli, fruit sorbet, and expresso. Go for yogurt with fruits for dessert if you feel like.

Friday: The day begins with a high-protein breakfast, the lunch adds fiber and protein, and a light dinner completes the day.

Breakfast: Boiled eggs, berries, and orange juice.
Lunch: Mediterranean bean salad, two slices whole-grain bread, and one pear.
Skip snack
Dinner: Baked potato and broccoli salad with two tablespoons low fat dressing of your choice.

An alternative is a breakfast of apple-crunch oatmeal, pineapple juice, six ounces green tea, carrot wheat germ muffin, hard-boiled egg, black grapes; a lunch of tortilla wrap, large banana, spicy shrimp, celery sticks, lime water; a snack of multifruit crumble and coffee; and a dinner of lemon grouper, vegetables, romaine lettuce salad, orange juice, blueberry-banana smoothie, and roasted cashews.

Saturday: High-soluble fiber breakfast complements a vegetable protein lunch with added vegetables, followed by a high-protein artichoke snack. A light dinner completes the day.

Breakfast: Sliced banana oatmeal.
Lunch: One cup lentil soup, two cups mixed salad greens, herbal tea.
Snack: Stuffed artichoke (bread crumbs, diced provolone cheese).
Dinner: Sweet and sour vegetables with tofu tossed green salad, orange.

An alternative is a breakfast of grapefruit juice, marbled eggs, whole-grain toast, fruit spread, green tea; a lunch of chicken salad, oat groat bread, iced green tea/lemon; a snack of popcorn, two cups fruit and spicy nut trail mix; and a dinner of avocado shrimp salsa, tomato dip/corn chips, chicken chili, pineapple. Go for a light dessert only if you have had a physically active day.

22.11 Food for Our Genes

The recipes above for each day of the week represent a nutrient-wise, complete and yet restricted-calorie methylation diet for our genes. Also, I have selected a few more recipes strictly for building variety that add a maximum of antioxidants, dietary fiber, probiotics, good fat, protein, essential minerals, and vitamins that we are generally deficient in. Each recipe gives a breakdown of per serving nutritional information, with notations of proven health effects and values in terms of nutrient deliveries. The recipes meet general criteria for a methylation diet, and they can be used as alternates in the weekly list of breakfast, lunch, snack, or dinner wherever possible without increasing total calories for the day. The criteria of diet design is to deliver high amounts of fiber, vitamins, and minerals by way of fruits, vegetables, seeds, and nuts.

22.12 Common and Easy Food Preparations

22.12.01 A Cold Drink of Vegetable Stock

(TBL is tablespoonful, tsp is teaspoonful, oz is 28.5 gram ounce.)

Ingredients: 1.00 TBL olive oil, two large chopped onions with skin on, 4 stalks of celery, 3 unpeeled coarsely chopped carrots, 2 coarsely chopped parsnips, 1 parsley sprig bunch of 10, 3 peppercorns, and 3 liters of water.

Preparation: Heat oil; add stir-fry vegetables just a little to make them crisp. Add all water and bring to boil. Simmer for one hour. Mash down the solids and strain. This can be frozen for three months for later use. Serve as a drink with ice for vitamins A and C and antioxidants.

Nutrition: Twelve servings of a very low-calorie beverage. Each one-cup serving provides only 18 calories, no cholesterol and saturated fat, only 20 mg sodium, 2 grams carbohydrates, and 1 gram sugar.

22.12.02 Recipes for High-Antioxidant and -Phytonutrient Dips, Salsa, and Spreads

Artichoke Bean Dip

Ingredients: 19 oz white kidney beans, 14 oz artichoke, one fourth cup of extra-virgin olive oil, 3 TBL lemon juice, 3 cloves of quartered garlic, 1 tsp coarse salt, 1 tsp cayenne pepper, ¼ tsp roasted red sweet pepper strips, ¼ cup snipped basil, and ¼ TBL sea or kosher salt.

Preparation: Use a food processor. Combine beans, artichoke, olive oil, lemon juice, garlic, salt, roasted red pepper, and cayenne pepper. Cover and process to a smooth paste. Transfer to serving bowl and chill for 24 hours.

Nutrition: This high-protein and high-fiber preparation makes four servings, each with 70 calories; 45 fat calories; 5 grams fat; 207 mg sodium; 8 grams carbohydrates; 3 grams fiber; 3 grams protein; % RDA of vitamin C: 18%.

Avocado and Edamame Spread

This preparation yields 14 servings of 1 oz size, each with 6.4 gram good fat, 4 gram fiber, and 8.97 gram protein.

Ingredients: 16 oz edamame, 1 avocado, 1 clove of garlic, 1 TBL olive oil, 1 TBL fresh lime juice, and salt to taste. Two tsp of mustard paste is optional.

Preparation: Cook edamame in boiling water for 5 minutes, strain, and rinse with copious amount of cold water to stop cooking. Combine all ingredients and make the pureed spread in a food processor. Blend in mustard paste if you like. This is a flavorful, high-protein, and good-plant-fat dip with minerals and fiber. My wife chooses to add mint flavor to it.

Spiced Carrot Spread for the Power of Spices

Ingredients: 6 thinly sliced carrots, 1 clove of garlic, ½ tsp grated and peeled ginger, ½ tsp ground cumin, ½ tsp ground cinnamon, a pinch of cayenne pepper, 1 TBL tahini, 2 tsp lemon juice, salt and pepper to taste.

Preparation: Set a steamer basket in a saucepan in 2 inches of simmering water. Steam carrots for 6 minutes to a tender texture. Transfer to food processor along with garlic, cumin, cinnamon, cayenne pepper, tahini, lemon juice, and ginger. Season with salt and pepper. Process until smooth, 1 minute. Add up to 2 TBL water if necessary. As an option ½ cup of baked and mashed butternut squash can be added.

Nutrition: This saturated fat-free recipe makes four servings, each with 59 calories; 2 grams fat; 9 grams carbohydrates; 3 grams fiber; 2 grams protein. Butternut squash adds vitamins A, B complex, C, and mineral manganese.

Powerfully Spiced Fruit Salsa

Ingredients: 2 peeled and chopped large kiwifruits, 2 cups of fresh chopped strawberries, 2 minced scallions, 2 TBL brown sugar, ¼ tsp cayenne pepper, and ⅓ cup lime juice.

Preparation: Combine all ingredients in a serving bowl. Refrigerate at least 1 hour. Serve with grilled seafood and poultry.

Nutrition: This fat-, cholesterol-, and sodium-free spiced salsa makes 10 half-cup servings, each with only 45 calories; 11 grams carbohydrates; 3 grams dietary fiber; 1 gram protein; and 7 grams sugar. In fact, this is a vitamin C salsa, with the health benefits of strawberries and scallions.

A Chutney or Dip of Celery Root with Garlic

Ingredients: 2 celery root or celery, 2 cloves of garlic, ½ tsp salt, ¼ tsp black pepper, ½ cup skim milk, and ¼ cup chopped parsley.

Preparation: Cut and peel celery. Add celery, garlic, salt, and pepper in a saucepan and cover with water. Bring to boil and cook until tender. Drain and mash. Stir in the milk and parsley.

Nutrition: Four servings of this fat- and cholesterol-free high-protein and high-fiber preparation, each of ¼ cup, provide 70 calories; 285 mg sodium; 14 grams carbohydrate; 3 grams fiber; 3 grams sugar; and 3 grams protein.

Power of Sautéed Spinach with Garlic

Ingredients: 16 oz baby spinach, 1 clove of garlic, and 1 TBL balsamic vinegar.

Preparation: Rinse spinach and add to skillet along with garlic. Cook on medium heat for 5 minutes. Sprinkle with drops of balsamic vinegar or lemon juice and serve.

Nutrition: This is a low-fat and cholesterol- and sugar-free preparation of four ¼-cup servings, each providing 25 calories; 4 fat calories; 0.4 grams fat; 0.1 g saturated fat; 90 mg sodium; 4 grams carbohydrates; 3 grams fiber; and 3 grams protein.

A Probiotic of Cabbage and Yogurt Slaw

Ingredients: 9 cups of medium green cabbage, 2 red bell peppers, 3 scallions, ⅔ cup Greek yogurt, 1 TBL cider vinegar, 1 TBL Dijon mustard, 2 tsp sugar, salt and pepper to taste.

Preparation: Toss together cabbage, bell pepper, and scallions in a large bowl. Whisk together yogurt, vinegar, mustard, and sugar. Season with salt and pepper in a small bowl. Pour dressing over greens. Toss to coat completely. Season to taste with salt and pepper. Refrigerate 20 minutes or up to two hours. This is a wonderful prebiotic fiber probiotic gut bacteria side dish.

Nutrition: Eight fat-free servings, each of 65 calories; 13 grams carbohydrates; 4 grams fiber; and 3 grams protein.

A Multinutrient Mayonnaise of White Beans, Greek Yogurt, and Sun-Dried Tomato

Ingredients: 15 oz white beans, ½ cup diced and dehydrated sun-dried tomato, ¾ cup nonfat Greek yogurt, 2 TBL fresh grated Parmesan cheese, ¼ cup chopped basil, salt and pepper to taste. This is a great replacement for mayonnaise.

Preparation: Add beans, sun-dried tomato, yogurt, and Parmesan cheese to the food processor and blend to creamy smoothness. Mix in basil; season with salt and pepper. Cover and refrigerate for 30 minutes before serving.

Nutrition: This fat-free preparation makes 16 servings, each of 2 TBL providing 35 calories; 55 mg sodium; 5 grams carbohydrates; 1 gram dietary fiber; 1 g sugar; and 3 grams protein. Also, it offers the benefits of high-protein, immunomodulating peptides from cheese, and eye health-promoting lycopene from tomatoes.

22.12.03 High-Fiber Grilled Fruit Kebab

Ingredients: ½ cup pineapple juice, 1 TBL brown sugar, juice of one lime, ½ tsp cinnamon, half a tsp allspice, and a pinch of clove.

Preparation: Soak wooden skewers in hot water for an hour. Heat the grill to medium. Combine basting-sauce ingredients of pineapple juice, brown sugar, lime juice, and cinnamon in a shallow bowl. Prepare the spice mixture of allspice and clove. Thread fruit cubes of kiwi, papaya, mango, and pineapple on the skewers. Sprinkle with spice. Grill and baste every few minutes for 10 minutes to brown color. Drizzle the remaining sauce over kebabs. Serve and enjoy.

Nutrition: Makes six cholesterol-free servings, each of 130 calories; only 5 fat calories; 0.5 gram fat, and only 0.1 gram saturated fat. 5 mg sodium; 33 grams carbohydrates; 4 grams fiber; 25 grams sugar; and 1 gram protein.

22.12.04 A Multinutrient Mash of Sweet Potato Turnip with Sage Butter

Ingredients: ¼ lb sweet potato, 4 oz turnip, 3 large cloves of garlic, 15 sage leaves, 1 TBL butter, ½ tsp salt, ½ tsp coarsely ground black pepper.

Preparation: In a medium saucepan, place sweet potato, turnip, garlic, and 6 leaves of sage and cover with water. Bring to boil. Cook at medium heat for 15 minutes to a tender soft texture. Drain and return vegetables to pan and cover. Heat butter to melt, and add remainder of sage. Let sage crackle to flavor for one minute. Pour sage butter over vegetables and mash. Stir in salt and pepper. Enjoy the nutritional benefits of sweet potato from vitamins A, C, B$_6$, choline, and cancer-fighting betaine.

Nutrition: Three servings, each serving with 88 calories; 4 grams fat; 3 grams saturated fat; 10 mg cholesterol; 224 mg sodium; 291 mg potassium; 12 grams carbohydrates; 2 grams fiber; and 1 gram protein.

22.12.05 Fruit-Based Smoothies and Drinks

Raspberry Smoothie

Ingredients: A very simple smoothie made from ½ cup low-fat milk, 3 TBL vanilla yogurt, ½ cup frozen raspberries, and optional skim milk.

Preparation: Combine all ingredients in a blender and puree to a desired consistency. It can be thinned by adding skim milk 1 TBL at a time.

Nutrition: I cup serving provides 210 calories, only 4.5 calories from 0.5 gram fat; no saturated fat and cholesterol; 85 mg sodium; and 45 grams total carbohydrates.

Strawberry-Kiwi Smoothie

Ingredients: One 2.6 oz bag of frozen strawberries, 2 fresh kiwi, 1 oz strawberry yogurt, 1 tsp almond extract, and optional skim milk.

Preparation: Combine all ingredients in a blender and puree to a smooth consistency. It can be thinned by adding skim milk 1 TBL at a time.

Nutrition: 1 cup saturated fat and cholesterol-free serving provides 11 calories; 9.00 fat calories from 1 gram fat;32 mg sodium; 26 grams total carbohydrates; and 3 grams dietary fiber. This is a high-vitamin C smoothie.

22.12.06 Soups and Salads

Asparagus and Scallion Soup with Almonds

Ingredients: 1 and ¼ cup sliced almonds, 2 medium sliced leeks, 6 thinly sliced scallions, 28 oz chicken or vegetable broth, ½ tsp dried thyme, salt and pepper to taste, 2 lb asparagus, 1 and ½ oz white beans, and 1 cup skim milk (optional).

Preparation: Place almonds in a saucepan over medium heat. Toast to golden brown color in 5 minutes. Set aside the nuts. Heat oil in the same pan, add leeks and scallions and cook for 5 minutes to tenderness. Add broth, skim milk, thyme, salt, and pepper and bring to boil. Add asparagus and beans. Bring to boil and then simmer for 15 minutes to soften vegetables. Cool and puree. Pour back into saucepan and heat to warm up. Serve in bowls garnished with scallions and toasted almonds.

Nutrition: Six servings, each providing 150 calories; 35 fat calories from 5 grams fat; 1 gram saturated fat; 290 mg sodium; 20 grams carbohydrates; 6 grams fiber; and 8 grams protein. This is a high-protein, high-fiber, high-folate, and high-magnesium soup. We should use it at least twice a week.

High-Fiber Spicy Black Bean Soup

Ingredients: 1 lb dried black beans, soaked overnight and drained; 2 diced large yellow onions; 2 minced jalapeño peppers; 1 TBL virgin oil; 4 garlic cloves; 2 tsp ground cumin; and salt and pepper to taste.

Preparation: Use medium saucepan. Add beans, water to cover 2 inches, and one quarter each of onion and jalapeño, and bring to boil over high. Simmer for 45 minutes. Refrigerate and use it later or next day.

Use a heavy pot. Heat oil. Add garlic and rest of onion; season with salt and pepper. Cook for 10 minutes. Stir in all remaining jalapeño, cumin, beans, and cooking liquid. Simmer for 20 minutes. Add water if necessary while cooking. Add beans and puree. Serve garnished with chopped onions.

Nutrition: Four servings, each providing 336 calories; 5 grams fat; 1 gram saturated fat; 62 grams carbohydrates; 20 grams fiber; and 20 grams protein. This is a truly high-protein and high-fiber dish.

Thick Tomato Soup

Ingredients: 1.25 oz diced celery, 1.25 oz diced onion, 16 oz diced carrot, 35 oz water, 5.5 corn bran fiber, 32 oz tomato juice, 14 oz chicken broth, 1 oz crushed tomatoes, half tsp oz extra-virgin olive oil, 1 tsp black pepper, 1 tsp oz salt, 1 tsp garlic powder, 1 tsp ground pepper, 2 tsp Worcestershire sauce, 1 tsp dried thyme, 2 tsp dried basil, and roasted red pepper puree and 2 TBL heavy cream optional for flavor and color. Red pepper is common in Indian recipes. This is a 6.5 lb recipe and 2 TBl heavy cream would not make much difference in nutrition per serving.

Preparation: Sweat the celery, onions, and carrots with the olive oil in saucepan. Add the chicken broth to the vegetables and simmer. Premix the corn-bran fiber with hot water using a blender, food processor, or immersion blender. Add corn-fiber slurry to the broth and vegetable mix. Add tomatoes, juice, Worcestershire sauce, ground black peppers, and all of the seasonings and continue to heat over medium heat. Once vegetables are tender and soup is heated throughout, use blender to puree until smooth. Place puréed soup over low heat and add the heavy cream; heat throughout. Garnish with basil or parsley and serve warm.

Nutrition: Use eight-fluid-oz serving that provides 100 calories; 54 fat calories from six grams fat; 25 mg cholesterol; 580 mg sodium; 8 grams carbohydrates; 3 grams dietary fiber; and 4 grams protein. This a high-sodium soup packed with protein and fiber.

High-Fiber Curried Sweet Potato and Lentil Soup

Ingredients: 2 TBL canola oil, 1 chopped medium onion, 3 cloves of garlic, 1 tsp fresh minced ginger, 2 TBL curry powder, 1 tsp salt, ½ tsp black pepper, 4 chopped plum tomatoes, 1 cup dry lentils, 1 quart vegetable broth, 14 oz can of light coconut milk, 1 lb half-inch chunks of sweet potatoes, 1 chopped zucchini, ½ cup cut green beans, ½ cup chopped cilantro or parsley.

Preparation: Heat oil in a medium-size pan over medium heat. Add onion and cook for 3 minutes to soft and translucent consistency. Add garlic and ginger and cook one more minute. Stir in curry powder, salt, and pepper and cook for 2 more minutes to proper color and fragrance. Stir in tomatoes, sweet potato chunks, and lentils, add vegetable broth and coconut milk, and bring to boil while partially closed. Reduce heat to simmering bubbles until lentils are soft in 15 minutes. Add stock or water for consistency. Cover and cook for 10 minutes. Stir in zucchini and green beans, adding more water for a good broth look. Cover and cook for 5 minutes to tenderness. Stir in cilantro; salt and pepper to taste.

Nutrition: Four servings, each of 1 cup that provides 446 calories; 15 grams fat; 6.5 grams saturated fat; 974 mg sodium; 63 grams carbohydrates; and 20 grams fiber. This is a high-fiber and high-vitamin A soup that also delivers 15% RDA of vitamin C.

Cool Cucumber Soup

Ingredients: 4 peeled and seeded small cucumbers, 2 cups Greek-style yogurt, 2 TBL lemon juice, ½ cup fresh-chopped mint, 1 clove of minced garlic, 1 TBL olive oil, and salt and pepper to taste.

Preparation: Add all ingredients to a food processor, puree for a minute, and pour into a container. Refrigerate for 3–4 hours. Serve with a garnish of minced dill.

Nutrition: Three servings of 1 cup each provide for 150 calories, including 45 calories from 5 grams fat; 0.5 gram saturated fat; 70 mg sodium; 11 grams carbohydrates; 2 grams fiber; 8 grams sugar; and 16 grams protein. This, in a sense, is high-protein Indian raiyata with Greek yogurt. This is a good way to increase daily vitamin K and molybdenum intake, along with anti-inflammatory and cancer-fighting cucurbitacins, lignans, and flavonoids.

Pea Soup

Ingredients: 2 lb fresh or frozen peas, ½ cup water, half-and-half (optional), vegetable broth (optional), goat cheese bits (optional), and croutons (optional).

Preparation: Puree, filter, and simmer for 1 minute. Add half-and-half or cream for creaminess Dilute with water/vegetable broth. Goat cheese bits can also be added with or without croutons.

Nutrition: 8 servings of 1 cup each provide 132 calories, including 1 gram saturated fat; 433 mg sodium; % RDA of folate: 20%; % RDA of vitamin A: 18%; % RDA of vitamin C: 77%. Pea and asparagus soups can be used once a week for a rather complete methylation diet.

Carrot Soup

Ingredients: 1.5 lb chopped carrots, 2 cups water, half-and-half (optional), and croutons (optional).

Preparation: Puree and filter. Simmer for 25 minutes. Can add half-and-half for creaminess. Can be served with or without croutons.

Nutrition: 8 servings of 1 cup each provide 77 calories with only 1 gram saturated fat; 484 mg sodium.

Asparagus Salad

Ingredients: ¾ tsp olive oil, ¾ lb asparagus, 1 TBL water, ½ lb sugar snap peas, 1 sliced scallion, ¾ tsp low-sodium soy sauce, and ¾ tsp honey.

Preparation: Heat oil in a pan over medium heat. Add asparagus and water. Cover and steam for 5 minutes. Add peas, scallion, soy sauce, and honey. Cover and cook for another 5 minutes. Season to taste with salt and pepper.

Nutrition: Four servings, each providing 34 calories, including 9 fat calories; 36 mg sodium; 5 grams carbohydrates; 2 grams fiber; and 2 grams protein. This is a high-folate salad to be used, if possible, twice a week.

President Clinton's Snow Pea Salad

Ingredients: 5 oz snow peas, 3 oz bean sprouts, ¼ medium julienned red bell pepper, 1 TBL soy sauce, 2 tsp sesame oil, 1 TBL toasted sesame seed, salt to taste.

Preparation: Blanch snow peas in boiling water for 10 seconds; then quickly soak in ice water and pat the peas dry. Combine peas, sprouts, and red pepper; toss with soy sauce, sesame oil, and sesame seed. Salt to taste. Serve chilled.

Nutrition: Makes four cholesterol-free servings, each of 57 calories; 2 grams protein; 5 grams carbohydrates; 2 grams fiber; 3 grams fat (1 gram saturated fat); and 304 mg sodium.

Artichoke Salad with Baby Greens

Ingredients: 1 cup baby field greens, 6 cherry tomatoes halved, 4 oz artichoke hearts, 1TBL red wine vinegar, 2 tsp extra-virgin olive oil, 1 finely minced garlic clove, salt and freshly ground black pepper to taste, and ½ TBL grated Parmesan cheese.

Preparation: Arrange greens, tomatoes, and artichoke in a bowl. Make dressing in a separate bowl by whisking in red wine vinegar, olive oil, garlic, salt, and pepper and toss into the greens for making two servings. Top with Parmesan cheese. Preparation time: Less than 15 minutes.

Nutrition: 91 calories per serving, including 45 fat calories from 5 grams fat, containing only 1 gram saturated fat; 318 mg sodium; total 9 grams carbohydrates; 2 grams fiber; and 4 grams protein.

Vegetable Salad

Ingredients—Pasta: 8 oz whole-wheat rotini noodle, 3 chopped medium tomatoes, 1 diced red pepper, 1 small yellow pepper, ½ cup minced parsley, 1 TBL minced fresh oregano, 1 diced cucumber, 1 can (15 oz) drained and rinsed chickpeas.

Ingredients—Vinaigrette: 3 TBL red wine vinegar, 1 TBL lemon juice, 1 tsp Dijon mustard, 2 and ½ TBL olive oil, and salt and pepper to taste.

Preparation: Combine the cooked pasta with tomatoes, red and yellow peppers, parsley, oregano, cucumber, and chickpeas. Combine the vinaigrette ingredients and mix. Toss it over pasta.

Nutrition: Six low-sodium and cholesterol-free servings, each providing 200 calories, including 65 calories from 7 grams fat, including 1 gram saturated fat; 120 mg sodium; 28 grams carbohydrates; 5 grams dietary fiber; 6 grams sugar; and 7 grams protein.

22.12.07 Lunch of Wild Mushroom and Lentil Burger

Ingredients: 6 oz Shitake mushrooms, 15.5 oz rinsed and drained lentils, ¼ cup whole-wheat bread crumbs, 1 large egg, ¼ cup chopped celery, 1.5 TBL fresh thyme, 1 tsp Dijon mustard, 1 cup chopped onion, 4 oz mild cheese, 6 TBL yellow cornmeal, 3 tsp olive oil, 4 whole-wheat buns, ¼ cup chopped roasted bell peppers, and ¼ cup watercress.

Preparation: Heat oven to 400°F. Coarsely mash ¾ of mushrooms and set aside. Add lentils, bread crumbs, egg, celery, thyme, roasted bell peppers, watercress, mustard, half of onion, and half of cheese. Pulse, and then form into patties. Coat both sides of patties with cornmeal. Heat 1 tsp oil and add remaining mushrooms and onion; heat to golden brown in about 5 minutes. Remove from pan. Add remaining oil to the pan and cook patties, turning once on each side. Transfer to baking sheet, top with remaining cheese, and bake for four minutes. Serve on buns.

Nutrition: Four servings, each providing 386 calories, including 125 fat calories from 13 grams fat; 5.5 grams saturated fat; 597 mg sodium; 51 grams carbohydrates; 12 grams fiber; 1 gram sugar; and 19 grams protein. This is a high-fiber and high-protein burger with mushroom and lentil nutrients.

22.12.08 A Multigrain Heart-Healthy Toss of Sweet Potato and Edamame

Ingredients: ½ cup shelled edamame, ¼ cup chopped onion, 1 cup mixed greens, 1 oz feta cheese, 1 medium sweet potato, salt and pepper, and 1 TBL vinegar.

Preparation: Boil edamame in water for 10 minutes. Drain and set aside. Heat 1 TBL vegetable oil in a pan; sauté onion and greens. Spread and flatten greens and cook at low heat. Add cheese and bind greens in bite size bits. Add chopped sweet potato, edamame, salt and pepper, and vinegar. Toss and mix. Serve warm.

Nutrition: Two servings each provide 200 calories, including 36 fat calories from 4 grams fat; 95 mg sodium; 47 grams carbohydrates; 1 gram fiber; and 3 grams protein.

22.12.09 A Lunch of Couscous with Chickpeas, Dried Fruit, and Cilantro

Ingredients: ½ cup Water, ¼ tsp ground allspice, 1 cup orange juice, ½ tsp salt, 1 cup whole-wheat couscous, 3 TBL olive oil, 1 sliced medium onion, 1 sliced green bell pepper, 3 cloves of minced garlic, 1 tsp curry powder, 1 can chickpeas, 15 oz sliced dried apricots, 1 and ½ cup sweetened cranberries, and 1 TBL cilantro.

Preparation: Combine water, allspice, half of the orange juice, and half of the salt in a saucepan. Heat at medium high and bring to boil. Add couscous and cook for 5 minutes. Remove from heat and let stand for 5 minutes.

Heat oil in a large nonstick frying pan over medium heat. Add onion, pepper, garlic, and curry powder. Cook while stirring 10 minutes. Add chickpeas, apricots, and cranberries, and cook while stirring. Pour in remaining orange juice and cook. Add cilantro and remaining salt. Serve over couscous.

Nutrition: Four servings of high-fiber and high-protein preparation, each of which provide 363 calories, including 76.5 fat calories from 8.5 grams fat, including 1 gram saturated fat; 320 mg sodium; 67 grams carbohydrates; 9 grams fiber; and 9 grams protein. A great lunch dish.

22.12.10 A Great Dinner of Cheesy Broccoli and Rice Casserole

Ingredients: Cooking spray, salt, 2.5 tsp olive oil, 1.25 cup quick-cooking brown rice, 4 cups vegetable or chicken broth, 1 can (12.5 oz) low-fat evaporated milk, 1 lb broccoli, one minced onion, 2 minced garlic cloves, ⅔ cup shredded cheddar cheese, a pinch of cayenne pepper, ¼ tsp dry mustard, pepper to taste, and 3 TBL Romano cheese.

Preparation: Coat 9 X 12-inch casserole with cooking spray. Boil a large pot of water 1 tsp salt. Heat 1 tsp olive oil in a large Dutch oven; add dry rice and sauté for two minutes. Add broth and evaporated milk and bring to boil. Simmer for 20 minutes until rice is tender. Add broccoli to boiling water in the pot, turn heat off, and let stand for 2 minutes.

Preheat oven to 400°F. Heat the remaining oil in a large skillet, add onion, and sauté for three minutes, then add garlic and broccoli and sauté for 2 minutes. Add broccoli-onion mixture to cooked rice; add cheddar cheese, cayenne pepper, dry mustard, and pepper. Pour the mixture in the casserole and sprinkle with Romano cheese. Bake for 15 minutes until bubbly.

Nutrition: 8 servings of 1 cup each provides 250 calories, including 55 fat calories from 6 grams fat containing 2.3 grams saturated fat but no trans fat; 10 mg cholesterol; 430 mg sodium; 38 grams total carbohydrates; 4 grams dietary fiber; 9 grams sugar; and 14 grams protein.

22.12.11 President Clinton's Herbed Quinoa with Green Onions

Ingredients: 1 cup quinoa, 2 cups vegetable stock, ½ cup disked cucumber, ½ cup diced tomatoes, 2 TBL diced red onion, 2 green onions finely sliced, 2 TBL chopped cilantro, 1 tsp chopped jalapeño, 3 TBL extra-virgin olive oil, and 1 TBL fresh lemon juice.

Preparation: Cook 1 cup of quinoa in 2 cups of vegetable stock, combine cooked quinoa with all other ingredients, toss well, and serve chilled.

Nutrition: Four cholesterol-free servings, each providing 268 calories; 34 grams carbohydrates; 13 grams fat including 2 grams saturated fat; 483 mg sodium; and 6 grams protein.

22.12.12 Brain Food for All: Hot Chocolate with Cinnamon, Peppermint, and Chili Powder: Take 2 cups of cocoa (hot chocolate) per day for enhanced memory (Harvard University, Farzaneh A. Sorond); it promotes blood flow to the brain. Cinnamon moderates blood pressure, and chili powder controls arthritis, headache, and insulin balance. Peppermint is a digestive aid.

22.12.13 Food for Eye Health and Stress Reduction: A butternut squash dish with rice contains three tablespoons olive oil, two garlic cloves, one small onion, one cup rice, one small butternut squash, one teaspoon paprika, one tablespoon maple syrup, four sage leaves, and one-half cup walnuts. To prepare the concoction, set oven to 400°F. In a saucepan, stir-fry garlic and onion to brownness in olive oil. Add rice, two cups water, salt, and pepper, and bring to boil. Simmer at low; stir-fry squash in oil. Add paprika, maple syrup, and salt, and roast for twenty minutes. Add walnuts to the cooked rice. The recipe adds the benefits of squash and walnuts to a common rice dish.

Nutrition: The recipe provides six servings, each providing 309 calories, thirteen grams fat, 45 g carbohydrates, 786 mg sodium, and three grams fiber. This is a rather high-sodium preparation.

Genes underlie all that happens in our body. A challenge in thought and action, a thought of a childhood food, a flavor appeal, memory of a fancy restaurant, or watching our spouse do the cooking can trigger gene expression. How many genes get lined for making in time and just in time delivery right after we think of eating or smell food is not precisely known but it does happen. We need all necessary nutrients or their metabolites for the right expression all through our lives for good disease-free health of body and mind.

Therefore we need methylation or gene-expression diets to (1) run our body's nano-factory with oxygen and glucose properly and efficiently with antioxidants, adaptogens, and anti-inflammatory foods; (2) maintain our second genome with probiotic foods like yogurt, cheese, sauerkraut, kimchi, and other fermented foods; and (3) maintain our second brain—the enteric nervous system—by consuming restricted calories and exercising appetite control by high-fiber (prebiotic) and high-protein foods.

22.13 Major Reference Institutions

The following institutions conduct research on food genomics, food gene interactions.

Institute of Health Metrics, University of Washington; Norwegian University of Science and Technology; Gene Expression Center of University of Wisconsin, Madison, Wisconsin; University of California, Davis; and Case Western Reserve University.

It is easy these days to use the Internet at home or the local public library and find out what great chefs are doing, preparing, and presenting. One can then concoct his or her own methylation diet from common food ingredients.. Well-known chefs such as Pierre Cognaire (French) and Ferran Adria can be our teachers for free. I add below a more extensive list of great chefs precisely for the purpose of our readers willing to google for their methylation dishes. These chefs are better disposed to come up with good dishes.

Heston Blumenthal, Fat Duck, London, UK	Thomas Keller, Yountville, California
Jose Rocha, Exec Chef, Hyatt Acapulco, MX	Jonathan Benn (NY), the rstaurant per se of Thomas Keller
Gary Farrell and Nigel Keen, Australia	Samuel Arnold, American Indian Mexican Cooking
Sanjeeva Kapoor and Neeraj Katyal, India	Paul McCabe, Exec Chef, Enchantment Resort, Sedona, Arizona
Antonio Carloccio, Italy	David Paulstich, Exec Chef, The Mark, New York
Florian Trento and Christof Syre, Hong Kong	Sandro Gamba, Chef de Cuisine, Park Hyatt, Chicago, Illinois
Christof Syre, Exec Chef, Reagent, Hong Kong	The Reagent, Wall Street, New York; Vietnamese food
Stephen Schreiber, Exec Chef, Hilton Hotel, Tokyo	Damon Baehrel, Earlton, New York
Patrick Lanes, Landmark, Bangkok	Cesar Ramirez, Brooklyn, New York

There is a global trend for nutrient-rich recipes, and a great chef knows all about it. One can access the best of culinary developments from all over the world and make a delightful methylation diet. The emphasis should always be on smoothie-like beverages based on fruits, and snacks based on high-protein dips (purees of seeds and nuts) and vegetables. Calories need to be restricted dish by dish- the breakfast, lunch, snack, dinner, beverages, and dessert.

CHAPTER 23

Summation

"Food and lifestyle choices can postpone death but not defeat it."
—Triveni P. Shukla

A half-million-years-long evolution has gifted us with a body with the best possible brain it could design in terms of its genetic construct.[599] Also, evolution made our second brain—the enteric nervous system and probiotic-based second genome in the colon, critically complementary to human immunity, metabolism, and physiology.[600,601] It used a system of seven octillion atoms and molecules for molecular coding, decoding, and patterns of recognition, all residing in DNA that lives in twenty-three pairs of chromosomes in the nucleus of every cell in our body. Irrespective of their type and function, all cells are endowed with the same DNA. They are programmed to die and regenerate at different times in our lifetime: nerve cells last for our entire life, certain immune cells regenerate every day, white blood cells regenerate weekly, red blood cells regenerate once every 120 days, and bone cells can last for years. With 10^{15} receptors and 100 billion neurons, our cells are collectively good at giving us a good life.

The cells have 25,000 to 35,000 genes that can code 173,000 proteins. Close to 10,000 of these genes are repaired each day with little or no error. Even the underlying processes and their efficiencies are encoded in our DNA, meaning human DNA has genes that repair it. Energy production from

---

599 http://www.ncbi.nlm.nih.gov/pmc/articles/PMC2715140/.
600 http://neurosciencestuff.tumblr.com/post/38271759345/gut-instincts-the-secrets-of-your-second-brain.
601 http://www.ncbi.nlm.nih.gov/pubmed/22703178.

glucose in the human body at 25 percent efficiency is much better than the 15 percent efficiency of our car. What an evolutionary construct!

All that happens in our body with oxygen, water, and nanoscale nutrients that enter our bloodstream during digestion and absorption. Oxygen reduces itself to water while it oxidizes glucose into energy in each of our cells, albeit at different rates. The work of half a liter of oxygen consumed per day by the human body depends on a huge number—1.29 X 10^{21}—of protein molecules involved in metabolic reactions, immune defense, and neurotransmissions. All 20 million stimuli from the brain running at 150 miles an hour require speedy proteins.[602] A mere one billionth of a gram of hormone or neurotransmitter in a 65,000-gram human body miraculously does it all. Our longevity rests on the success of these fast, and sometimes furious, events, mediated always by proteins.[603] When ubiquitin fails to do its job in generating telomere, we begin to age faster, and longevity begins to fade away[604] with the decay of mitochondrial activity. No doubt our life and health depend on a perfect set of uncorrupted genes. A perfect set of genes demands good choices of food and lifestyle, without which prolonging longevity and even partially defeating death become impossible. What is possible, as long as oxygen and water are there for our body, the temple, is to sustain and regulate itself with appropriate molecules of food that keep our genome stable, the epigenome free of any corruption, the second genome maximally functional, and the second brain fit for ideal digestion of what we eat.

This book is about you, the conscious you. It is about your body, with the most highly developed brain that Darwin's evolution could afford. It is about humans as an *assembly of evolved molecules* and about the human body as a nanoscale molecular factory designed to replicate and reproduce with a relatively error-proof defense system, a human body that runs an information-processing network that connects to all the 210 varieties of 10 trillion cells with no more than 250 millivolt, 100 watt/hour electrical power. It is about a human body that knows how to exchange matter and energy at a speed known only to electrons and protons. It is about a human body that, as a universe of electrons, protons, atoms, and molecules, takes in another universe of electrons, protons, atoms, and molecules in a single breath. The

---

602 http://www.che.utexas.edu/research/biomat/PDFReprints/JDrugDelSciTech_16_11-18(2006).pdf.

603 http://www.sciencedaily.com/releases/2013/09/130930152741.htm.

604 http://www.ncbi.nlm.nih.gov/pubmed/21324320.

energy that our body depends on comes from the vibrational bond energy of molecules we eat. *In effect, then, the energy of vibration is life.* We hear by vibration, our cell receptors sense by vibration, and the villi in our intestine maximize absorption by vibration. The ear converts sound vibrations to electrical signals, the retina deals with transitions of electrons with each absorption of photons, protein receptors on cell surfaces vibrate for specific topology, and the vibrational energetics of contraction of heart muscles is vital to circulation of the blood.

More specifically this book is about how the human body manages molecules of oxygen, water, and foods we eat to be what it is and about the speed at which it maintains its being. This book is about the dynamics of interaction of food molecules in sustaining as complex a system as our body. The body begins as a fetus with DNA from the two parents, but then it grows and develops by expression of genes (the little DNAs) controlled largely by the epigenome. All this depends on the foods the body gets, whether from the mother's blood and milk or from foods and beverages we consume later in our life, and the environments that it is subjected to. Therapeutics by food thus involves mechanisms unrelated to DNA itself, but it does involve the biology that DNA is designed to control.

23.01 The Human Genome and DNA

The human body contains 600 muscles for movement and motor actions. Muscles of the heart, bladder, and intestine have unique functions. Each of the 210 types of cells in our body makes 173,000 proteins per day. This ability comes from a 2.5-nanometer (one billionth of a meter) DNA in the genome. This is possible because the genome, packed in a protein globule, has a trillion times more information than any computer known today. Maybe the quantum computers will match this capacity. The DNA, with all twenty-three pairs of chromosomes, has 6 billion base pairs and some 25,000 to 35,000 genes capable of coding 75,000 enzymes. Enzyme-catalyzed reactions that occur at a given time in the human body are incalculable. Electrical and chemical processes control mechanisms are there in all eleven systems of our body, with ten to 150 impulses per second in a circuit diagram larger than today's global telephone network.[605] This is part of the grand organization.

---

605 http://www.nimh.nih.gov/health/educational-resources/brain-basics/brain-basics.shtml.

What matters to the network are electrons, ions, photons that impinge on our retina, and special proteins.

All 2.5 trillion red blood cells are busy carrying four oxygen molecules on the backs of each of the 280 million hemoglobin molecules in them.[606] This is a grand oxygen-carrying strategy designed by evolution. The red blood cells are without DNA and thus are exempt from routine corruption by bacteria and viruses. The remaining 7.5 trillion cells have nuclei with genome-containing DNA in all twenty-three pairs of chromosomes. Human beings evolved with a proteome of protein molecules for a maximally developed brain and a reliable immune system. The DNA on all twenty-three pairs of chromosomes has 6 billion base pairs on its genes. Unravel your DNA, and it would stretch from here to the moon.

How 25,000 genes encode 173,000 proteins is a dilemma. ATP, the grand energy molecule for all that lives, as a molecular motor generates a circular motion along the double helix of DNA to make it click. The DNA has evolved to like and love this dance by ATP. Gene expression, which underlies all metabolism, homeostasis, and health, follows. Some 2,831 genes are involved in inflammation processes of genetic decoding and protein making[607] in relation to various diseases. This busy schedule causes ten thousand changes in DNA of each cell to be repaired daily by foods that we eat. Such foods must be devoid of chemicals and toxins responsible for undesirable methylation and epigenetic effects.

23.02 The Human Body: An Orchestrated Ensemble of Molecules

Bacteria came 3.5 billion years ago, then came plants some 3 billion years ago, followed by a conscious human being with a developed mind of free will coming rather recently, only 0.5 million years ago. There are common denominators in the DNA of living plants and animals. The current human DNA, it stands to reason, is fashioned from food we have been eating over half a million years. It is kept stable by the foods in our lifetime, and its products, the genes, are expressed rightly only with the right nutrients and under the right environmental conditions of additives, chemicals, toxins, and lifestyle, inclusive of exercise, meditation, faith, and spirituality.

---

606 http://faculty.etsu.edu/currie/hemoglobin.htm.
607 http://www.nature.com/srep/2013/131205/srep03426/full/srep03426.html.

The human body grows as a fetus from a zygote by nutrition from the mother's blood and then at least for a few months from her milk after birth. For the rest of life, a human being develops and grows on the basis of what he or she consumes and digests. Daily diet, therefore, makes us into what and who we are; it supports all other systems in the human body: skin, muscles, skeleton, brain and neurons, endocrine glands, heart and circulatory system, respiratory system, immune and lymphatic system, digestive system, excretory system, and finally the reproductive system, which prepares us for the next generation. Excretory system, I thought, should be listed here in summary.

Ten percent of our weight is due to bones, all 206 of them. Replaced completely every three months, human bone is five times stronger than mild steel. The smallest bone is in the middle ear, about 25 mm long. We need twelve hours to digest food—actually much longer considering digestion by probiotics in the appendix. Nutrients go to the blood and finally to every cell in the body. The blood returns, but only after the kidneys have done their hard work of balancing and purifying it. Waste removal and electrolyte balance are critical to the body. It's no wonder that kidneys are designed to filter 1.5 gallons of blood every forty-five minutes, or forty-five to forty-eight gallons per day. Daily food is the basis for all these works of growth, repair, maintenance, and waste disposal. A key food molecule, if missing, can have dramatic effect on our health. So what, how much, and when we eat matters in our daily biology.

23.03 Efficiency of Our Body

The human body does all this work with an efficiency of 25 percent, better than the 15 percent efficiency of a car.[608] Among plants, a cornfield is only 1.5 percent efficient in performing as a solar collector. An average 150-pound man spends 80 to 100 watts or 1,800 calories (kilocalories to be exact) per day just to stay alive. This translates to twenty-six calories per kilogram of body weight that is required to run the heart, lung, and brain and to maintain body temperature. The human body is rather frugal in spending energy even when we exercise; it stores it as fat if one consumed excess calories. This is why we gain weight. This efficiency and fat storage are not only encoded

---

608 http://retremblay.net/Text_book/2nd_edition_files/4.5.pdf.

in DNA,[609,610,611] but can be influenced by environment—say exercise and meditation.[612]

23.04 The Human Body as a Molecular Machine

The evolution of the human body and mind took place over five thousand years, and it was based on energy production by the oxidation of glucose and simultaneous reduction of oxygen to water by mitochondria in our cells. Molecules made from food are at the core of all experiences of the human body's organs of sensory perception. An army of such molecules includes 10^{15} receptors on 7.5 trillion cells. Receptors are everywhere—in the brain, eyes, ears, heart, and kidneys. The human body has to control and communicate incessantly with 100 billion motor and sensory neurons as the metabolism purposefully exchanges matter and energy involving electrons and protons. In theory, although we are far from knowing the details, all 7 octillion or 10^{27} atoms of our body are involved in making our life possible.

Protein molecules rule the human body chemistry. The nervous and immune systems are very well-coordinated. Mind and body are connected by chemicals, neurotransmitters, hormones, and receptors. Emotions are controlled by neuropeptides, giving feelings an ever-changing chemical and electrical dimension. Actually there are protein enzymes that can destroy, mend, and amend neurotransmitters; they can change our feelings.[613] The protein ribosome, as a 7,400 amino acid structure in our cells, is also common to lives of bacteria and plants as well. But we are human because of a much better equipped ribosome that transcribes DNA into protein. It can produce 2,000 proteins per second, or 173 million proteins per day per cell. All body cells combined can produce an astounding 1.29 X 10^{21} molecules per day if they have to.

The parasympathetic CNS (central nervous system) controls 25 percent of digestion, and GI hormones control the remaining 75 percent of it. Such a control system is in line with intermittent delivery of oxygen, water, and food molecules to the body. Our body gets in trouble below the 95 percent oxygen saturation level (percent of hemoglobulin bound to oxygen), just

609 http://www.ncbi.nlm.nih.gov/pmc/articles/PMC3287999/.
610 http://www.sciencedaily.com/releases/2008/05/080521171808.htm.
611 http://www.eurekalert.org/pub_releases/2008-05/gi-gsr052108.php.
612 http://www.sciencedaily.com/releases/2013/07/130703101344.htm.
613 http://www.ncbi.nlm.nih.gov/books/NBK21521/.

as it gets in trouble even with 2 percent dehydration below the optimum of 58 percent water comprising body mass. There can develop all sorts of digestive, cardiovascular, immune, and musculoskeletal problems if dehydration exceeds 10 percent. The body can get up to 10 percent water from metabolic exchange, but the rest has to come from foods and beverages. Low oxygen is worse than low water because a majority of life-sustaining metabolic reactions are oxidation-reduction reactions. Oxygen that we breathe in and oxygen that can come from the 58 percent water in the body have got to be at their optimum. The cells in the human body live with ATP, the molecular motor that uses hydrogen ions for its power. Food thus not only runs the body for building, growth, energy, antibodies, hormones, neurotransmitters, and receptors, but it also keeps the genome stable and the epigenome uncorrupted. We need to eat, exercise, and drink water to stay alive in good health.

23.05 The Timescale

The number of reactions that occur at a given time in the human body is incalculable. The brain can process 20 million stimuli per second. To repeat for emphasis, the tiny 2.5-millimeter DNA in the human genome, packed in a chromatin protein globule, helps each of 7.5 trillion cells make 173,000 proteins per day. All 7.5 trillion cells make an astounding 1.3×10^{21} proteins per day. Our nose has 10 million cells for detecting scent, and it has hairlike cilia that flick fifteen times each second in order to push bacteria and fungi away and out from the throat. The brain has a trillion times more information-processing power than any computer known today. There are 10,000 changes in DNA of our cells per day that are repaired daily. This repair and control depends on foods we eat. The details of the design and speed of events in our body are real, but their enormity is beyond our comprehension. The molecules of food we eat maintain this enormity. The electrons and ions rule, and the protons give us the power to live.

The retina of our eye has rhodopsin as a photoreceptor, and the change of rhodopsin to bathorhodopsin happens in a quadrillionth of a second involving a stretch of an unsaturated carbon-to-carbon bond. Our eyes blink ten times a minute in order to irrigate the entire eye with tears, with an average blink taking no more than 100 milliseconds. Beta-carotene, vitamins C and K, lutein, zeaxanthin, and omega-3 fatty acids are of great help for maintaining eye health. I am a living example of curing night blindness by consuming carrots for only three weeks. I wish my mother had managed to add sweet potatoes in my regular diet also. Carrots and sweet potato together would have done a better job.

The involuntary convulsion called sneezing is an automatic neurologic reflux involving high ATP production. The lower brain stem knows all about sneezing. Chest muscles compress, pushing a burst of air upward at a speed of eighty to one hundred miles per hour in order to eradicate incoming bacteria and viruses. We have no idea how many protons are involved in this extravagant but necessary expenditure of ATP.

The brain can process 20 million stimuli per second, each traveling at 150 miles an hour. The myelin around nerves helps accelerate such transmissions by a factor of fifty. We don't know how the sheath of proteins accelerates the process. May be there are built-in transient electronics in the myelin. We know there is a transmembrane potential.[614] Vasodilator nitric oxide, the actual messenger behind nitroglycerin and Viagra, has a half-life of only ten seconds, but it can communicate from cell to cell as a tiny one-nanometer hormone at the speed of modern e-mail. This gas is critical to our nervous, respiratory, circulatory, immune, and reproductive systems. We don't inhale it, but endothelial cells of our body know when and how much of it to produce from amino acid arginine on very short notice when we begin to exercise and run.

Most ultrahigh speed works in our body use no more than a billionth of a gram of hormones or neurotransmitter. But that is high traffic among the 1.29×10^{27} atoms and molecules in our body. Just to make the point clearer, let us take a kitchen example. Bromelain, a protein enzyme from pineapple in our kitchen, can locate, secure, chop, and tenderize 30,000 proteins per second in a piece of steak right in our kitchen. That is exactly what happens in our stomach. The incredible speed at which our body operates depends on our daily food. Food selection, therefore, is critical for sufficiency and balance.

23.06 Day-to-Day Survival

The day-to-day survival of human body systems depends on the nervous and endocrine system. We must note that the digestive, endocrine, and circulatory systems are closely interdependent. Furthermore, the circulatory system is connected to the respiratory and lymphatic systems. As a matter of fact,

---

614 http://stbb.nichd.nih.gov/pdf/sti_mye_ner91.pdf.

the lymphatic and immune systems can't function without the circulatory system. It is wise to examine the dependency of these critical systems in terms of foods we eat because this dependency is critical to maintaining our vital signs and everyday health. The gatekeeping recognition experts are the membranes enveloping our cells, including the nuclear membrane that houses all our genes. The energy-producing mitochondria have their own molecular mysteries.

23.07 The Lung and Heart for Oxygen-Based Energy Production

We need our daily oxygen. Our body consumes six liters of air or 0.443 liter of pure oxygen per day. It knows how to dispose of more than half a gram of dirt that comes with air every day. The lung has a surface area of seventy square meters. Just imagine that one breath takes in as many molecules of air as there are stars in all our galaxies. Red blood cells, 2.5 trillion of them with 280 million hemoglobin molecules, are meant to carry 1.12 billion oxygen molecules to the remaining 7.5 trillion body cells. Carbon dioxide and nitric oxide are equally important, and hemoglobin carries them also. We breathe out 1.5 kilograms of carbon dioxide per day; there is 3.96 percent carbon dioxide in each exhalation. The circulating blood still can have up to half a gram of carbon dioxide as carbonic acid, which is used to neutralize stomach acid and help maintain acidity. Our life depends on three gases— oxygen, carbon dioxide, and nitric oxide—all transported by hemoglobin. Food molecules that bring in iron become an almost obligatory daily food.

The heart, with its unique muscular constitution capable of squirting blood up to thirty feet, beats ten thousand times a day and pumps two thousand gallons of blood by its own electricity. Blood circulation is fast enough to allow every drop of blood to pass through the heart once every minute, in line with the oxygen need of all cells in the body. The heart muscles are striated for continued rhythmic self-contraction. Low-cholesterol foods, which improve blood-flow rate and prevent atherosclerosis along with anti-inflammatory vitamin D and omega-3 fatty acids, are good for the heart. High-dietary fiber foods lower cholesterol for heart health.

23.08 Command and Control by the Brain

Twenty-five percent of all oxygen and 10 percent of glucose is used by the brain, the most energy-intensive organ in our body. The left side of the body is controlled by the right brain and vice versa. Neurons are largely minerals, cholesterol, and phospholipids capable of electrical excitation. Sensory and motor neurons are more directly controlled genetically, and they make a lot of proteins and small neuropeptides known as miniproteins.

Chemicals and protein receptors are the basis of the mind-body connection. Most control comes from our unconscious mind. We see a world model only as our brain interprets it. Feeling is both electrical and chemical. Neurotransmitters can be modulated by enzymes, and emotions are controlled by neuropeptides. Whereas steroidal hormones work with genetic material DNA, protein and peptide hormones work with the cellular receptors. Unlike the ultrafast nervous system, the endocrine system works slower and lasts longer. Neurons and the immune system are well-coordinated. The foods we eat are involved every step of the way in all kinds of coordination and communication in our body. Let us look at some details.

1. Neurons with long axons carry information along long neural pathways directly between the brain and spinal cord or between different regions of the brain. There is no alteration in the transmitted information.

2. Multineuronal or multisynaptic pathways are made up of many neurons or synapses. Such pathways can integrate new information into the transmitted information. We need to pay special attention to the balance between the sympathetic and parasympathetic peripheral systems: the sympathetic for stress control, and the parasympathetic for control of exercise and digestion. Let us note that evolution has designed us for exercise; otherwise there would be no need for its control by the parasympathetic nervous system.

3. The central nervous system and immune system work together via hormone and neuropeptide interrelationships.[615,616] Nerve cells talk to immune cells, and food influences the conversation.

23.09 Cellular and Molecular Defense in Our Body

Different types of cells of our body are tabulated in Table 23.01. This book has been emphatic all along that we are creatures of oxygen and that evolution

---

615 http://thyroid.about.com/library/immune/blimm28.htm.

616 http://www.niaid.nih.gov/topics/immunesystem/pages/nervoussystem.aspx.

has so accommodated us over thousands of years. It is no wonder that 2.5 trillion tiny, six- to eight-micrometer red blood cells that circulate in our blood are dedicated to oxygen transport to all the other 7.5 trillion cells every twenty seconds. Oxygen transport is their main function.

Table 23.01 Characteristics of Human Blood Cells

Cell Type	Diameter (microns)	% in Blood	Target	Lifetime
Red blood cells	< 8.50	90% of blood	Transport oxygen	120 days
		% of WBC		
Neutrophiles, first responders against bacteria and viruses.	< 12	62.00%	Bacteria & fungi	6 hours to few days
Eosinophiles against asthma, hay fever, hives	< 12	2.30%	Parasites & allergies	8–12 days
Basophiles (T cells, B cells)	< 15	0.40	Inflammation	A few hours
Lymphocytes, antibody-producing B cells, T Cells	7–15	30.00	Adaptive & regulatory response, NK cells	Weeks to years
Monocytes & denditric cells	12–70	5.30	Can become microphages	Hours to days
Platelets; vitamin K plays key role in clotting	2–3		Produce growth factors; affect clotting function	A few days

The red blood cells don't have mitochondria and do not produce ATP by the electron-transport mechanism of the Kreb cycle; they make energy by glycolysis aenerobically. They have a relatively longer life of 120 days dedicated to the purpose of oxygen transport because evolutionary design found it useful and necessary. This relatively long life is by evolutionary design. Red blood cells are designed without DNA, nucleus, and other cellular components just in order to make room for as many hemoglobin molecules as possible for carrying as much oxygen as possible.[617] All else in the human body depends on red blood cells, but only when connected to the information-carrying longest-living neurons or nerve cells.[618] Homeostasis requires constant communication between the brain, spinal cord, and nervous system, and the grand infrastructure of the circulatory system. Whereas there are 2.5 trillion red blood cells, the number of white blood cells is only 5 million (see table 23.01).

White blood cells, some 9,000 of our sanitary engineers in every microliter of blood, are cells with nuclei; they can glide on and around walls of tissues, go out of the bloodstream, and chew and eat foreign bacteria and fungi in our body's defense. High-protein foods of balanced amino acids are a must for white blood cells and our immune system.

23.10 The Power of Perception

The retina is the best television set, and our eyes are photon detectors. No wonder it uses considerable energy to do its job.[619] It is in constant vibration for signal transfer to the brain.[620] Stimulation of opsin—for example, rhodopsin—by photons takes place in a billionth of a second.[621] Exposed to the external world, eyes have to have their own immune system.[622]

The complex circuitry of human brain processes color of things we see at thirty different places, and a given region participates in a variety of ways. Brain cells, the neurons, are unique in connecting thought, emotion, and attitude; they are connected with the world of energy and the energy within us. Thoughts as energy and consciousness are busy constructing reality.[623] Consciousness is the sum total of perceptions by all our sense organs capable of deciphering the grand holographic view of the universe.[624]

Extremely well-coordinated with the brain, our eyes are the most

---

617 http://www.pbs.org/wgbh/nova/genome/dna_sans.html.

618 http://www.mrsdupar.com/teachers/specializedcells.pdf.

619 httphttp://serinet.meei.harvard.edu/faculty/peli/papers/Garcia-Perez_JSV_2003.pdf.//www.fi.edu/learn/brain/carbs.html.

620 http://www.perceptionweb.com/abstract.cgi?id=v020413.

621 http://www.pnas.org/content/72/1/381.full.pdf.

622 http://www.ncbi.nlm.nih.gov/pmc/articles/PMC2948372/.

623 http://www.perceptionweb.com/abstract.cgi?id=v020413.

624 http://www.marxists.org/reference/archive/spirkin/works/dialectical-materialism/ch03.html.

developed sense organ. They monitor motion, contrast, quality, and color on a nano-timescale. Color is a brain response, and we learn to define ourselves *with and in color* because we can color our mood and personality. We can color our karma and our day. Photoreceptors on the retina collect light and send it to the neurons; the neurons convert light to electrical impulses that reach the brain, which processes it all into recognizable and meaningful information—an information pattern. The brain collects information from both eyes using only the mild part of the sun's radiation and puts it all together in a 3-D format. Our color preferences change as we grow and as we learn to improvise the best output along with other senses. Behind the color of foods we eat are the molecules that can manage cancer, diabetes, and eye health. Food color can help us manage appetite.

Hearing is our second most advanced sense organ. Sound waves cause the tympanic membrane (eardrum) to vibrate. Vibrations go to the nerves through vibrating small bones. The three bones in the ear (malleus, incus, and stapes) pass these vibrations on to the cochlea, a snail-shaped, fluid-filled structure in the inner ear. Inside the cochlea is another structure, called the organ of Corti. Hair cells are located on the basilar membrane of the cochlea. The cilia of the hair cells make contact with another membrane, called the tectorial membrane. When the hair-cell proteins are excited by vibrations, a nerve impulse is generated in the auditory nerve. These electrical impulses are then sent to the temporal lobe of the brain. The question "Do daily foods affect the travel of this vibrational energy?" is not answered very well, and neither is loss of hearing as we gain longevity. Humans can hear sound waves with frequencies between 20 and 20,000 Hz. At a temperature of 68°F (20°C), sound travels at 1,125 feet/sec (343 meters/sec). This amounts to traveling at 756 miles/hr or 1,217 km/hr. Our ear can sense one one-hundredth of PSI wave difference. We sense more by hearing than what we generate by speaking. Our sense of smell is critical to the quality of breathing and selecting the quality of food we eat. What we breathe and eat daily is critical to our sensory existence, including hearing.[625]

Vapor molecules reaching the nose through the mucous membrane reach receptors. These receptors for camphor, musk, flower, mint, ether, acrid, and putrid smells connect to the olfactory nerve, which transmits signals to the brain. We know from daily experience that dogs have a better

625 http://www.audicus.com/blogs/hearing-aids-blog/6071762-you-hear-what-you-eat-5-foods-that-can-prevent-hearing-loss-and-hearing-aids.

sense of smell than humans. Although our senses of smell and taste often go together, the sense of smell is ten thousand times more sensitive than that of taste.

We have taste buds on our tongue. The top senses salt and sour, the tip senses sweet, and the back of the tongue senses bitter. The umami taste of amino acids of food proteins can also be sensed. Buds signal taste to nerves and then to the brain, which coordinates taste and smell. We need, therefore, to maintain the health of our nose and tongue for selective and beneficial breathing and eating for good nutrition and health. An ordinary influenza tells us a lot about this relationship. We lose our taste, and we lose our sense of what we eat.

Skin is embedded with nerve endings that can sense cold, hot, contact, and pain. The hairs on the skin magnify sensitivity. Fingertips and private parts have maximum sensitivity due to endocrine stimulation. Others can tickle you, but you can't tickle yourself.

The key to perception is coordination of senses. We can detect pressure, pain, and temperature by coordinating more than one sense. Balance is managed by the eye, gravity sensors, and stretch sensors of the muscles, joints and skin, and ears. Without this balance we get dizzy. Muscle and joint movements help us walk and talk.

We are electromagnetic beings, and we manage vibration energy very well. Very few of us, less than 0.1 percent, have the ability to associate sound and color or shape and smell. Sound is the energy of vibration. Light is electromagnetic energy. Pressure and temperature are other forms of energy. Smell molecules have their own chemical energy. Our sense organs, in coordination with the brain, manage various energy forms in order to help us observe the world we live in. They are the basis for our daily observation and learning. We respond by our organs of action: feet, hands, reproductive organs, and speech apparatus. Speech can represent music, anger, surprise, emotion, and other tones in between. I am convinced that our ability to perceive is tied to the foods we eat. To eat, therefore, is to perceive and manage various forms of energy to stay alive and well. Our senses of smell, taste, and even hearing are synchronizable; they are very relevant to food choice.

23.11 Food Conversion to Nanoscale Nutrients

Unlike voluntary skeletal muscles, smooth and proteinaceous muscles of the digestive tract are involuntary, like the heart muscle. Digestion, the key to functioning of all other body systems, is involuntary, without conscious control. The details of digestion are intriguing in that nutrient absorption uses a surface area of two hundred square meters, high-calcium foods empty slowly, stem cells of six million microvilli in the ileum are busy regenerating themselves second by second, zinc-based lysozyme is there to sanitize beyond acid sterilization in the stomach, and absorption of vitamins B_{12} and K takes place in the colon, not any earlier in the intestine. People from different cultures have the same physical brain but differently conditioned minds for foods that digest differently. It is well-established that people of different cultures have varying propensity to epigenome-based diseases,[626] but the molecular mysteries remain unknown.

23.12 Common Foods That We Must Consume

Common foods have all the nutrients we need to live in good health. Trace minerals, major minerals, water-soluble vitamins, and fat-soluble vitamins are found in everyday common foods (see the tables in chapter 22). A good combination of foods can be full of antioxidants and phytochemicals for maintaining eye, brain, gut, heart, and kidney health; foods can even manage allergies, pain, and stress. A complete compilation of fiber foods, fruits, vegetables, grains and nuts, and fish is available in appendixes I through V. Daily foods if properly selected can help avoid use of food supplements like unnecessary vitamins and L-carnitine, excessive amounts of calcium, and unproven chemicals. Meta-analysis of available data shows that the value of artificial vitamin supplements not part of daily food items is weak and questionable./[627]

The brain is all fat, made of omega-3 fatty acids. Protein is second only to water in our daily requirements, a must for our immune system. We need to select foods for protein and good plant-based fat. The often promoted no-fat diets are bad advice. Water balance in our daily life is a must, and that is why we have the thirst signal.

As repeatedly pointed out in this book, necessary gene-expression nutrients are beta-carotene, vitamin A, vitamin B complex including B_6, folic acid, B_{12}, and vitamins C, D, E, and K, mineral antioxidants like

---

626 http://groups.anthropology.northwestern.edu/lhbr/kuzawa_web_files/pdfs/Kuzawa%20and%20Sweet%20AJHB%20early%20view.pdf.
627 http://www.bmj.com/content/330/7496/871.

selenium, omega-3 fatty acids, and trace minerals boron, copper, chromium, magnesium, and zinc. Common foods that can provide all such nutrients are almonds, green beans, beans, Brazil nuts, broccoli, cheese, egg whites, egg yolk for choline, flaxseed, ginger, kale, nuts, pineapple, papaya, pumpkin seed, red bell pepper, sesame seed, spinach for riboflavin and magnesium, soy foods, sweet potatoes, sprouts, sunflower seeds, tomatoes, trout, Swiss chord, salmon, walnuts, and wasabi, which delivers isothiocyanates. The collectivity of these common foods makes a methylation diet.

Most important of all, we need symbiotic bacteria in our colon in order to maintain a proper ratio of anti- and pro-inflammatory cytokines. Also, they provide us with vitamins and short-chain fatty acids, talk to our immune cells, and help maintain homeostasis and immunity. We need to avoid bad bacteria and boost the population of good ones by consuming yogurt, kefir, kimchi, and sauerkraut and strengthen our second genome.

23.13 Malpractice of Food Science and Nutrition

There is no doubt that the post-World War II period has witnessed malpractice of food science and technology. We invented hydrogenated trans fat and then high omega-6 fatty acids and commercialized them widely without sufficient clinical trials, not to the extent we do today. The very idea of cereal milling and throwing away nutrients in bran and germ has been counterproductive to human health. These blind innovations caused serious public health problems. Food became plentiful with chemicals that corrupted our epigenome, and to add more insult to our body, we began overeating more or less in a culturally and socially default mode. Actually we are now accustomed to eating too much refined sugar and high glycemic carbohydrates. Food additives and hazardous chemical contaminants as part of our drive for mass food production and processing caused considerable collateral damage to public health. Common consequences we face today are overweight conditions and obesity among children and adults alike. These effects, it is widely believed, are epigenetic because the genes simply don't change that fast. Good foods can manage protons in our stomach at almost no extra cost, for which we consume $2.2 billion worth of antacids. Day-to-day eating behavior is affecting the brains of 25 percent of college students who are now on pills against attention-deficit disorder.[628] Food and

628 http://totallyadd.com/adhd-medication-abuse-amongst-college-students/.

its contaminants have changed gene expression in a majority of Americans, and junk food has now become a national problem.[629] We need to reverse this trend by preparing our restricted calorie daily dish with common foods and not exotic ones talked about in the nmedia. Managing eating, manging daily sleep, and following dictates of evolution's diurnal design of our body can balance our daily physiology. Daily exercise that we did for centuries can normaluize our metabolism and meditation and mindfulness can help us connect our body and mind. Things go wrong when mind is not well connected to the body.

23.14 Special Foods in Retail Stores

Retail stores are now proactive with bulk sales of tree nuts, seeds of pumpkin and sunflower, rice bran, flaxseed meal, and probiotic yogurt and other fermented foods. They are available in megastores and specialty health food stores alike. Foods of the old cultures of India, China, and Japan are becoming popular. Antiallergy 82 percent dark chocolate is a famous daily treat now. Adaptogenic food items like India's aswagandha, China's astragalus, Korea's ginseng, and gugal from India have now crept into the common vernacular of diet and nutrition in the United States. The sales of spices has almost quadrupled during the last ten years. These are good indicators of practical food therapeutics right in the American kitchen, and to some extent, even in fast-food establishments and dining places.

23.15 Essential Molecules of Food

Fruits, vegetables, roots, herbs, and spices with antioxidants, minerals, and vitamins are what is implied by old-fashioned good foods that carry good molecules of food. They have been part of our diet for centuries. It is a physiological fact that we can't live long without protein and fat, a fact that most diet programs fail to emphasize upfront. They fail to tell us that our DNA has evolved with these common foods.

Brain Foods: The methylation diet provides brain foods. Foods for brain cells include avocadoes, beans, berries, cauliflower, coffee, dark chocolate, flaxseed, grapes, lentils, mustard greens, sesame seeds, sweet potatoes, spinach, and vitamin D from sunlight. Vitamin B_{12} deals with DNA synthesis,

---

629 http://www.scientificamerican.com/article.cfm?id=courchesne-gene-expression.

regulation of our cells, brain function, the nervous system, and blood formation. No one can afford to be deficient in vitamin B_{12}. Nutrient vitamins B_6, B_9 (folic acid), and B_{12}, vitamin E, carotenoids, flavonoids, antioxidants, magnesium, selenium, omega-3 fatty acids, tyrosine, phenylalanine, and tryptophan are provided by almonds, avocadoes, berries, broccoli, beets, chocolate, cinnamon, curry, lentil, pumpkin seed, pistachio, spinach, sunflower seed, sage, salmon, Brazil nuts, walnuts, and whole grains. These nutrients increase alertness, cognition, and mental health. They fight ADHD, schizophrenia, and depression.[630]

Foods for Immunity: Food for white blood cells include good protein, beta-carotene, folic acid, and B vitamins. Other immune-boosting foods are vitamin E, carotenoids, flavonoids, omega-3 fatty acids, zinc, and selenium. Common foods to eat then are barley, broccoli, crab, kiwi, oranges, green tea, salmon, mushrooms, carrots, garlic, spinach, sweet potatoes, oysters, and yogurt. Coenzyme Q_{10} from broccoli is a must to incorporate in our daily diet.[631]

DNA Methylation Nutrients: DNA methylation nutrients like vitamins B_2, C, K, and B_{12}, amino acids methionine, lecithin for neurotransmitter choline, and minerals copper and iron are all there if our daily foods contain a combination of asparagus, beer, eggs, pomegranate, soy foods, sesame seed, orange juice, beans, cereals, lentils, green vegetables, peanuts, nuts, and yeast. Such foods are good for red blood cells, genome stability, and upkeep of the epigenome.

Antistress Foods: B vitamins (B6, folic acid, B_{12}) and stress-controlling adaptogens like ginseng, astragalus (China), aswagandha (India), maitake mushrooms (Japan), vitamin C, coenzyme Q_{10}, and antioxidants in general improve the mind-body connection. Black pepper, cinnamon, and turmeric-like spices we use in preparing our daily dishes are considered to be adaptogens that modulate the mind-body connection. Selected foods for specific purposes include[632]:

---

630 http://www.cfwellness.com/health-topics/add-adhd.

631 http://www.health.harvard.edu/flu-resource-center/how-to-boost-your-immune-system.htm.

632 http://www.readersdigest.ca/food/healthy-food/5-anti-stress-foods.

1. Use foods to maintain HDL in a range of 40 to 60 mg/dl for heart and brain health.
2. Use good foods for magnesium and calcium (300–600 mg/day). Avoid too much calcium.
3. Use black pepper, cinnamon, and turmeric as readily available adaptogens for stress control.
4. Use omega-3 fatty acids from walnuts and flaxseed regularly.
5. Use the methylation diet for DNA stability and a full complement of 10,000 proteins made by our cells each day.
6. Use a variety of proteins. They are next to water in importance to our daily health.
7. Use broccoli for coenzyme Q_{10}. It improves the immune system.
8. Use vegetables as salads and a sumptuous dose of probiotics. A good proof is the Indian diet. People of Indian origin have more bacteria in their stool than people in the West.
9. Avoid L-carnitine as a food supplement.
10. Talk to your physician for specific tests, such as (1) hypocretin or orexin tests for diagnosis of narcolepsy-type sleep disorders; (2) amyloid proteins for diagnosis of Alzheimer's disease and dementia; (3) stress hormone cortisol, which affects DNA methylation; (4) sugar in the brain, which suppresses digestion and the immune system and causes sleep disorders, heart diseases, obesity, eczema, and other skin problems. In the near future, brain mapping may assist in diagnosing autism, anxiety, stress, and even depression.

23.16 Gene Expression

Our 25,000 to 35,000 genes have to be expressed in our cells to make proteins that make things happen in our body as antibodies for defense, enzymes to run our metabolism, receptors, and neurotransmitters. In order to make these proteins, DNA needs to be transcribed into messenger RNA, which has to be tailored and pruned before it can line up amino acids to make proteins on the back of another protein called ribosome. Proteins so translated from DNA often need modification to perform their specific functions. This entire protein-making machinery of gene expression is under tight control before a genotype human being becomes a phenotype human being. Foods affect DNA's expression, and thus our phenotypic being in good routine development and health. Diet, digestion, and the immune system are intricately connected, and diets of high carbohydrates make genes work overtime, causing inflammation, type 2 diabetes, cardiovascular diseases, cancer, and even dementia.[633,634] Key controllers of genetic susceptibility are nutrients we consume, like folic acid, B_{12}, other B vitamins, S-adenosyl methionine, choline, omega-3 fatty acids, and betaine (a specific ampiphilic zwiterion), Betaine is approved by the FDA for reducing homocysteine.[635] Betaine can also come from whole-wheat bread, rolled oats, quinoa, and other whole grains.

Choline is a must because it is the donor for the methyl group.[636] Folic acid and carbohydrates affect methylation indirectly. Dietary protein, fat, fatty acids, zinc, biotin, vitamin A, and vitamin C are also known to affect gene expression. Diet composition, total calorie intake, and food contaminants influence the methyl tagging of DNA, although genes have differential methylation patterns. Some of these tags can pass on to the next generation. *Most noticeable is the silencing of the tumor-suppressor gene, a phenomenon that causes cancer.*[637] A major industrial contaminant to food and water is bisphenol A, which decreases DNA methylation.[638]

Cancers are definitely under epigenetic control, and proper selection of foods for our daily diet can help. Nutrients and foods that supply them are relisted in tables 22.02, 22.03, and 22.07. The information makes it clear that some nutrients coming from similar common foods are clinically well-established remedies for pains, ailments, allergies, and chronic diseases. Just as suggested in chapter 22, we can make our selections of high-value foods and create new dishes, drinks, and snacks for healthful living and even enhance longevity.

"Can we monitor epigenetic changes in real time for disease diagnosis and prevention?" is the key question to be answered in the near future. The answer is yes and it becomes clear from summaries of health promoting nutrients in tables 23.02 and 23.03.

---

633 http://www.ivlproducts.com/Health-Library/Health-Concerns/Weight-Management/The-Relationship-Between-Food-and-Genes/.

634 http://www.independent.co.uk/life-style/health-and-families/cut-the-carbs-and-eat-frequent-meals-to-optimize-health-finds-study-2357956.html.

635 http://umm.edu/health/medical/altmed/supplement/betaine.

636 http://www.ncbi.nlm.nih.gov/pubmed/12163687.

637 http://www.ncbi.nlm.nih.gov/pubmed/18185590.

638 http://www.ncbi.nlm.nih.gov/pubmed/22699882.

Table 23.02 A Comparison of Health-Promoting Nutrients

Disease	Mechanism of Action	Curative Foods
Freedom from osteoporosis	Avoid inflammation by reducing prostaglandins and COX-2 activity. Vitamin B_6 prevents homocysteine formation, thus permitting proper collagen formation. Other helpful nutrients are B_6, folic acid, vitamin K, vitamin D, coenzyme Q_{10}, boron, calcium, selenium, magnesium, and zinc. Bosvellia extract (Bosvellic acid) switches off pro-inflammatory cytokines.	Green vegetables, roasted nut snacks, snacks of seeds of pumpkin and sunflower, and omega-3 fatty acids from flaxseed, walnuts, black pepper, turmeric, black currant, ginger, oranges, sweet potatoes, and prunes. Bosvellia extract (Bosvellic acid), black pepper, and turmeric are good food cures.
Pain-free life	Glutamine, Flavonoids (apricots, avocadoes),omega-3 fatty acids, vitamin D, glutathione, tryptophan; calcium-to-magnesium ratio of 600:300 mg per day; chromium, selenium, and zinc can help reduce pain. So can lack of anxiety and a stress-free life, meditation, yoga, and exercise.	High-glutamine foods (cabbage, beets, beef, chicken, fish, beans, and dairy products) contribute to gamma amino butyric acid—a pain inhibitory neurotransmitter. Trace minerals of chromium from garlic and onion, zinc from pumpkin seed, and selenium from Brazil nuts are involved in elastin formation.

Disease	Mechanism of Action	Curative Foods
Skin and beauty	Vitamin D is a beauty vitamin. Vitamins A, B complex, C, D, E, and K, choline, minerals copper, selenium, and zinc, and DMAE (dimethyl aminoethanol) are good for skin and overall health. The objective is to prevent DNA damage, reduce stress, and maintain the estrogen-progesterone ratio. Also, a good night's sleep is good for skin health.	Zinc from mustard and pumpkin seed, selenium and copper from sunflower seeds and Brazil nuts, and chromium from garlic and onion are great for skin health, and so are phytoceramides 3 and 6 and bromelain from pineapple. Reducing stress by exercise and meditation also helps. Genes p53 and Satb1 are involved in chromatin remodeling and gene expression.
Diabetes-free life	Control sugar intake. Consume omega-3 fatty acids, foods containing chromium, B vitamins, vitamins D, C, E, and K, chromium, selenium, zinc, and antioxidant vitamins C and E.	Betatropin hormone therapy is on the horizon. Glutamic acid decarboxylase encoded by two genes is seen by immune cells as an antigen; the T-cells with autoantibody begin to destroy beta cells. Gene therapy may become possible in this case.
Heart health	Foods containing natural statins (gugal in India, red rice yeast in Japan), coenzyme Q_{10}, phytosterols, omega-3 fatty acids, and prebiotic dietary fiber reduce risks of cardiovascular diseases. Prevention of dense small LDL particles and keeping homocysteine below 15 micromoles/liter is a medical need. Test it periodically.	Maintain the health of 2.5 trillion red blood cells. Get enough exercise. Do yoga. We need 6 molecules of oxygen for every molecule of glucose we burn in our cells by exercise and yoga. More research is needed about genes governing cholesterol receptors, genes behind homocysteine metabolism, and low oxygen-tolerant robust CA3 neurons of the hippocampus.

Disease	Mechanism of Action	Curative Foods
Cancer-free life	Consume niacin, vitamin C, folic acid, protein, vitamins A, B_6, and B_{12}, carotenoids, D, E, and K_2. Consume antioxidants, leutin, lycopene, kaempherol, and 3, 3 diindolylmethane from kale, endive, spinach, cauliflower, and broccoli, and minerals selenium and zinc. Use low-calorie diet. Protect against DNA damage by a full complement of methylation nutrients.	Use twice the daily recommended allowance of antioxidants; use common fruits, dark green vegetables, and tomato products for leutin and lycopene. Minerals selenium and zinc from tree nuts and seeds of pumpkin and sunflowers must be part of the anticarcinogenic dietary arsenal. Consume anticarcinogenic Gouda and Emmentaler cheeses.
Longevity	DMAE methylates choline to DNA; it can go across the blood-brain barrier. Other key nutrients are B_6, B_9, B_{12}, amino acids arginine, tryptophan, tyrosine, minerals magnesium, selenium, zinc, and antioxidant ubiquinine. Also needed are ribose, 3-phenyl-3-acetamineindol, high proteins, and high antioxidants. Dimethyl aminoethanol works with neurotransmitter acetylcholine. Don't allow stress to shorten telomere.	Out of 44 genes that can go wrong as we age, 12 are involved in growth, metabolism, and fat and cholesterol processing. Alpha-synuclein gene, IGF gene, telomerase gene, FOXO 3 gene that upregulates genes, and daf2 gene are very important in this regard. The antiaging molecule dimethyl aminoethanol (DMAE) is an antioxidant present in anchovies and sardines. Neurotransmitter acetylcholine is involved in heart function, breathing, and sleep.

Table 23.03 A Comparison of Health-Promoting Therapies and Choices: Role of Nutrients and Therapeutic Agents

Gene therapy	Mature technology has yet to evolve.	Possible germ-line and somatic gene therapies are in the making.
Enzyme therapy	Low-lactose milk treated with lactase is already sold.	Sprouted grains and legumes are other routine sources of enzymes.
Vitamin D	Vitamin D is a blessing for beauty and skin health.	Vitamin D is a multifunctional steroidal anti-inflammatory hormone. It is necessary for bone and heart health and protection from infections.
Antioxidants	Antioxidants are managers of electrons. The redox potential of hydroxyl is + 0.23 volt, of hydrogen peroxide is + 0.36, and of superoxide is + 0.07 volt. Thus, different antioxidants are suited for dealing with different radicals.	Whereas enzyme antioxidants such as superoxide dismutase, catalase, and glutathione peroxidase prevent initial free-radical attack, vitamin C, vitamin E, glutathione, and coenzyme Q_{10} repair oxidizing radicals. Vitamin C cooperates with vitamin E, and it actually repairs and recycles vitamin E.
Exercise	Consume high-arginine proteins for nitric oxide. Exercise for more ubiquitin and brain-derived neurotrophic factor. Manage hypothalamic response by exercise, which can potentially affect DNA methylation of 7,663 genes and thus promote gene expression. It can prevent resistance of insulin and leptin. It prevents even Alzheimer's disease. Exercise promotes genes PGC-1a, DPK4, PPAR δ, and CIDE.	The genetics of exercise is different for different people. Exercise can change blood flow, gene expression, activity of nitric oxide synthase from arginine, serotonin transfer protein 5HTT, and mutated genes in white fat cells for small fat globules. Common foods that reduce the number of white fat cells are not known. Exercise can burn white fat cells off.

Meditation	Meditation works with the parietal lobe and increases production of nitric oxide and GABA (gamma amino butyric acid). It reduces stress and thus influences gene expression.	Can help with modulation of homeostasis, helps renew cells, and helps control of somatic nervous system.
Stress	Causes imbalance of calcium, sodium, and potassium ions; it damages cells, DNA, and the immune system. It shortens telomere. ATF3 gene, expressed in the central nervous system, is silenced by stress.	Reduce stress by adaptogenic foods, exercise, meditation, and yoga. Cortisol, serotonin transporter gene, and neuropeptide Y (NPY) induce corticoprotein-releasing hormone in response to stress.
Sleep	Deals with brain and chronobiology	Consume foods high in tryptophan for serotonin production.
Belief and faith	Ubiquitin generates telomerase and influences gene expression for a good immune system. Consume methylation foods for cell membranes.	Belief is critical to neurobiology and physiology; faith deals with cells and not DNA, though. Epigenetic effects can be inheritable.

Our very important health-care practices should include reducing inflammation and enhancing the immune system. Eudaemonics (those who believe in a higher purpose) among us, it appears, have a better genetic profile than hedonics.[639,640] Mood matters. Serotonin transporter gene 5 HTTLPR may be involved. How well do nerve cells manage serotonin fixes our mood.

Those with a functional variant of 5 HTT are happier people, says Jan Emanuel de Neve of the London School of Economics[641]. Volunteering can create a better gene profile (less inflammation, better antibodies and antiviral activity (Dr. Steven W. Cole, UCLA[642]). Mood for sure affects gene expression. Commitment to simple things like investing time with grandchildren can

---

639 http://well.blogs.nytimes.com/2013/08/23/what-our-genes-reveal-about-true-happiness/?_php=true&_type=blogs&_r=0.

640 http://rethinked.org/?tag=eudaemonic.

641 http://personal.lse.ac.uk/deneve/

642 http://newsroom.ucla.edu/releases/don-t-worry-be-happy-247644

make one happy. Mental health and dementia[643] are serious health problems of aging people; 44 million in the world today have dementia and mental problems, and the figure may jump to 125 million by 2050. Even today, there are 450 million people worldwide with mental problems (WHO[644] and Mental Health Problem Australia[645]). A methylation diet can improve the situation, and bioactive ingredients present in it for sure are known to reduce inflammation and boost the immune system. Such a diet, I believe, works by using our body's metabolic processes for pharmacological effects.

23.17 Health in the Near Future

Health of the body and body cells, its genome, molecularity of physiological events and DNA-encoded efficiency, speedy defense against molecular corruption, speed of brain works, digitization of perceptions, and conversion of daily food to other critical nanoscale molecules can improve health and longevity. However, the idea of defeating death is not plausible yet on the basis of what we know today. Good foods can no doubt postpone death by a combination of (1) a disease-free life and (2) a long life. For this effect, each and every cell in our body has to remain healthy. A good understanding of what goes on in our cells every second of the day can better aid us in selecting our foods for preventing damage to mitochondrial and nuclear DNA. What follows is an account of cellular health.

23.17.01 Mitochondria—The Thread of Life: Our power to live comes from mitochondria, and mitochondria in our cells, acquired or inherited, are small, 0.5-to-10 micron-in-diameter organelles. They are our power plants, providing 90 percent of the body's energy. Their number depends on the energy-requirement cells in a given organ. Heart, liver, and muscle cells can have up to 7,000 to 10,000 of them. Twenty percent of liver volume is all mitochondria. Sperm cells have only sixteen, whereas the ovum has a thousand mitochondria. The mitochondria change in shape constantly and use NADH as the main source of high-energy electrons. The power comes from ten protons produced for every two electrons. We should recognize that production of protons by five enzyme molecular protein machines of

---

643 http://www.newscientist.com/topic/mental-health.
644 http://www.who.int/whr/2001/media_centre/press_release/en/
645 http://www.mentalhealthvic.org.au/index.php?id=112

the electron-transport chain working in combination and molecular synergy depend critically on the nature, completeness, and general composition of the food we consume.[646] Actually, production of energy molecule ATP happens by way of a coupled proton circuit.[647] The mitochondria in our cells with thirty-seven genes are a purely maternal contribution, and on them rest largely our basic metabolic rate, energy production, weight, obesity, and general health.[648] The new research under the mitochondrial proteome initiative is revealing a lot.[649] Some two hundred mutations of mitochondrial DNA found to date are involved in diseases such as type 2 diabetes, hypertension, and neurodegenerative diseases like Parkinson's, Huntington's, and Alzheimer's. Nutrigenomic works need to be more specific in this area.

The sequencing of the 16.6 KByte mtDNA genome is now complete. There can be up to a thousand copies of mitochondrial genome in a given cell. Very prone to mutation due to constant exposure to free radicals, the mtGenome is involved in many inherited diseases because oxidation of DNA, lipids, and proteins is the underlying cause for many of them. On a daily basis, mitochondria help make twenty-four essential polypeptides for electron transport, two RNAs and twenty-two mRNAs. Our longevity depends greatly on mtDNA, and so do cell death, cardioprotection, and daily metabolism. We must choose foods as part of our daily methylation diet and preserve mitochondrial health for a longer life.

Mitochondria in our cells produce the energy molecule ATP via the electron-transport chain, where electrons and protons are produced from hydrogen[650] and the electronic energy of glucose is converted into chemical energy molecule ATP for our life processes. Four electrons are finally used to reduce oxygen into water. Superoxide, hydrogen peroxide, and hydroxyl radicals are produced as reactive oxygen species when the reduction process uses only one electron.

Mitochondria are a bit alkaline and negatively charged. Electron transport off glucose carbons to oxygen takes place in steps. In the process, hydrogen and electron currents simply extrude protons for power through

646 http://www.sciencedaily.com/releases/2013/06/130627142404.htm.

647 http://www.life.illinois.edu/crofts/bioph354/lect11.html.

648 http://www.ncbi.nlm.nih.gov/books/NBK7587/.

649 http://acceleratingscience.com/proteomics/the-mitochondrial-human-proteome-initiative/.

650 http://www.ncbi.nlm.nih.gov/books/NBK26904/.

the ATP synthase enzyme. Energy is released in small increments as the sum of electrical potential and proton concentration through a spatially separated oxidation-reduction system. This is no more than 200 mV at a given time, and the process is miraculously 90 percent efficient. The mechanical work that we perform is actually proton power in our cells' mitochondria, actually proton pumps in their inner membrane. The membrane makes the difference as it picks up electrons on one side and transfers protons on the other. It is no wonder that exceedingly mobile water protons thus become the basis of our life. In other words, water itself is life.

Mitochondrial DNA is not coiled around any spool of histone protein.[651] It has its own loop, like old-fashioned DNA. As such, it is subject to damage by electrons that leak off the electron-transport chain.[652] The situation is worse for mitochondria in skeletal muscles, lungs, liver, neurons, and heart cells with numerous mitochondria,[653,654] causing diseases of the brain and heart. Other problems due to mtDNA mutation include ATP production problems, cytochrome C oxidase deficiency, aging, cancer, cardiac hypertension, dementia, deafness, peripheral neuropathy, diabetes, migraines, epilepsy, optic neuropathy, and constipation. Therefore, mitochondria have to be protected at all costs throughout our lives.

Calorie-restriction diets simply imply less use and longer life of mitochondria for longevity.[655] Asian medicines and mineral micronutrients keep the mitochondria active and efficient.[656] I call mitochondria the *thread of life,* and they are already a pharmacological target[657] for chronic disease control. They can be trained to kill cancer cells, act as vectors for chemotherapy, and receive and promote drug functions and effects. This area, I believe, will deliver much help for maintaining human health. Their role in neurodegenerative diseases such as Parkinson's, Alzheimer's, and Huntington's is now more apparent than a decade ago. Use of antidiabetic sulfonyl ureas, immune suppression and antitumor drugs, anti-lipidemic

651 http://www.leica-microsystems.com/science-lab/mitochondrial-dna-molecules-are-packaged-individually/.
652 http://nar.oxfordjournals.org/content/35/22/7505.full.
653 http://www.nature.com/nrn/journal/v9/n7/fig_tab/nrn2417_F1.html.
654 http://www.ncbi.nlm.nih.gov/books/NBK26894/.
655 https://www.fightaging.org/archives/2011/04/calorie-restriction-increases-mitochondrial-biogenesis.php.
656 http://www.nature.com/news/screen-uncovers-hidden-ingredients-of-chinese-medicine-1.10430.
657 http://pharmrev.aspetjournals.org/content/54/1/101.long.

medicines, antiviral agents, and drugs for potassium channel opening are good examples. ATP per se regulates potassium channels, angina, and blood pressure. Low blood and low oxygen supply to the brain are mitochondrial problems.

Foods can prevent peroxidation and damage of mitochondrial DNA. Quircetin, a flavonoid antioxidant with anti-inflammatory and antihistamine power, seems to be a promoter of mitochondrial biogenesis.[658] We must include berries, red grapes, purple fruits and vegetables, licorice, and onions in our daily diet. To maintain good health is to maintain good mitochondria by the methylation diet.[659] We eat not only to get energy from electron-bearing food molecules but also to quench reactive oxygen species and prevent routine damage to our DNA.[660] The methylation diet provides all necessary nutrients (magnesium, manganese, zinc, iron, vitamin C, coQ_{10}, vitamin K_2, vitamins C, E, and beta-carotene, pyrroquinoline quinone, alpha-lipoic acid, and even antioxidant carnosine from poultry, fish, and beef. N-acetyl cysteine and curcumine from turmeric are necessary for antioxidant glutathione, ubiquinone is necessary for coQ_{10}, and pantothenic acid (B_5) is necessary for enzyme cofactor CoA or acetyl CoA. Even gut hormones affect cognition via mitochondrial function. Ubiquinol acts very much like vitamin C and beta-carotene. Diet, exercise, and mindfulness promote mitochondrial health and enhance homeostatic controls.[661] There have been two hundred mutations in mitochondrial DNA since our beginning.[662,] All such DNA passes to the offspring from the mother without any recombination.

23.17.02 Homeostasis: The human body operates under a regulated equilibrium called homeostasis. It keeps the body stable and constant by controlling internal environment acidity, temperature, glucose concentration in blood, blood concentration, and even the equilibrium of the digestive system.[663] Foods we eat and the lifestyle we choose must help maintain homeostatic mechanisms. We now know that exercise, meditation, yoga, and prayer

---

658 http://www.ncbi.nlm.nih.gov/pubmed/19516153.

659 http://www.mitoresearch.org/treatmentdisease.html.

660 http://www.ars.usda.gov/News/docs.htm?docid=17382.

661 http://www.mindfulhealth.biz/store/#!/~/product/category=3151016&id=13557208.

662 http://hihg.med.miami.edu/code/http/modules/education/Design/Print.asp?Course Num=2&LessonNum=4.

663 http://www.biology-online.org/dictionary/Homeostasis.

do help maintain homeostasis[664] and homeostasis is involved in pain perceptions.[665,666]

23.17.03 Stress: All 7.5 trillion cells in our body have mitochondria, the energy-producing power house. Energy is produced by burning glucose in an electron-proton furnace with the help of oxygen. The process is efficient, but some reactive oxygen species break off the electron-transport chain. These species are very damaging to DNA and cell constituents of proteins and lipids. Stress due to free radicals is bad for mitochondrial health. The damage can cause cancer, Parkinson's and Alzheimer's diseases, heart failure, autism, and chronic fatigue. This is homeostasis gone unregulated and electrons gone out of control.[667,668] Glutathione, the powerful endogenous antioxidant, and antioxidants such as vitamins C, E, D, and K and carotenoids in daily food can help maintain homeostasis, and so can exercise, meditation, and faith. They can prevent leakage of electrons and quench them in the event they do leak. Genes can help activate the ATF3 gene (that codes AMP-dependent transcription factor), which is silenced by stress. Neuropeptide NPY can calm us down by helping reduce stress (ibid).

23.17.04 Nitric Oxide: With a life of no more than ten seconds, nitric oxide (NO) is a great vasodilator.[669] It binds to hemoglobin, the carrier molecule in red blood cells. Nitric oxide is produced from arginine in endothelial cells by nitric oxide synthase at an in vivo concentration of no more than 5 nanomoles. Nitric oxide has a life of only a few seconds, but it does the job of nitroglycerin and Viagra rather effectively. As a free radical, nitric oxide can be dangerous and cause neurodegenerative diseases. We should be very careful with arginine supplement overdose.

23.17.05 Homocysteine: Homocysteine above 10 micromoles/liter is known to cause congestive heart failure.[670,671] It comes from high-methionine foods

664 http://www.asm.org.au/lifestyle/articles/overcoming-stress-with-meditation/.
665 http://www.ncbi.nlm.nih.gov/pubmed/21533708.
666 http://www.ncbi.nlm.nih.gov/pubmed/12798599.
667 http://www.ncbi.nlm.nih.gov/pubmed/10863530.
668 http://livasperiklis.files.wordpress.com/2013/06/free-radicals-and-antioxidants-in-normal-physiological-functions-and-human-disease.pdf.
669 http://www.ncbi.nlm.nih.gov/pubmed/9390947.
670 http://www.ncbi.nlm.nih.gov/pubmed/21828948.
671 http://eurjhf.oxfordjournals.org/content/8/6/571.full.

such as meat, poultry, eggs, and pork. Keeping a beans and legume diet is a good strategy in order to maintain a good homocysteine level in the plasma. Drinking, smoking, and deficiencies of vitamin A and carotenoids, B_2, B_6, B_{12}, and folate can cause high homocysteine. These are the same nutrients that are required for DNA methylation and repeated in tables 23.02 and 23.03; their deficiency causes pains, stress, and chronic diseases.

We have a dilemma in that methional, which makes S-adenosyl methionine for the good work of DNA methylation, is bad when it produces homocysteine. Specific genes and homeostatic control may be the answer. Nonetheless, we should ask our physician to test our homocysteine regularly.

23.17.06 Brain-Derived Neurotrophic Factor (BDNF): The BDNF gene encodes a protein necessary for survival of stratial (dorsal root ganglion) neurons.[672] Largely present in the central nervous system, it affects cells in the eyes and the sensory and motor neurons of the peripheral nervous system. Expression of this gene is reduced in both Alzheimer's and Huntington's disease patients. It is believed that the BDNF gene plays a role in the regulation of stress response and in the biology of mood disorders.[673] Multiple variants of this gene have been discovered.[674]

23.17.07 Hormone Peptide Cholecytokinin: Produced in the small intestine, the peptide hormone helps digestion of fat and protein by inducing discharge of bile acid and enzyme secretions in the pancreas. Also, it acts as a hunger suppressor for bringing about satiety. High-fat and high-protein chyme in the stomach stimulates CCK secretion. What nutrients specifically induce CCK secretion is not very clear, and neither are the genes involved in the process.

23.17.08 Human Microbiome (the Second Genome) and Fecal Transplant: A new clinical practice in treating clostridium difficile, Crohn's disease, colitis, and diabetes has emerged. The genome attributable to our gut bacteria can be analyzed for treating yet unknown other diseases. We know that the gut bacteria convert allagitannins and gallic acid in strawberries and pomegranates to urolithin, which could be a colon cancer treatment.[675,676]

---

672 http://www.nature.com/mp/journal/v17/n6/full/mp2011107a.html.
673 http://www.nature.com/mp/journal/v7/n6/full/4001211a.html.
674 http://www.nature.com/mp/journal/v5/n5/full/4000749a.html.
675 http://www.hindawi.com/journals/ecam/2013/247504/.
676 http://www.faqs.org/patents/app/20120264819.

Also, colon bacteria convert soy isoflavone into equol, an estrogenic metabolite that is therapeutic to cardiovascular diseases, osteoporosis, and even some cancers.[677] As the underlying science evolves to perfection, we can sure benefit from consumption of soy foods, strawberries, and pomegranates. Considerable research is underway to understand the positive effect of probiotics on cytokine signaling and possible prevention in the case of colon cancer.

23.17.09 The Future of Gene Therapy: At the present time, germ-line therapy is the least possible. However, somatic gene therapy may become a standard healthcare activity in the near future. Somatic gene therapy is better because the molecules of therapy are not involved in reproduction, and none will be passed on to sperm or egg cells.[678] Currently at an infant stage, this may be the greatest advance in cancer treatment in the near future. Noteworthy gene therapies are for choroideremia eye disease[679] and HIV vaccine with a synthetic gene(ibid). A lot will have to be investigated and learned about genes p53, BRCA1, and BRCA2.

The gene for p53 protein[680] is very critical in activating DNA repair, arresting growth, helping express a few mitochondrial genes, and even initiating cell death in all multicellular organisms. The protein is also known as phosphoprotein 53, tumor-suppressor protein, and cellular tumor antigen. The gene that encodes p53 is the guardian of our genome. We know very little about managing this gene by the foods we eat. BRCA1 and BRCA2 are also tumor-suppressor genes, more directly in the case of breast and ovarian cancers. Consumption of restricted-calorie, low-fat, high-fiber foods, including fruits and vegetables, helps breast cancer patients, but scientific information on the mechanism of action of food molecules is uncertain.

Antistress Genes: Our immune cells have stress genes that spread breast cancer. Gene ATF3 on chromosome 1 is a link between stress and cancer.[681] Expressed when we are under stress, ATF3 helps cancer cells (ibid). The Swedish twin study concludes that our personality type and propensity

677 http://onlinelibrary.wiley.com/doi/10.1002/mnfr.200600262/abstract.
678 http://www.iptv.org/exploremore/ge/features/somatic.cfm.
679 https://www.blindness.org/index.php?view=article&id=2974%3Agene-therapy-for-choroideremia-university-of-oxford-uk-&option=com_content&Itemid=99.
680 http://www.bioinformatics.org/p53/introduction.html.
681 http://researchnews.osu.edu/archive/ATF3.htm.

to stress are attributable to our genes, but once produced, stress can cause further genetic damage.[682]

The gene for membrane protein VMAT2 (vesicular monoamine transporter 2), which transports dopamine, is a spirituality gene that makes stress-free life possible.[683] While the idea of a God gene is debatable, release of neurotransmitters is essential for life itself. That VMAT2 reduces stress is great news.

Genes for Sleep: The twenty-four-hour circadian rhythm is a matter of human evolution, and it depends on the coexpression of at least five clock genes. The details of coexpression are not yet known well. Gene DBQ-0602 keeps open the potassium channel during the wake state and closes it during sleep.[684] What mechanism there is to manage neuronal oscillations in seconds, minutes, hours, days, months, and years is not yet known. It is clear, though, that temporal organization (solar and lunar rhythms) are just as important as cellular organization, meaning that our body reacts to the sun and moon in terms of sleep and wake states. A good proof is jet lag, which we experience during long-distance travel.

Cytokines are peptide, protein, or glycoprotein molecules. They are used for signaling by our immune system. Infections during sickness induce cytokine production, which once transmitted to the brain, causes a change in physiology. Interleukin 1 in particular changes the brain circuit that alters sleep-wake behavior.[685] Many other cytokines interact with hormones and neurotransmitters and regulate sleep indirectly, and many of them are dependent on nutritional status. Pro-inflammatory cytokines can create or terminate immuno-response and oxidative stress. Omega-3 and omega-6 fatty acids present in the proper ratio of 1:2 in our daily diet are essential for managing the effect of cytokines.[686] Cytokines are also produced by the immune system to control sleep. They help maintain a good serotonin-melatonin ratio for sleep. The serotonin transporter gene (SERT) encodes

682 http://www.cnn.com/2012/09/19/health/health-genes-stress/.

683 http://www.medicalnewstoday.com/releases/16378.php.

684 http://books.google.com/books?id=xyjwJWfuNhQC&pg=PA168&lpg=PA168&dq=Gene+DBQ-0602&source=bl&ots=XyuHPI8kRH&sig=JJh-ey9p5G1S1EX2A9NCsWFINuo&hl=en&sa=X&ei=JEDUUpyqGYSZrgGpq4C4CQ&ved=0CDwQ6AEwAg#v=onepage&q=Gene%20DBQ-0602&f=false.

685 http://www.jneurosci.org/content/18/16/6599.abstract.

686 http://www.ncbi.nlm.nih.gov/pubmed/17600534.

integral membrane protein necessary for transporting and recycling serotonin. The SERT gene is thus involved in circadian rhythm and sleep, and so is the neuropeptide NPY.[687]

Gene Therapy by Exercise, Meditation, and Yoga: A very well-planned study at Lund University in Sweden reveals exercise-driven methylation of 1,800 sites on 7,663 genes, and many of these genes are known to be involved in fat storage.[688] The genome analysis company *23andMe* has also concluded that exercise induces genome-wide changes[689]. The gene FTO is responsible for high BMI, but its control by diet is not clear yet. We do know, however, that exercise, meditation, and yoga can induce production of nitric oxide for cardiovascular health, change gene response to stress, and induce monoaminooxidase A gene. Such changes may reverse the effects of neurotransmitters of anger and violence. Exercise, we must take to our hearts, manages mitochondrial performance.

Skin-Care Genes: Gene therapy for melanoma cancer in patients who have tumor-fighting specialized cells is now within reach. Adaptive immunotherapy involving a patient's own immune cells for treating metastatic melanoma skin cancer may be the first case of indirect gene therapy (National Cancer Institute[690]). The underlying technology is about genetic engineering of the patient's own antitumor immune cells.

Gene p16 involved in cell senescence is a tumor-suppressor gene.[691] It is inactivated in the case of melanoma families. Another skin-care gene, p53, also a tumor-suppressor induced with UV light, works with it in carcinogenesis.[692] This highly investigated gene with 64,000 publications is known to regulate more than one hundred other genes, including those for mitochondrial function, cell division, and cell death. Gene p53 is known to have suffered 2,500 mutations, with consequences not identified and fully understood.

Green tea has been shown by workers at the university hospital in Zurich

---

687 http://www.ncbi.nlm.nih.gov/pubmed/15867649.

688 http://www.lunduniversity.lu.se/o.o.i.s?id=24890&news_item=6074.

689 http://blog.23andme.com/wp-content/uploads/2012/11/Brian_Naughton_TruVar_ASHG2012poster.pptx-1.pdf

690 http://www.cancer.gov/newscenter/newsfromnci/2006/melanomagenetherapy

691 http://www.ncbi.nlm.nih.gov/pubmed/9508208.

692 http://www.ncbi.nlm.nih.gov/books/NBK22268/.

to have photoprotective effects on skin because of its anti-inflammatory and antioxidative effects[693]. Loss of p53, it appears, is a definite cause of carcinogenesis. Stress causes mutations, and mutated forms of p53 are present in half of human cancers. We may be able to postpone skin-cell senescence by regular use of green tea and use of other antioxidants.

Genes That Kill Pain: Gene SCN (sodium channel norciceptor) gives instructions for making sodium channels, which convey pain sensations. SCN9A has been implicated in congenital insensitivity to pain.[694] This gene has suffered from thirteen known and documented mutations as they relate to variation in pain tolerance. Pain-relief drugs may be designed in the near future for blocking sodium channel proteins on cell membranes and thus inducing pain relief.

There may not be a treatment of rheumatoid arthritis soon, but FDA recently approved an antibody against nerve growth factor for controlling joint pain in the case of osteoarthritis.[695] The antibody, a kind of vaccine, works best with nonsteroidal painkillers like aspirin and ibuprofen. The relief comes by preventing NGF from doing its job of transmitting pain through the sympathetic and sensory nervous system.

The benefit of arthritic knee-pain reduction by chondroitin sulfate and glucosamine is very individual. The best approach is to try for two months. My wife and I tried it for more than two months, and it seemed to work when used twice a day.

Gamma aminobutyric acid (GABA), an inhibitory neurotransmitter, regulates anxiety and stress and controls muscle tone by opening pores in ion channel proteins on cell membranes.[696] It is a zwiterion (both positive and negative), with considerable ease of flexibility in its shape, and therefore, it can bind with a variety of receptors and help form pores for passage of ions. GABA can put an end to pain, and we can get it by exercise, meditation, and yoga and also from foods like kefir, hummus, tomatoes, shrimp, and Darjeeling tea. A good Darjeeling tea can give 15 mg of GABA per serving of 15 grams of tea. There is no Recommended Daily Allowance for GABA, but 250 mg per day can reduce stress and pain. Since GABA doesn't cross

---

693 http://www.madhippie.com/green_tea_skin_benefits.pdf

694 http://www.jci.org/articles/view/33297.

695 http://www.fda.gov/downloads/AdvisoryCommittees/CommitteesMeetingMaterials/Drugs/ArthritisAdvisoryCommittee/UCM295202.pdf.

696 http://www.med.nyu.edu/content?ChunkIID=222543.

the blood-brain barrier, it is wiser to have it produced in the brain from glutamic acid present in our dietary proteins. The value of excessive GABA supplement is very questionable.

The Gene That Could Cure Parkinson's Disease: Northwest Perkin's Foundation, founded by Michael J. Fox, is very active in the area of research on this inherited disease. The genes involved may be SNCA and Parkins.[697] The last fifteen years of research are about to pay huge dividends because Parkinson's disease may be the first successful case of regenerative medicine. High-protein foods, being popularized as the dopamine diet, can be of some help, but for a sure cure against Parkinson's disease at the present, the best choice is dopamine drugs.

Controlling Genes of Multiple Sclerosis: One American in every seven hundred suffers from multiple sclerosis, a disease largely attributable to genes that make IL7R-alpha proteins that affect our immune cells and signal transmission. The company *23andMe* does genetic profile testing that may diagnose the extent of mutative damage. The best nutrients in the case of multiple sclerosis are omega-3 fatty acids and vitamin D_3.

23.17.10 Food and Neurotransmitters: Neurotransmitters such as adrenalin for reaction time and reflex, dopamine and norepinephrine for alertness, gamma amino butyric acid for pain relief, glutamate for learning, histamine for focus, and serotonin for good sleep come from foods we eat. We should use eggs, tree nuts, fruits, vegetables, and whole grain in our daily methylation diet. This should help us keep our neurotransmitters intact and efficient and match the speed with which our body uses its signaling molecules.

23.17.11 Phytoceramides 3 and 6: These are skin-proofing sphingolipid molecules claimed to improve skin health with respect to skin hydration and wrinkle-free softness.[698] Good food sources for ceramides are sweet potatoes, rice, wheat, baker's yeast, and virgin plant oils. Ceramides have been cleared by the FDA, but the activating effect of gene PPARs on natural ceramides in

---

697 http://ghr.nlm.nih.gov/gene/SNCA.
698 http://phtytoceramidesantiaging.com/.

our body is still in early stages of research.[699] Diet-based routine intake of magnesium and B$_2$ vitamin riboflavin are known to be good for skin health also.[700]

23.17.12 Vitamin D$_3$ Therapy for Nerve Growth Factor: Vitamin D is key to gene expression and extremely necessary for those who take calcium supplements. A daily dose of 500 mg is sufficient for proper calcium absorption, and a blood test showing 2 mg/dl is fine. The intake is too much if it is causing nausea, vomiting, and constipation. A diet that includes fortified milk, fish, eggs, and special varieties of mushrooms supplies sufficient amounts of vitamin D. This can be supplemented, but with no more than an added 400 mg/day.

23.17.13 Vitamin C: Our body regulates vitamin C extremely well. Total intake of vitamin C in a range of 75 to 90 mg per day is attainable from fruits and vegetables such as citrus fruits, green peppers, berries, broccoli, and brussels sprouts. Optimal absorption and full interaction of vitamin C with other nutrients like vitamin E is possible when it comes from natural sources. It is worth noting that no more than 6 percent of the US population exhibits vitamin C deficiency.

23.17.14 Vitamins A and E: We don't need more than 0.8 to 0.9 mg per day of vitamin A and no more than 22.4 IU of vitamin E per day. A daily diet based on fruits and vegetables for vitamin C and virgin plant oils and whole grains for vitamin E is sufficient.

23.17.15 Omega-3 Fatty Acids: Two servings of fish every week does the job of meeting our omega-3 fatty acid need of 300 mg. Vegetarians and vegans may like to boost omega-3 fatty acids by taking a 1 gram fish oil tablet per day or design their daily dishes with flaxseed and walnuts. They should snack on one ounce of walnuts every day. Omega-3 fatty acids are anti-inflammatory and help reduce risk of cardiovascular disease, improve mental health, and help us fight against chronic diseases. Omega-3 fatty acids are brain foods.

699 http://www.sigmaaldrich.com/img/assets/6780/PPARS.pdf.
700 http://www.webmd.com/vitamins-supplements/ingredientmono-957-RIBOFLAVIN%20(VITAMIN%20B2).aspx?activeIngredientId=957&activeIngredientName=RIBOFLAVIN%20(VITAMIN%20B2).

23.17.16 Dietary Fiber: Snacking on fresh fruits and roasted tree nuts, having *crunchy* salad every day, preparing foods with whole grains and legumes, having fruit-based smoothies, and eating high-fiber bars and cereals can easily provide more than 25 grams of dietary fiber per day. This is very good for blood-sugar control in the case of type 2 diabetes and digestive health, including appetite control. Daily dietary fiber consumption of 35 grams per day is fine for those trying to reduce weight.[701]

23.17.17 The Myth Behind Vitamin and Mineral Supplements: Current science about vitamin RDAs is available in the FDA guidelines. It is excerpted from a variety of sources in chapter 6 in this book. What is not established is possible overdosing in the case of people who take supplements on top of a good diet. This is obviously risky. There is little need for vitamin supplements if one were to use the methylation diet regularly. Meta-analyses find costly vitamin supplements of questionable health value.[702,703]

23.18. Food Supplements

Calcium: Skip calcium pills unless you are suffering from osteoporosis. To keep your bones strong and let muscles, heart, nerves, blood, and hormones function optimally, elect to get 600 mg calcium from calcium-rich foods like milk, yogurt, broccoli, and fortified orange juice and cereal. This approach assures the complementary and synergistic benefit of other minerals. Too much calcium causes kidney stones, heart problems, and constipation.

23.19 Enzyme Therapy

Enzyme therapy is either for digestive enzyme supplements targeted to curing ulcers or pancreatic cancer or metabolic enzymes targeted to cell damage repair. There are no well-designed studies supporting the concept of regular enzyme therapy in this area, but a few developments are noteworthy.

The enzyme asparginase converts asparagines to aspartic acid so that

---

701 http://www.mayoclinic.org/fiber/art-20043983.

702 http://www.thelancet.com/journals/lancet/article/PIIS0140-6736(13)61647-5/abstract.

703 http://www.ncbi.nlm.nih.gov/pubmed/23255568.

production of the carcinogen acrylamide is reduced.[704] It has also been found effective in treating lymphoblastic leukemia.[705] Another commercial practice is production of lactose-reduced milk for those who are lactose-intolerant and suffer from a deficiency of the enzyme lactase. At the kitchen level, use of meat tenderizers, including papain from papaya or bromelain from pineapple, is common. Green salads and sprouted grains and legumes add a lot of digestive enzymes to our daily diet, and many of them survive the acidity in the stomach and continue working in the intestine. Enzyme therapy is not a routine clinical practice at the present time.

23.20 The Methylation Diet and Gene-Expression Diet

Chapter by chapter, it has been the major objective in preparing this book to demonstrate that B vitamins (B_1 thiamine, B_2 riboflavin, B_3 niacin, B_5 pantothenic acid, B_6 pyridoxine, B_7 biotin, B_9 folic acid, and B_{12} cyanocobalamine) and methionine can cure pain and many chronic diseases. We could call these mood and methylation nutrients. Such nutrients are known to improve high-density lipoprotein, cure acne, improve reaction time, and reduce risks of type 2 diabetes, Alzheimer's disease, colon cancer, and migraine headaches. They positively affect ATP production, prenatal development, and level of neurotransmitters, and reduce susceptibility to cancer, autism, food allergies, and asthma. They are involved in suppressing by turning off tumor genes P16, MGMT, DAPK, RASSF1A, GATA4, PAX5a, and PAX5β. Our diet must secure these nutrients from foods, not supplements, for day-to-day feeding of our genes for health and long life. The methylation diet in simple terms is to include fruits, vegetables, tree nuts, seeds of pumpkin, sunflower, and sesame, and whole grains for dietary fiber, minerals, and vitamins.

704 http://www.google.com/url?sa=t&rct=j&q=&esrc=s&frm=1&source=web&cd= 2&ved=0CDgQFjAB&url=http%3A%2F%2Fwww.researchgate.net%2Fpublication %2F258402636_Impact_of_L-asparaginase_on_acrylamide_content_in_potato_products %2Ffile%2F9c96052824a7bea856.pdf&ei=iEzUUpWYF8T-qAHmooHYDQ&usg=AFQjCN FvjFsJqBeX1uY6BS5yFaCyzG3DIA&sig2=55hR6ol15mYcvQeCuNI6cg.
705 http://www.cancer.org/treatment/treatmentsandsideeffects/guidetocancerdrugs/ asparaginase.

A restricted-calorie methylation diet inclusive of omega-3 fatty acids, cucurmin from turmeric, flavonoids from cocoa and oranges, vitamin B complex and vitamin D, choline, and minerals calcium, zinc, copper, selenium, and iron improves cognition via their effect on mitochondrial performance and energy production.[706]

23.21 Defeat of Death

Death is much more than our neurons gone wrong; often it is malnutrition leading to senescence or biological aging. Such is the case attributable to 90 percent of the deaths in the United States and other advanced industrialized societies. It happens when lung and heart stop and when oxygen and glucose to brain cells is dangerously depleted. Death means the destruction and absence of electric impulse or voltage differential in our cells. It means no energy, no mitochondrial function, no signals, and no communication. Our food can postpone death, but defeating it altogether is impossible. The best way to postpone death and enhance longevity is to manage the health of our cellular power plant—the mitochondria.

How do proteins travel and get around from the Golgi apparatus in the cell? How is senescence programmed by genes? What is the lifelong accumulative damage to gene expression before the final hour of death is decided physiologically? What is behind the biological limitation underlying the life of a male sperm of more than five days compared to only one day for the female egg's life? Why is there a structured nucleus in sperm, and at best only a loose nucleus in the egg? Why is there a need for millions of sperms to impregnate a single egg? Why is the sperm the smallest and the ovum the largest cell in the human body? How does a centenarian or even an octogenarian evolve from the combining of two short-lived human cells? What is behind the programmed life span of different cells? Why do neurons last for our lifetime and not regenerate on demand? How could we ever defeat death without answering these questions? Many of these questions more aptly center around ubiquitin and telomere.

Ubiquitin: Proteins are the basis for metabolism, and the ubiquitin/proteosome pair manages protein homeostasis.[707,708] Behind it is a

---

706 http://www.ncbi.nlm.nih.gov/pmc/articles/PMC2805706/.
707 http://www.nature.com/nature/supplements/insights/ubiquitin/.
708 http://homepages.bw.edu/~mbumbuli/cell/ublec/.

longevity gene, but whether it can be tweaked to prevent death is not known. We do know that it does help generate telomere.

Telomere: Telomere is also a longevity gene. The idea of unlimited longevity and immortality seems absurd in view of our current knowledge of telomere, which shortens after every cell division. From what we know now, our longevity simply means the longevity of telomere. On an average, our cells divide maybe 10,000 trillion times during our lifetime. While telomere, the end cap of a chromosome, is there to protect DNA from an end-replication problem, it does shorten at each cell division. Although it is remade again quickly, it gets shorter and shorter due to being consumed at each cell division and also due to oxidative damage. A short telomere leads to a poor immune system, chromosome fusion, gene deterioration, and even death of the cell. A way to defeating death then would be to prevent telomere shortening, and the answer may be in the genes of immortalized cancer cells or the jellyfish Turritopsis nutricula.[709] It appears that at fault is the very biology of sexual reproduction. Our DNA can't do what the DNA of jellyfish does, or even the DNAs of less-immortal tortoises and bristlecone pine trees do. For now we humans are mortals, and a century long life span would do fine.

We can't go back to the stage of a zygote that we started with. We do know that human height, weight, strength, and longevity vary because of genes. Therefore, the answer for how to postpone death may lie in our own genes.[710] We should learn how to feed our genes like the centenarians in Okinawa, Japan, do. The goal of defeating death should always keep us on guard for added research in the biology of death. For now it appears that keeping entropy at a minimum forever and permitting biology to proceed ad infinitum is not possible. The methylation diet can for sure postpone death and enhance longevity. There are many research efforts whose outcome has revealed ways to postpone death:

1. Maintain dehydroepiendrosterone, a hormone linked to aging. Supplements have been available for ten years. Dehydroepiendrosterone

---

709 http://www.nytimes.com/2012/12/02/magazine/can-a-jellyfish-unlock-the-secret-of-immortality.html?pagewanted=all&_r=0.

710 http://www.positivefuturist.com/archive/273.html.

peaks at the age of 20 years; by age eighty-five, it is only 5 percent of normal.

2. Maintain the work speed of key genes even if they have to work overtime.

3. Don't let genes IGF 1, FOXO3, and daf 2 slow down.

4. Genes p53 and BML1 are molecular clock components. They must remain functional forever.

5. Use the methylation diet to repair DNA efficiently, in particular the mitochondrial DNA.

6. Aging is only 24 percent affected by genes. The rest is lifestyle. Sleep for eight hours, exercise, reduce stress, and maintain an optimal weight.

7. The genetic clock slows down and telomere shortens with age. Actually telomerase is switched off in the case of humans. We run into cancer when we switch it on! Can we activate the telomerase gene safely? Cytoastrogenol, a saponin from astragals roots, boosts telomerase. Resveratrol, astragaloside, and ginkgo biloba have similar effects (UCLA work[711]).

8. Consume B vitamins, vitamins C, D, E, and K, omega-3, and polyphenols. Learn about genes for antioxidant defense. This may help postpone death.

Let's learn to use the methylation diet for a disease-free long life and call it postponing death with happiness.

To conclude, I need to have my readers pay attention to quantum biology, more rightfully nutrigenomics, bioinformatics and DNA computers that may identify a single nucleotide polymorphism on our DNA and relate it to a specific disease in milliseconds. I should speculate along with the readers that effects of foods may thus be detected in a very short time. Why not, when we know that no more than 0.15 gram hydrogen cyanide can uncouple electron transfer from citochrome C oxidase in our cells' mitochondria and kill us in a hurry? Just to reemphasize: photosynthesis, DNA mutation, Brownian motors in our cells, magneto-reception in animals, and human vision present routine examples of quantum biology, a phenomenon beyond chemical bonds, vibration, and ionization. Energy conversion in our body

711 http://letsgothin.com/health-articles/how-to-optimize-your-health-through-telomerase-activation/

into molecules that are of quantum mechanical nature is another example of quantum biology operating in our body. Our mind and consciousness may be the highest-form quantum biological system.

There is already a beginning of what we are speculating about here. Genetic programming and quantum computation[712] in regard to cancer diagnostics are being used by D-Wave's DNA-SEQ company using a 128-qbit quantum computer that is to be used for cancer diagnostics.[713] On the remediation side, there is the possibility of flexible and stretchable transient electronics sheathed in silk protein that can act as sensors for monitoring performance of stents in our arteries, epidermal electronics for vital signs, electroceuticals for fighting pain, and wireless electronics for monitoring functions of body organs and even monitoring our time-dynamic brain as it accommodates its motion during its routine electrochemical functions.[714,715] Such protein-based transient circuits will work in the body's aqueous environment for monitoring food functions for days and weeks. Quantum computers and bio-based transient electronics may help us detect cancer earlier, develop more selective drugs, and study tumor treatments by discrete combinatorial optimization.

Evolution will continue no matter what. The speed of evolution in terms genomic change will depend on when and what molecules we subject our body to via our foods and drinks. "Can research institutions beat evolution?" is an unanswered question. It is possible though to manage our health by managing our genes, mitochondrial genes in particular, with foods and beverages we consume and lifestyles we choose to live by. With easy access to our genetic profile and cost effective blood tests in response to therapeutic nutrients, sleep cycle maintaince, proper exposure to sunlight, and a choice of lifestyles that includes physical activity and day to day mindfulness there is hope that public health shall improve decade by decade. "How does what you eat" and "what lifestyle you opt for routine living" affects your body and mind will be deciphered right through your smart phone once applications become available for personal genomic assistant, and for routine blood chemistry testing. You will be able to monitor your vital signs and in relation

---

712 http://faculty.hampshire.edu/lspector/pubs/qc-aigp3-prepress.pdf.

713 http://www.diagnomics.com/Diagnomics_D-wave.php.

714 http://www.smithsonianmag.com/innovation/electronics-that-can-melt-in-your-body-could-change-the-world-of-m.

715 http://rogers.matse.illinois.edu/files/2012/annurevbioeng.pdfedicine-180947638/.

to what you eat and drink much more frequently and you will become your own physician.

Live well with the choices of calorie restricted methylation diet made of common foods and live well and long by a lifestyle that includes daily exercise, mindfulness, social interaction, and a good night sleep.

Major Research Institution
23andMe, Mountainview, Ca
µBiome of San francisco, CA
Pathway Genomics, San Diago, CA
Theranos, Inc, Palo Alto, CA
Illumina, San Diego, CA

APPENDIX I

Fiber-Rich Common Foods

Ingredient	Amount	G Fiber	Ingredient	Amount	G Fiber
Raw oat bran	1 oz	12	Wheat berries	¼ cup	5
Raw wheat bran	1 oz	12	Wild rice	1 cup	3
Raw corn bran	1 oz	22	Brown rice	1 cup	4
Raw rice bran	1 oz	6	Bulgur	1 cup	8
Fiber One bran cereal	½ cup	14	Bread	1 slice	2
All-Bran cereal	½ cup	10	Rye crackers/wafers	1 oz	6
Fiber One chewy bars	1 bar	9	Cooked spaghetti	1 cup	6
Fiber from beans			Frozen green peas	1 cup	14
Cooked lima beans	1 cup	14	Cooked peas	1 cup	5
Adzuki beans	1 cup	17	Cooked Swiss chard	1 cup	4
Cooked broad beans	1 cup	9	Cooked beet greens	1 cup	4
Cooked black beans	1 cup	15	Cooked spinach	1 cup	4
Cooked garbanzo beans	1 cup	12	Cooked collard greens	1 cup	5

Ingredient	Amount	G Fiber	Ingredient	Amount	G Fiber
Cooked lentils	1 cup	16	Cooked mustard greens	1 cup	5
Cooked cranberry beans	1 cup	16	Cooked turnip	1 cup	5
Black turtle beans	1 cup	17	**Fiber from nuts**		
Kidney beans	1 cup	16	Almonds	1 oz	4
Navy beans	1 cup	19	Pistachios	1 oz	3
White beans	1 cup	19	Cashews	1 oz	1
Cooked French beans	1 cup	17	Peanuts	1 oz	2
Cooked mung beans	1 cup	15	Walnuts	1 oz	2
Cooked yellow beans	1 cup	18	Brazil nuts	1 oz	2
Pinto beans	1 cup	15	Sunflower seeds	¼ cup	3
Fiber from berries			Pumpkin seeds	½ cup	3
Raspberries	1 cup	8	Sesame seeds	¼ cup	4
Blueberries	1 cup	4	Flaxseed	1 oz	8
Red raw currants	1 cup	5			
Strawberries	1 cup	3			
Boysenberries	1 cup	7			
Gooseberries	1 cup	6			
Loganberries	1 cup	8	Cooked Hubbard squash	1 cup	7
Elderberries	1 cup	10	Zucchini squash	1 cup	3
Blackberries	1 cup	8	Acorn squash	1 cup	9
			Cooked spaghetti squash	1 cup	2

Ingredient	Amount	G Fiber	Ingredient	Amount	G Fiber
Fiber from grains			Kale	1 cup	3
Amaranth grain	¼ cup	6	Cauliflower	1 cup	5
Barley pearled	1 cup	6	Kohlrabi	1 cup	5
Buckwheat groats	1 cup	5	Savory cabbage	1 cup	4
Popcorn	3 cups	4	Broccoli	1 cup	5
Oats	½ cup	4	Brussels sprouts	1 cup	6
Dry rye flour	¼ cup	7	Red cabbage	1 cup	4
Cooked millet	1 cup	2	Russet potato	1	4
Quinoa	1 cup	5	Red potato	1	3
Teft grain flour	¼ cup	6	Sweet potato, fresh and skinned	1	4
Tricale flour	¼ cup	5			

APPENDIX II

Composition of Fruits

	Vit A	Vit C	Vit E	Folate	Vit K	Iron	Potassium	Calcium	Phosphorus	Selenium	Zinc	chomiumhr	Mg	CopperU
Units	IU	mg	IU	mcg	mcg	mg	g	mg	mg	mcg	mcmg	mcg	mg	mcg
RDA	1000	15	15	50	30	7	3	600	460	20	3	11	80	300
Apple	98	8.4				0.22	195	11	20				9	
Avocado	293		4.16	163	42	1.11	975	24	105		1.29		58	0.38
Banana						0.31	422		26	1.2				
Blackberries	308	30.2	1.68	36	29	0.89	233	42	32		0.76		29	0.24
Black currants	238	202.7	1.12			1.27	361		66		0.30		27	
Boysenberries	88	202.7	1.15		10.3	1.12	183		36		0.29		21	
Breadfruit		63.8		31	1.1	1.19	1,078	37	66		0.26		55	0.18
Cantaloupe	2,334	25.7			1.7	0.14	184		10		0.12			
Cherries	88	9.7			2.9	0.50	306	18	29				15	
Chinese pear		10.4	0.33		12.4		333	11	30				22	0.14
Cranberries	60	13.3	1.2		5.1	0.25	85		13					
Dates				28	4		964		91	4.4			63	0.30
Figs	91				3	0.24	148	22	9	1			11	
Gooseberries	435	41.5	0.56	9			297	38	40	0.9	0.18		15	
Grapefruit	2,132	79.1	0.30	23		0.21	320	28	18		0.16		18	
Grapes		16.3	0.29		22	0.54	288		30		0.11		11	0.19
Guava	1,030	376	1.2	81	4	0.43	688	30	66	1	0.38		36	
Kiwi	60	64	1.01	17	27.8	0.21	215	23	23				12	
Lemon	18		0.13			0.50	116		13	0.3			7	
Lime	34	19.5					68		12	0.3				
Lychee		135.8	0.13			0.59	325		59	1.1	0.13		19	0.13
Mango	1,584	57.3		29		0.27	323		23				19	
Mulberries		51	1.22		10	2.59	272		53		0.17		25	
Nectarine	475					0.4	287		37					0.12
Olives						0.28								
Orange		69.7	0.24	39			237		18				13	
Papaya	1532	86.5	1.02	53	3.6		360		7	0.8			14	
Passionfruit	3,002	70.8				3.78	821		160		0.24			0.20
Peach	489		1.09		3.9	0.38	285		30		0.26		14	0.10
Pear					8	0.3	212		20		0.18		12	0.14
Pineapple		78.9		30	1.2	0.48	180		13		0.2			0.18
Plum	569		0.43	8	11	0.28	259		26		0.17		12	

	Vit A	Vit C	Vit E	Folate	Vit K	Iron	Potassium	Calcium	Phosphorus	Selenium	Zinc	chomiumhr	Mg	CopperU
Units	IU	mg	IU	mcg	mcg	mg	g	mg	mg	mcg	mcmg	mcg	mg	mcg
Pomegranate		28.8	1.69	107	46	0.85	666		102		0.99		34	0.45
Raisins					1.5	0.81	332		43				14	0.14
Raspberries			1.07	26	10	0.85	186		36		0.52		27	0.11
Starfruit	81	45	0.2	16			176		16					0.18
Strawberry		84.7	0.42	35	3	0.59	220	23	35	0.6	0.2			
Tomato			0.66	18	10	0.33	292	12	30		0.21		14	
Watermelon	1,627	23.2				0.69	320		31		0.29		29	0.12

Eat avocado for vitamin K, fiber, potassium, folic acid, vitamin B_6, and vitamin C; grapefruit, cantaloupe, blueberries, lemon/lime, kiwi, orange, papaya, pineapple, strawberries, and watermelon for vitamin C. Use condiments and spices like cayenne pepper for vitamin A; and cinnamon, cloves, cumin, oregano, parsley, thyme, and turmeric for trace minerals such as manganese, Fe, and selenium.

APPENDIX III

Composition of Vegetables

	Vit A	Vit C	Vit E	Folate	Vit K	Iron	Potassium	Calcium	Phosphorus	Selenium	Zinc	Chromium	Magnesium	CU
Per 100 Gram Product														
Units	IU	mg	IU	mcg	mcg	mg	g	mg	mg	mcg	mg	mcg	mg	mcg
RDA	**1000**	**15**	**15**	**50**	**30**	**7**	**3**	**700**	**460**	**20**	**3**	**11**	**80**	**300**
Asparagus	905	8	1.35	134	45.5	0.82	202	21	49	5.5	0.54			0.15
Bamboo shoots						0.29	640		24	0.50	0.56			
Beetroot				68		0.67	229		32	0.60	0.3		20	
Bok choy	7,223		0.15	70	58		631	158	49				19	
Broccoli	1,207		1.13	84	110		229		52		1.2	0.35	16	
Brussels sprouts	1,209		0.67		219	1.87	495	56	87		0.51		31	
Butternut squash	22,868	31	2.64	39	2	1.23	582	84	55	1	0.27		59	
Cabbage		28.1	0.1	22	81.5	0.13	147	36	25				11	
Carrots	13,286		0.8	11	10.7		183		23					
Cauliflower				27	8.6									
Celery	782		0.53	33	56.7		426		38	1.5				
Chinese broccoli	1,441		0.42	87	74.6		230	88	36	1.1			16	
Chinese cabbage	1,151			63		0.36	268	38	46				12	
Corn	310					0.53	257		91		0.73		31	
Cucumber	55				8.5		76	8	12					
Daikon radish				25			419	25	35				13	
Eggplant			0.41	14	2.9	0.25	122							
Fennel	177			23		0.64	360		44				15	
French beans				133			655	112	181	2.1			99	
Green pepper	274		59.5	7	5.5		130		15					
Kale	17,707	53	1.1	17	1,062	1.17	296		36		1.2		23	
Leek	1,007		0.62		31.5	1.36	108	37	21				17	
Lima bean			0.36	156	3.8	4.49	955	209			1.79		81	0.45
Mushroom				6			11		30	3.3				
Okra	453		0.43	74	64				216	51			58	
Onion,				9										
Parsnip			1.56	90	1.6		573	58	108		0.41			
Peas	1,282		0.22		41.4	2.46	434	43	187	3000	1.9		62	
Pumpkin	12,230		1.96	22	2.0	1.40	564		76		0.56		22	
Spinach	2,183	8.40	0.61	58	145		167						24	
Spaghetti squash	170		0.19		1.2	0.53	181	33			0.31			

	Per 100 Gram Product													
	Vit A	Vit C	Vit E	Folate	Vit K	Iron	Potassium	Calcium	Phosphorus	Selenium	Zinc	Chromium	Magnesium	CU
Units	IU	mg	IU	mcg	mcg	mg	g	mg	mg	mcg	mg	mcg	mg	mcg
Summer squash	2,011		0.22	41	7.9		319	40	52		0.40			0.11
Winter squash	10,707		0.25	41	9.0		494	45	39		0.45			
Sweet potato	21,909		0.81		2.6		542	43	62		0.36		31	0.18
Taro	79		2.48	23	1.0		615		87		0.24		34	
Turnip		18.10		14	1.0		276	51	41				14	
Yellow squash	190	24.50	2.48		4.1		282	27	41				25	

Eat asparagus for vitamin K, folic acid, tryptophan, and vitamin C; mushrooms and mustard seed for selenium, copper, and vitamins B_2 and B_3; barley, brown rice, buckwheat, millet, oat, quinoa, rye, spelt, and whole wheat for manganese, selenium, tryptophan, and magnesium; lettuce, garlic, and onion for chromium; cashews for copper, magnesium, and tryptophan; peanut, pumpkin seed, sesame seed, sunflower seed, and walnuts for copper, magnesium, tryptophan, and manganese; black beans, dried peas, garbanzo beans, kidney beans, lentils, lima beans, navy beans, pinto beans, soy protein, and tofu for molybdenum, manganese, tryptophan, folic acid, copper, and fiber.

APPENDIX IV

Composition of Grains and Nuts

	Vit A	Vit C	Vit E	Folate	Vit K	Iron	K	Calcium	Phosphorus	Selenium	Zinc	Chromium	Mg	CU
Units	IU	mg	IU	mcg	mcg		g	mg	mg	mcg	mg	mcg	mg	mcg
RDA	1000	15	15	50	30	7	3	600	460	20	3	11	80	300
Almonds			7.43	14			200	75	137		0.87		76	
Amaranth				22	2.1		135	47	148	5.5	0.86		65	
Barley				16	1.33		93		54	8.60	0.82			
Brazil nuts			1.62	6			187	45	206	543	1.15		107	0.50
Buckwheat				30	2.20		460		347	8.80	2.40		231	1.10
Cashews			0.26	7	9.70	1.89	187		168	5.6	1.64		83	0.62
Chestnuts	22		0.42	59	0.66	0.76	497		90		0.48		28	
Coconut				21		1.94	285		90	8.10	0.88		26	0.34
Flaxseed						0.59	84		66					
Hazelnuts			4.26	32	4		193		82				46	0.49
Macadamia nuts			0.15				104		53				37	0.21
Millet				19		0.63	62		100				44	
Oats				56		4.27	429	54	523		3.97		177	0.62
Peanuts						0.64	187		101		0.94		50	
Pecans							116		79				34	0.34
Pine nuts				10	15.30		169		163				71	0.37
Pistachios	74		0.55	14	3.70		295		137					
Pumpkin seeds				16			223		233		2.17		156	0.36
Quinoa			0.63	42			172		152		1.00		64	
Rice, brown							79		77				44	
Rice, wild						0.60					1.34		32	
Sesame seed				9		1.31	42	88	57					0.36
Spelt			0.26				143		150		1.25		49	0.21
Sunflower seeds			7.40	67	0.80		241	20	327				37	

| | Per 100 Gram Product | | | | | | | | | | | | | |
	Vit A	Vit C	Vit E	Folate	Vit K	Iron	K	Calcium	Phosphorus	Selenium	Zinc	Chromium	Mg	CU
Units	IU	mg	IU	mcg	mcg		g	mg	mg	mcg	mg	mcg	mg	mcg
Walnuts			0.20	28	0.20	0.82	125	28	98				45	0.45
Wheat durum						3.5	431	34	508	89	4		144	0.55
Wheat, red						3.6	340	25	332	70	3		124	0.41
Wheat,				38			432	32	355					

APPENDIX V

The Value of Common Fish in Our Diet

Fish	Calories	Fat, g	Protein, g	Omega-3, mg	B_{12} mcg	Vit D IUs	Selenium mcg
Salmon	144	6	21	840	4.8	447	31
Canned tuna	73	1	17	196	2.2	154	60
Trout	143	6	20	905	3.5	645	24
Pollock	100	1	21	484	3.1	43	40
Catfish	122	6	16	165	2.4	81	8
Sardines	177	10	21	1.259	7.6	164	45
Mackerel	223	15	20	1,208	16.2	311	44
Baramundi	70	1	35	480	n/a	n/a	n/a

Glossary and Definitions

Adaptogen: Helps maintain homeostasis and decrease cellular sensitivity against stressors. Good examples are Maitake mushroom, ginger, blueberries, garlic, and aloe vera, but none are approved by the FDA.

Bile acid: Bile acid is produced in liver abnd it passes to duodenum via gall bladder. It is an ampiphilic, meaning both positively and negatively charged, steroid produced in the amount of 0.5 to one liter per day.

Blood: A total of five liters in an adult contains 55 percent plasma. It contains red blood cells, white blood cells,, hormones, oxygen, vitamins, and minerals. It is a liquid organ.

Blood-brain barrier (BBB): A mechanism that creates a selective barrier between brain capillaries and tissues and circulating blood; serves to protect the central nervous system. It is a physical barrier due to high electrical resistance and permeability of endothelial cells of blood capillaries, and a transport mechanism at the same time. Even sodium and potassium need a carrier-mediated mechanism to get in. Alcohol and caffeine can get in, but bacteria, viruses, and other harmful chemicals are stopped from getting in or else we would not survive too long. Sugar, oxygen, and amino acids can enter the cerebrospinal fluid.

Blood pressure: 115/76 millimeter is the best reading, representing systolic and diastolic events.

BMI (body mass index): BMI is a statistical measure of adiposity. It is calculated as weight in kilograms divided by the square of height in meters. A number below 15 represents starvation, from 15 to 18.5 represents underweight, greater than 25 represents an overweight condition, greater than 30 is obese, and close to 40 means morbid obesity. A good range is 20.0 to 24.9.

Body: As a molecular factory, it is a nanoscale energy machine.

Calories: Food calories in effect are kilocalories. Two thousand kilocalories per day is 7,920 BTU per day or 330 BTU per hour = 96.7 watt hour = 3.5 horsepower.

Capillary system: The 60,000-mile-long blood capillary system connects to every cell in our body.

CAT (computerized axial tomography) SCAN : Unlike two-dimensional X-ray imaging, tomography is cross-sectional imaging of, say, a 3 cubic mm slice of tissue or organ (*tomos* in Greek means slice).

Cells: There are 100 trillion cells in our body. These are highly coordinated units of our existence. Each has DNA.

Chromatid: Each threadlike strand of chromosome with double-helical structure intact.

Chromosome: A self-replicating genetic structure, composed primarily of proteins and DNA. Our genes reside in our chromosomes. We have twenty-three pairs of them, each with a strand from both parents. Chromosome is DNA on a protein spool. We see chromosomes more vividly during cell division.

Creatinine: A measure of muscle breakdown. Normal level is 0.6 to 1.2 mg/100 gram.

Digestion: Digestion is an eighty-hour process. It starts with the thought and smell of food. It is controlled by the autonomic nervous system. Digestive hormones are the key to our growth, reproduction, and existence. Food stays two hours in the stomach, five hours in the small intestine, and seventy-three hours in the large intestine.

DNA (deoxyribonucleic acid): DNA is composed of two antiparallel strands that wind about a common axis to form a double helix. Each strand of DNA is composed of a linear array of nucleotides bonded in such a way that the bases extend toward the central axis of the molecule, while the two backbones are composed of alternating sugar and phosphate subunits. The bases of the two strands are weakly bonded to each other in a complementary fashion. DNA is the master molecule that encodes genetic information.

Energy: Energy as ATP is the chemical currency for living systems.

Enzyme: Enzymes are made out of proteins. They are our molecular motors.

Esophagus: Delivers food from mouth to stomach by peristalsis.

FAD, FADH, and FADH$_2$: Flavin adenine dinucleotide is a molecule of vitamin riboflavin bound to phosphate of adenosine diphosphate (ADP). FAD can be reduced to FADH$_2$, gaining two electrons and two protons. FADH$_2$ is an energy-carrying coenzyme, and when oxidized, it delivers energy.

Genes: An ordered sequence of nucleotides that act as the functional subunit of hereditary information. The collection of genes in an organism determines the characteristics of that organism. We have 20,000 to 25,000 genes. An ordinary roundworm has 10,000 genes.

Health metrics: A rated index based on personal environment, lifestyle, diet, and social interaction. Like a credit rating, we must meet an index in excess of 800, 1000 being the maximum, and we should value it more highly and respectfully than a credit card. A rating of 800 health metrics implies restricted calorie diet, exercise, good sleep, meditation and mindfulness, and an optimal social and spiritual interaction.

Human body: The human body emerges from a single cell from Mom and Dad to a zygote to an embryo and finally to a walking and talking human being. The human being is a highly organized and ordered molecular factory, and its existence largely depends on the food he or she consumes.

Kidney: With 1 million nephrons, the two kidneys filter forty-five to fifty-two gallons of blood per day and adjust its pH; return good blood and a half gallon of waste fluid per day; measure and maintain levels of sodium, potassium, and phosphorus; and create the right balance for life. Also our kidneys produce three major hormones—erythropiotin, renin, and calcitrol (active vitamin D, which maintains calcium).

Large intestine: Houses E. coli, bifidobacteria, and lactobacilli bacteria, produces short-chain fatty acids from prebiotic dietary fibers, and reabsorbs and recycles part of the water in colonic residues.

Lung: With a surface area of a tennis court, our lungs are gas-exchange organs.

Meiosis: The process of cell division in which a single cell produces four daughter cells, each of which contains half of the number of chromosomes of the parent cell. For example, a single diploid spermatogonium (primordial germ cell) will divide meiotically to produce four haploid sperm cells.

Metabolome: The total number of metabolites present in a cell, tissue, and organ.

Microbiome or our second genome: Probiotic bacteria in the human colon and their genome.

Microbiome: The gene pool of intestinal bacteria that is integral to our health, immune system, and longevity.

Mitochondria: Mitochondria are the energy-production factory in each cell in our body. Evolved from bacteria, they are 70 percent efficient, compared to the hydrogen-oxygen fuel cells in spacecraft, which only 40 percent efficient. We age when mitochondria lose efficiency and too many electrons leak out and become free-radical super oxide.

Mitosis: The process of cell division in which a single cell produces two daughter cells that are identical to one another and to the original parent cell.

MRI: Our body has hydrogen atoms. The protons of water in any biologic tissue are mini-magnets. They can deflect the radio frequency waves from a powerful external magnetic field through which the test individual travels. The difference in incoming and outgoing radiofrequency radiation is used to produce cross-sectional images of organs and tissues (heart, lung, liver, breast, blood vessels, spleen, kidney, tumors, etc). The abnormal image is then compared with the normal.

NAD and NADH: Nicotinamide adenine dinucleotide, respectively in oxidized and reduced forms. NAD is an oxidizing agent, and NADH a reducing agent. NAD accepts electrons, and NADH donates electrons. These coenzymes are present in every cell such that we can use electrons in our metabolic oxidation-reduction reactions.

Neurons: There are 100 billion of these cells in our brain, and they connect in a billion ways for creating, keeping, and enhancing cognition and memory.

Nucleosome: Unit of DNA coiled around a spool of positively charged eight histone proteins; the coil is two-meter-long DNA. A nucleosome becomes a chromosome.

Omega-3 fatty acids: Decosahexaenoic acid (DHA) and eicosapentaenoic acid (EPA). These are essential fatty acids involved in critical life processes. They are critical to the functioning of brain, eyes, and nerves.

Pancreas: A gland that produces sodium bicarbonate, hormones secretin and cholecytokinin, intrinsic factor for vitamin B_{12}, and enzymes amylase, lipase, trypsin, trypsinogen, chymotrypsin, carboxypeptidase, elastase, and nuclease for routine digestion.

Perspiration: The brain has centers for temperature, heating, and cooling. Our energy machine, the body, produces heat from burning food, which is sufficient to raise the temperature of twenty-five liters of water to boiling. Our body would never allow it to happen, though. We perspire and lose 540 calories/g water.

Proteome: The entire complement of proteins that can be expressed in the human cell.

Proteosome: Protein complex present in nucleus and cell cytoplasm that is responsible for destruction of proteins produced with faults and imperfections. These are largely proteases.

Pulse rate: Our heart beats 3 billion times in our lifetime, an average of seventy beats per minute.

Reducol: Cholesterol-lowering plant sterol (ADM and Forbes Meditech are major manufacturers).

Ribose: A five-carbon sugar common to the structures of DNA, RNA, ATP, NAD, and FAD.

Saliva: We produce 1 to 1.5 liters of saliva per day, delivering 140 IU of enzyme amylase per liter. The water is recycled.

Skin: The largest organ of our body. Around 90 percent of heat is lost through our skin.

Small intestine: A twenty-foot-long section of digestive tract that includes the duodenum, jejunum, and ileum; capable of processing one liter fluid/day, 90 percent of which is reabsorbed in the large intestine as we complete the process of digestion.

Sneeze: We sneeze because of irritation of the respiratory walls due to foreign substances. It is a high-speed (102 miles per hour) protective response.

Sodium bicarbonate: pH buffering in our body is due to bicarbonate. In the digestive system, bicarbonate is produced by epithelial cells of the pancreas in order to counter acidity of chyme coming to the duodenum of the small intestine. The key enzyme is carbonic anhydrase for its production from carbon dioxide. Let us always remember that up to 75 percent of carbon dioxide is converted into bicarbonate. The air we inhale has only 0.04 percent carbon dioxide, but a lot more is produced by our cells. Its level in our blood is controlled, including that which is used in making bicarbonate by the pancreas.

Stomach: A one-liter hopper and acid reactor; it kneads, mixes, squeezes, and pushes chyme to the small intestine when it is not busy. Food resides in the stomach for around two hours. Hydrochloric acid is produced by the ATPase proton pump of G cells. Gastric juice has a pH in the range of 1 to 2.

Telomerase: An enzyme that lengthens telomere. Tumor cells get their immortality from telomerase.

Telomere: Repetitive nucleotide unit of up to 15,000 base pairs that act as a cap on the ends of chromosomes. They are there to protect chromosomes from faulty duplication, and they get shortened after every cell division. They are short in the case of our cells that become malignant.

Urea: A major daily waste of the human body to the extent of 20 mg/100ml or 0.02 percent in the urine. It is a measure of protein breakdown in the body.

List of References for Additional Reading

Chapter 1: Introduction

Adventist Health Study in California. *A Multiyear Long Study of 154,000 Vegetarian Male and Female Seventh-Day-Adventists, Vol. 5.* City of Publication: Publisher's Name, 2008.

Atkins, Robert C. *Atkins for Life.* New York: St. Martin's Press, 2003.

Deutsch, Ronald M. *The Family Guide to Better Food and Better Health.* New York: Bantam Books,, 1971.

Dolinoy, D. C., D. Huang, and R. L. Jirtle. "Maternal Nutrient Supplementation Counteracts Bisphenol A-Induced DNA Hypomethylation in Early Development." *Proceedings of National Academy of Sciences* 104 (2007): 13056–13061.

Dolinoy, D. C., J. R. Weidman, R. A. Waterland, and R. L. Jirtle.. "Maternal Genistein Alters Coat Color and Protects Avy Mouse Offspring from Obesity by Modifying the Fetal Epigenome." *Environmental Health Perspectives* 114 (2006): 567–572.

Kay-Tee Khaw and Nick Wareham (19930. The EPIC (European Perspective Investigations into Cancer), Norfolk Cohort, MRC Epidemiology Unit, University of cambridge (www.srl.cam.ac.uk/epic/

Evans, William, and Irwin H. Rosenberg. *Biomarkers.* New York: Fireside/ Simon and Schuster, 1992.

Willet, Walter C. (1991). Health Professionals Follow-up Study (HPSF) . Prospective studies of diet and cancer in men and women, www.hsph. harvard.edu/hpfs Iowa Women's Health Study, Epidemiology of cancer in a cohort of women (1985). NCI extramural program, Epidemiology and Genomics Research Program, Division of cancer control and population sciences. http://www.cancer.umn.edu/research/prevention-and-etiology/research-studies/index.htm

Kaati, G., L. O. Bygren, M. Pembrey, and M. Sjostrom. "Transgenerational Response to Nutrition, Early Life Circumstances and Longevity." *European Journal of Human Genetics* 15 (2007): 784–790.

Katz, David L. *The Flavor Point Diet*. New York: Rodale Press, 2005.

Kucharski, R., J. Maleszka, S. Foret, and R. Maleszka. "Nutritional Control of Reproductive Status in Honeybees via DNA Methylation." *Science* 319 (2008): 1827–1830.

Kurzweil, Raymond, and Terry Grossman. *Fantastic Voyage: Live Long Enough to Live Forever*. New York: Rodale, Inc., 2004.

McDougall, John A. *Maximum Weight Loss*. New York: Penguin Books USA, 1985.

McGowan, P. O., M. J. Meaney, and M. Szyf.. "Diet and the Epigenetic (Re) Programming of Phenotypic Differences in Behavior." *Brain Research* 1237 (2008): 12–24

Meyers, S. "Use of Neurotransmitter Precursors for Treatment of Depression." *Alternative Medicine Review*. 5(1) (2000): 64–71. http://umm.edu/health/medical/altmed/supplement/tyrosine.

Spreizer Frank (1976) and Willet Walter C. (1989). The Nurses' Health Study (1976–1989). Funded by National Institute of Health. (http://www.channing.harvard.edu/nhs/) Ornish, Dean. *The Spectrum*. New York: Ballantine Books/Random House, 2007.

Physicians' Health Study (1982). Bringham and Women's Hospital, Boston, MA (http://phs.bwh.harvard.edu/) Roizen, Michael F. *The Real Age Makeover*. New York: Harper Collins, 1999.

Roizen, Michael F., and Mehmet Oz. *You on a Diet*. New York: Free Press/Simon and Schuster, 2006.

Siri-Tarino, P. W., Q. Sun, F. B. Hu, and R. M. Krauss. "Meta-Analysis of Prospective Cohort Studies Evaluating the Association of Saturated Fat with Cardiovascular Disease." *The American Journal of Clinical Nutrition* 91 (2010) (3): 535–46. doi:10.3945/ajcn.2009.27725. PMID 20071648.

Siri-Tarino, P. W., Q. Sun, F. B. Hu, and R. M. Krauss. (2010) Metaanalysis of prospective cohort studies evalualting the assocoation of Saturated Fat with Cardiovascular Disease., *The American Journal of Clinical Nutrition* 91 (3): 535- 546. doi:10.3945/ajcn.2008.26285. PMID 20089734.

Dietary guidelines for Americans, 2005. U.S. Department of Health and Human Services (HHS) and *U.S. Department of Agriculture(USDA)*. http://www.health.gov/dietaryguidelines/dga2005/document/

Willett, Walter C. *Eat, Drink, and Be Healthy*. New York: Simon and Schuster, 2001.

Joint WHO/FAO Expert Consultation (2003). Technical Report series 916, Diet. *Nutrition and the Prevention of Chronic Diseases, Geneve, (http://whqlibdoc.who.int/trs/who_trs_916.pdf)*

World Health Organization, Milestones in health Promotion, First International Conference on Health Promotion, ottawa, November 1986http://www.who.int/healthpromotion/Milestones_Health_Promotion_05022010.pdf

World Health Organization (2007). Prevention of cardiovascular disease, a pocket guide for assessment and management of cardiovascular Risk,, Geneve (http://ish-world.com/downloads/activities/PocketGL_ENGLISH_AFR-D-E_rev1.pdf).

Chapter 2: Nutrition, Genomic Stability, and Epigenetics

Clayton T.A., Lindon J.C., Cloarec O, Antti H, Charuel C, Hanton G, Provost J.P., Le Net J.L., Baker D, Walley R.J., Everett J.R., Nicholson J.K. (2006) "Pharmaco-Metabonomic Phenotyping and Personalized Drug Treatment." *Nature* 440 (4): 1073–1077.

Dolinoy, D. C., D. Huang, and R. L. Jirtle. "Maternal Nutrient Supplementation Counteracts Bisphenol A-Induced DNA Hypomethylation in Early Development." *PNAS* 104 (2007): 13056–13061.

Dolinoy, D. C., J. R. Weidman, R. A. Waterland, and R. L. Jirtle. "Maternal Genistein Alters Coat Color and Protects Avy Mouse Offspring from Obesity by Modifying the Fetal Epigenome." *Environmental Health Perspectives* 114 (2006): 567–572.

Kaati, G., L. O. Bygren, M. Pembrey, and M. Sjostrom. "Transgenerational Response to Nutrition, Early Life Circumstances and Longevity." *European Journal of Human Genetics* 15 (2007): 784–790.

Kucharski, R., J. Maleszka, S. Foret, and R. Maleszka. "Nutritional Control of Reproductive Status in Honeybees via DNA Methylation." *Science* 319 (2008): 1827–1830.

Mansour, J. C., and R. E. Schwarz. "Molecular Mechanisms for Individualized Cancer Care." *J. American . College of Surgions.* 207 (August 2008) (2): 250–8.

McGowan, P.O., M. J. Meaney, and M. Szyf.. "A Diet and the Epigenetic (Re) Programming of Phenotypic Differences in Behavior." *Brain Research* 1237 (2008): 12–24.

Oldenburg, J., M. Watzka, S. Rost, and C. R. Müller. "VKORC1: Molecular Target of Coumarins." *J. Thrombosis and Hemostasis* 5 (July 2007) Suppl 1: 1–6.

PricewaterhouseCoopers' Health Research Institute(2009). Diagnostics 2009: moving towards Personalized Medicine. (http://www.pwc.com/us/en/healthcare/publications/diagnostics-2009-moving-towards-personalized-medicine.jhtml)

Saglio, G., A. Morotti, G. Mattioli, Mattoli, G., Messa, A., Guigliano, E., Volpe, G., Giovanna, R., and Celloni, D. "Rational Approaches to the Design of Therapeutics Targeting Molecular Markers: The Case of Chronic Myelogenous Leukemia." *Annals. New York Academy of Scencesi.* 1028 (December 2004): 423–31. (http://onlinelibrary.wiley.com/doi/10.1196/annals.1322.050/abstract)

Shastry, B. S. (2006). "Pharmacogenetics and the Concept of Individualized Medicine.", *Pharmacogenomics Journal* (1): 16–21.

van't Veer, L. J., and R. Bernards. "Enabling Personalized Cancer Medicine through Analysis of Gene-Expression Patterns." *Nature* 452 (April 2008) (7187): 564–70.

Chapter 3: Our Body, Our Mind, and the Body-Mind Connection

Chopra, Deepak. *Magical Mind, Magical Body: Mastering the Mind-Body Connection for Perfect Health and Total Well Being, An Audio CD.* City of publication: A Nightingale Conant Production, 2003.

Eagleman, Daniel 2011). Incognito: *The Secret Lives of the Brain.* City of publication: Pantheon Books, New York.

Freeman, Scott. *Biological Science.* Upper Saddle River: Prentice Hall, Inc., 2002.

Gray, Hnery.(1918). *Anatomy of tHE Human Body,* 20th ed., Lea & Febiger, pHILADELPHIA

Logan, Alan C. (2007). *The Brain Diet: The Connection between Nutrition, Mental Health, and Intelligence.* City of publication: Cumberland House Publishing Nashville, TN

O'Rahilly, R., Muller, F., Carpenter, S., and Swenson, R. (2004). *Basic Human Anatomy.* City of Publication: Online version of Dartmouth Medical School.

Pinkler, Steven. *How the Mind Works.* City of Publication: W W Norton and Company, 1999.

Ramchandran, V. S. *Phantoms in the Brain: Probing the Mysteries of the Human Mind.* City of Publication: Harper Perennial, 1998.

Schmidt, Michael A., and Jeffrey Bland. *Brain Building Nutrition: How Dietary Fats and Oils Affect Mental, Physical, and Emotional Intelligence.* City of publication: Frog Books, 1997.

Seaward, Brian Luke. *Achieving the Mind-Body-Spirit Connection—A Stress Management Workbook.* City of publication: Jones and Bartlett Publishers, 2004.

Smith, Penny, sr. ed. *First Human Body Encyclopedia.* City of publication: D K Publishing, Penguin Group, 2005.

Walker, Richard. *The Kingfisher First Human Body Encyclopedia.* Boston: Kingfisher, 1999.

Chapter 4: Our Second Brain and Food Processing in Our Body

Brody, Tom. *Nutritional Biochemistry.* San Diego: Academic Press, 1999.

Davenport, H. W. *Physiology of the Digestive Tract: An Introductory Text.* City of publication: Publisher, 1971.

Furness, John B. *The Enteric Nervous System.* New York: John Wiley & Sons, 2008.

Gershon, Michael. *The Second Brain: A Groundbreaking New Understanding of the Stomach and Intestine.* New York: Harper Collins Publishers, 1999.

Kong, F., and R. P. Singh. "Disintegration of Solid Foods in Human Stomach." *Journal of Food Scencei.* 73 (2008) (5): R67–80.

Maton, Anthea (1993). *Human Biology and Health.* Englewood Cliffs, NJ: Prentice Hall..

Morton, J. E. *Guts: The Form and Function of the Digestive System.* City of publication: Publisher, 1967.

Mullin, Gerald E. and Kathie Madonna Swift. *Inside Tract: Your Good Gut Guide to Great Digestive Health.* New York: Rodale, Inc., 2011.

Nicholos, Trent W., and Nancy Faass (2005). *Optimal Digestive Health.* Healing Arts Press, Rochester. Saladin, Kenneth S. *Anatomy & Physiology: The Unity of Form and Function,* 2nd ed. New York: McGraw-Hill, 2001.

Smith, M. E. (2001). *The Digestive System: Systems of the Body Series.* Churchil Levingstone/Elsevior, New York.

Chapter 5: Probiotic Foods and Our Second Genome

Adolfsson, O, S. N. Meydani, and R. M. Russell (2004). "Yogurt and Gut Function." American Journal of Clinical Nutrition. 80 (August (2): 245–56.

Corthesy, B., Gaskins, H.R., and Mercenier, A. (2007) "Cross-Talk between Probiotic Bacteria and the Host Immune System." Journal of Nutrition, 137 (2007) (3) 7815–7905.

Fedorak, R. N., and K. L. Madsen. "Probiotics and Prebiotics in Gastrointestinal Disorders." Current Opinion in Gastroenterology. 20 (March 2004) (2) 146–55.

Guarner, F., and J. R. Malagelada. "Gut Flora in Health and Disease." *Lancet* 361 (February 8, 2003) (9356): 512–9.

Mackie, R. I., A. Sghir, and H. R. Gaskins. "Developmental Microbial Ecology of the Neonatal Gastrointestinal Tract." American Journal of Clinical Nutrition. 69 (May 1999) (5): 1035S–1045S.

Montrose, D. C., and M. H. Floch. "Probiotics Used in Human Studies." Journal of Clinical Gastroenterology. 39 (July 2005) (6): 469–84.

Peltonen, R., W. H. Ling, O. Hanninen, and E. Eerola. "An Uncooked Vegan Diet Shifts the Profile of Human Fecal Microflora: Computerized Analysis of Direct Stool Sample Gas-Liquid Chromatography Profiles of Bacterial Cellular Fatty Acids. Applied Environtal Microbiology. 58 (November 1992) (11): 3660–6.

Salminen, S. J., M. Gueimonde, and E. Isolauri. "Probiotics that Modify Disease Risk." Journal of Nutrition. 135 (May 2005) (5): 1294–8.

Sartor, R. B. "Probiotic Therapy of Intestinal Inflammation and Infections." Current Opinion in Gastroenterology. 21 (January 2005) (1): 44–50.

Wheeler, J. G., S. J. Shema, M. L. Bogle, M. A. Shirrell, A. W. Burks, A. Pittler, and R. M. Helm. "Immune and Clinical Impact of Lactobacillus Acidophilus on Asthma." Annals of Allergy, Asthma, andImmunology. 79 (September 1997) (3): 229–33.

Chapter 6: Food Functions

Groff, James,L., Gropper, Sareen S., and Hunt Sara, M. (2009). *Advanced Nutrition and Human Metabolism*, 5th ed. Belmont CA: Wadsworth..

Insel, Paul. *Nutrition*. City of publication: Jones & Bartlett Learning, 2004.

McGee, Harold. *On Food and Cooking: The Science and Lore of the Kitchen*. New York: Scribner, 2004.

Point, Fernand. (2008). *Ma Gastronomie*. Groupe Flammrion, Paris,France:

Ross, A. C., Cabarello, Benjamin, Cousins, Robert, J., Tucker, K.L., and Zeigler, T.R. (2012).. *Modern Nutrition in Health and Disease*. City of publication: Lippincott Williams and Wilkins, Walterskluwer Health..

Chapter 7: Food and Disease Connection

Alzheimer's Association. *Alzheimer's Disease Facts and Figures 2007*. Chicago: Publisher, Date of publication.

American Diabetes Association. *The Dangerous Toll of Diabetes*. Alexandria: American Diabetes Association, 2005. May 18, 2007.

American Diabetes Association. *Direct and Indirect Costs of Diabetes in the United States*. Alexandria: American Diabetes Association, 2005. Available at http://www.diabetes.org/diabetes-statistics/cost-of-diabetes-in-us.jsp. Accessed September 20, 2007.

Barker, D. J. P. *Mothers, Babies and Health in Later Life*, 2nd ed. Edinburgh, UK: Churchill Livingstone, 1998.

Centers for Disease Control and Prevention. *Chronic Disease Overview: Costs of Chronic Disease*. Atlanta: Centers for Disease Control, 2005. Available at http://www.cdc.gov/nccdphp/overview.htm. Accessed July 24, 2007.

Centers for Disease Control and Prevention. *Overweight and Obesity*. Atlanta: Centers for Disease Control, 2007. Available at http://www.cdc.gov/nccdphp/dnpa/obesity/trend/index.htm. Accessed July 24, 2007.

Joseph, K. S., and M. S. Kramer. *Review of the Evidence on Fetal and Early Childhood*. City of publication: Publisher, 1996.

Mensah, G. A., and Brown,D.M. (2007). "An Overview of Cardiovascular Disease Burden in the United States." *Health Affairs* 26(1)): 38–48.

Murphy, Annie. "The Womb. Your Mother. Yourself." *Time*, October 4, 2010, p. 51.

National Heart, Lung, and Blood Institute. *Morbidity and Mortality: 2004 Chart Book on Cardiovascular, Lung, and Blood Diseases.* Bethesda, MD: National Institutes of Health, 2004.

Partnership for Solutions. *Chronic Conditions: Making the Case for Ongoing Care, September 2004 Update.* Baltimore: Johns Hopkins University, 2004. Available at http://www.partnershipforsolutions.org/DMS/files/chronicbook2004.pdf. Accessed July 24, 2007.

Partnership to Fight Chronic Disease. *The Implications for Individuals.* Washington, DC: Partnership to Fight Chronic Disease, 2007.

Prentice, A. M. "Intrauterine Factors, Adiposity, and Hyperinsulinaemia." British. Medical. Journal. 327 (2003): 880–881.

World Health Organization/Food and Agriculture Organization. *Diet, Nutrition and the Prevention of Chronic Diseases.* WHO Technical Report Series 916, 2003. World Health Organization, Geneva, Switzerland.

Yajnik, C. S. "Interactions of Perturbations in Intrauterine Growth and Growth During Childhood on the Risk of Adult-Onset Disease." Proceedings of Nutrition. Society. 59 (2000): 257–265.

Chapter 8: Foods That Build Immunity

Daizo, A., Y. Egashira, and H. Sanada. (2005)"Suppressive Effect of Corn Bran Hemicellulose on Liver Injury Induced by D-Galactosamine in Rats.", Nutrition, 10:1044-1051

Fehervari, Z., and Shimon Sakaguchi. "Mysteries of the Immune System." *Scientific American* October 2006, pp. 57–63.

Galisteo, M., J. Duarte, and A. Zarzuelo. "Effects of Dietary Fibers on Disturbances Clustered in the Metabolic Syndrome." *The Journal of Nutritional Biochemistry* vol. 19, issue 2, pp. 71–84.

Gorsek, Wayne F. Cancer Immune Composition for Prevention and Treatment of Individuals. US Patent 6582723, 2003.

Nagata, Jun-ichi, Yutaka Higashiuesato, Goki Maeda, Isao Chinen, Morio Saito, Kazuya Iwabuchi, and Kazunori Onoē. (2001). "Effects of Water-Soluble Hemicellulose from Soybean Hull on Serum Antibody Levels and Activation of Macrophages in Rats." *Journal of. Agriculture and Food Chemistry.* 49 (2001) (10), 4965–4970.

Chapter 9: Foods That Fight Osteoporosis and Osteoarthritis

Anderson, J. J. B. (2004). Nutritional Assessment: Analysis of relations between nutritional factors and bone health, in "Nutrition and Bone Health, (Hollick, M.F. and Dawson-Hughes, B. eds), Humana Press, Totowa, NJ Bonaluti, D., Shea,B., Lovine,R., Negrini, S., Robinson, V., Kemper, H.C., Wells, G., Tugwell, P, and Crannery, A. (2002). *Exercise for Preventing and Treating Osteoporosis in Post-Menopausal Women*, Cochrane Database of Systematic Review, 7:CD000333

Dawson Highes, B. "Effect of Calcium and Vitamin D Supplementation on Bone Density in Men and Women 65 Years and Older." *New England Journal of Medicine* 337 (1997) (10): 670–676.

Holt, E. H. Diseases of Calcium Metabolism and Metabolic Bone Disease, ACP Medicine Section 3, Chapter 6, Hamilton, ONT, 2008.

National Osteoporosis Association. *Prevention Exercise for Healthy Bone*. City of publication: Publisher, 2008.

Rosen, C. J. "Restoring Aging Bone." *Scientific American* 288 (2003): 70–77.

Zhang X, Shu XO, Li H. (2005). Prospective cohort study of soy food consumption and risk of bone fracture among postmenopausal women. Archives of International Medicine, 165(16):1890-1895.

Chapter 10: Foods That Fight Chronic Pain

Bowe, Whitney P., and Alan C. Logan. "Acne Vulgaris, Probiotics and the Gut-Brain-Skin Axis—Back to the Future?" *Gut Patholog.* 3 (2011): 1.

Denk, F., and S. B. McMahon.. "Chronic Pain: Emerging Evidence for the Involvement of Epigenetics." *Neurons*, 73 (2012) (3): 435–444.

Dubner, Richard, and Michael Gold. "The Neurobiology of Pain." *National Academy of Sciences,* 96 (1999) (14): 7627–7630.

Harper, Jean. *Food: Your Miracle Medicine.* New York: Harper Perenials, A Div of Harper Collins, Publishers, 1993.

Hunr, Stephen, and Martin Kolzburgh. *The Neurobiology of Pain.* City of publication: Oxford University Press, 2005.

Institute of Medicine. *Relieving Pain in America, Committee on Delivering Pain Research, Cases, and Education.* Washington DC, Publisher, 2011.

Levingston, William K. *Pain and Suffering.* City of publication: Publisher, 1996.

Murray, Michael. *Encyclopedia of Natural Medicine,* 2nd edition. New York: Random House, Inc., 1997.

Pitchford, Paul. *Healing with Whole Foods.* Berkeley, CA: North Atlantic Books, 2002.

Ren, K., and R. Dubner. "Interaction between the Immune System and Pain." *Natural Medicine* 16 (2010) (11): 1267–1276.

Silverthorn, Dee. *Human Physiology: An Integrated Approach,* 2nd ed. Upper Saddle River,, NJ: Publisher, 2001.

Vandenkerkhof, Elizabeth G., McDonald, H.M., Jones, G.T., Powers, C., and mcfarlane, G.J. (2011).. "Diet, Lifestyle, and Chronic Widespread Pain— 1958 British Birth Cohort study." *Pain Research Management,* 16 (2): 87–92.

Wall, Patrick. *The Science of Suffering.* London: Waldenfeld and Nicolso, 1999.

Wall, Patrick, and Donald Melzack. *Textbook of Pain.* Edinburgh: Churchill Levingston/ Elsevier, NY, 2000.

Young, E. E. "Genetic Basis of Pain Variability—Research Advances." *J. Medical Genetics* 49 (2012) (1): 1–9.

Chapter 11: Foods for Skin and Hair Health

Chang, L. "Top Ten Foods for Healthy Hair.". WebMD, 2009., http://www. webmd.com/beauty/hair-styling/top-10-foods-for-healthy-hair

Denda, M, K. Takei, and S. Denda. "How Does Epidermal Pathology Interact with Mental State?" *Medical Hypotheses* 80 (February 2013) (2):194–6. doi: 10.1016/j.mehy.2012.11.027. Epub 2012 Dec 12.

Fessing MY[1], Mardaryev AN, Gdula MR, Sharov AA, Sharova TY, Rapisarda V, Gordon KB, Smorodchenko AD, Poterlowicz K, Ferone G,Kohwi Y, Missero C, Kohwi-Shigematsu T, Botchkarev VA. (2011). p63 regulates Satb1 to control tissue-specific chromatin remodeling during development of the epidermis. Journal of Cell Biology, 194(6): 825-839

Gee, S. N., L. Zakhary, N. Keuthen, D. Kroshinsky, and A. B. Kimball. "A Survey Assessment of the Recognition and Treatment of Psychocutaneous Disorders in the Outpatient Dermatology Setting: How Prepared Are We?" *Journal of American Academy of Dermatology.* 68 (January 2013) (1): 47–52. doi: 10.1016/j.jaad.2012.04.007. Epub 2012 Sep 3.

Ghajarzadeh, M, M. Ghiasi, and S. Kheirkhah. "Associations Between Skin Diseases and Quality of Life: A Comparison of Psoriasis, Vitiligo, and Alopecia Areata." *Acta Medica Iran.* 50 (2012) (7): 511–5.

Gray, John. *The World of Hair—A Scientific Companion.* City of publication: McMillan Press, 1977.

Kurtzwell, P., and T. A. Young. "Vitamin of the Month—Biotin." *FDA Consumer* 25 (1991) (8): 34.

Leslie, Mitch. "p63 Delegates during Skin Development." *Journal of Cell Biol0gy.* 194 (2011) (6): 808.

Magin, P., J. Adams, G. Heading, D. Pond, and W. Smith. "Psychological Sequelae of Acne Vulgaris: Results of a Qualitative Study." *Can Family Physicians.* 52 (2006): 978–9.

Mallon, E., J. N. Newton, A. Klassen, S. L. Stewart-Brown, T. J. Ryan, and A. Y. Finlay. (1999). "The Quality of Life in Acne: A Comparison with General Medical Conditions Using Generic Questionnaires." *British Journal of Dermatoogyl.* 140 : 672–676.

Rabin, F., S. I. Bhuiyan, T. Islam, M. A. Haque, and M. A. Islam. "Psychiatric and Psychological Comorbidities in Patients with Psoriasis—A Review." *Mymensingh Medical Journal.* 21 (October 2012) (4): 780–6.

Rapp, D. A., G. A. Brenes, S. R. Feldman, A. B. Fleischer, G.; F. Graham, M. Daily, and S. R. Rapp. "Anger and Acne: Implications for Quality of Life, Patient Satisfaction and Clinical Care." *British Journal of Dermatology.* 151 (2004): 183–9.

Stokes, J. H., and D. H. Pillsbury. "The Effect on the Skin of Emotional and Nervous States: Theoretical and Practical Consideration of a Gastrointestinal Mechanism." *Archives of Dermatology and Syphilology.* 22 (1930): 962–93.

Thorslund, K., B. Amatya, A. E. Dufva, and K. Nordlind. (2013). "The Expression of Serotonin Transporter Protein Correlates with the Severity of Psoriasis and Chronic Stress." *Archives of Dermatology Research.* 305 (2): 99–104.

Yaghmaie, P., C. W. Koudelka, and E. L. Simpson. (2013). "Mental Health Comorbidity in Patients with Atopic Dermatitis." *Journal of Allergy and Clinical Immunology.* 131 (2): 428–33.

Chapter 12: Foods for Diabetes Therapy

American Diabetes Association. "Nutrition Recommendations and Interventions for Diabetes." *Diabetes Care,* 31 (2008) (Suppl 1): S61–S78.

American Diabetes Association. "Prevention of Type 1 Diabetes Mellitus: Clinical Practice Recommendations 2004." *Diabetes Care*, 27 (2004) (Suppl 1): S133.

Brownley, Kimberley A., Light, K.C., Grewen, K.M., Bragdon, E.E., Hendenliter, A,L., and West, S.G. (2005). "Postprandial Ghrelin Is Elevated in Black Compared to White Women." *Journal Clinical Endocrinology and Metabolism*, 89 (9): 4457–4463.

Davis, T. M., I. M. Stratton, C. J. Fox, Holman, R.R., and Turner, R.C (1997). "U.K. Prospective Diabetes Study 22. Effect of Age at Diagnosis on Diabetic Tissue Damage during the First 6 Years of NIDDM." *Diabetes Care* 20 (9): 1435–1441.

Meslier, N., T. Gagnadoux, P. Giraud, Person, C., Ouksel, H., Urban, T., aqnd Raceneux, J.L. (2003). "Impaired Glucose-Insulin Metabolism in Males with Obstructive Sleep Apnoea Syndrome." *European Respiratory Journal* 22 (1): 156–160.

Purnell, J. Q. . "Obesity." In D. C. Dale and D. D. Federman, eds. *ACP Medicine*, section 3, chap. 10. Hamilton, ON: BC Decker, 2008.

Shobhana, R., Rama Rao, P., Lavanya,,A., Williams, R., Vija, V., and Ramachandran A. (2000). "Expenditure on Health Care Incurred by Diabetic Subjects in a Developing Country—A Study from Southern India." *Diabetes Research in Clinical Practice* 48 (1): 37–42.

U.S. Department of Health and Human Services. *2008 Physical Activity Guidelines for Americans* (ODPHP Publication No. U0036). Washington, DC: U.S. Government Printing Office, 2008. Available online: http://www.health.gov/paguidelines/pdf/paguide.pdf

World Health Organization. *Prevention of Diabetes Mellitus. Report of a WHO Study Group.* Geneva: World Health Organization, 1994. No. 844.

Yannakoulia, M., Yannakourius, N., Meristas, L., Kontogianni, M.D., Melagiris, I., and Montzoros, C.S. (2008). "A Dietary Pattern Characterised by High Consumption of Whole-Grain Cereals and Low-Fat Dairy Products and Low Consumption of Refined Cereals Is Positively Associated with Plasma Adiponectin Levels in Healthy Women." *Metabolism Clinical and Experimental,* 57 : 824–830.

Chapter 13: Foods for Fighting Cardiovascular Diseases

German, J. B., R. A. Gibson, R. M. Krauss RM, Nestel, P., Lamarche, B., and van Staveran, W.A.(2009). "A Reappraisal of the Impact of Dairy

Foods and Milk Fat on Cardiovascular Disease Risk." *European Journal of Nutrition* 48 (4): 191–203. Glew, R. H., M. Maragaret, W., Conn, C.A, Cadena, S.M., Crossey, M., Okolo, S.N, Vanderjagat, D.J. (2001). "Cardiovascular Disease Risk Factors and Diet of Fulani Pastoralists of Northern Nigeria." *The American Journal of Clinical Nutrition* 74 (6): 730–736.

Halton, T. L., W. C. Willett, Liu, S. (2006). . "Low-Carbohydrate-Diet Score and the Risk of Coronary Heart Disease in Women." *The New England Journal of Medicine* 355 (19): 1991–2002.

Lichtenstein, A. H., Appel, L.J., Brands, M. Carnethon M, Daniels S, Franch HA, Franklin B,Kris-Etherton P, Harris WS, Howard B, Karanja N, Lefevre M, Rudel L, Sacks F, Van Horn L, Winston M, Wylie-Rosett J. (2006). "Diet and Lifestyle Recommendations Revision 2006: A Scientific Statement from the American Heart Association Nutrition Committee." *Circulation* 114 (1): 82–96.

Mente, A., L. de Koning, H. S. Shannon, and S. S. Anand. (2009). "A Systematic Review of the Evidence Supporting a Causal Link between Dietary Factors and Coronary Heart Disease." *Archives of International. Medicine.* 169 (7): 659–69.

Mozaffarian, D., E. B. Rimm, and D. M. Herrington. "Dietary Fats, Carbohydrate, and Progression of Coronary Atherosclerosis in Postmenopausal Women." *The American Journal of Clinical Nutrition* 80 (November 2004) (5): 1175–84.

Prior, I. A., F. Davidson, C. E. Salmond, and Z. Czochanska. "Cholesterol, Coconuts, and Diet on Polynesian Atolls: A Natural Experiment: The Pukapuka and Tokelau Island Studies." *The American Journal of Clinical Nutrition* 34 (August 1981) (8): 1552–61.

Smith, S. C., Jackson, R. . Pearson, T. A., Fuster, V., Feargeman, O., Wood, D.A, Alderman, M., Morgan, J., Home, P., Hunt, M., and Grundy, S.M. (2004). "Principles for National and Regional Guidelines on Cardiovascular Disease Prevention: A Scientific Statement from the World Heart and Stroke Forum." *Circulation* 109 (25): 3112–21.

*Austin, G.L., Ogdon, L.G., and Hill, J.O. (2011).*Trends in carbohydrate, fat, and protein intakes and association with energy intake in normal-weight, overweight, and obese individuals: 1971–2006, American Journal of Clinical Nutrition, 93(4): 836-843.

Chapter 14: Foods for Fighting Cancer

American Institute of Cancer Research Science Now.(2010). "Fat: Finding the Cancer Link." *ScienceNow* 32, Spring 2010. http://preventcancer.aicr.org/site/DocServer/ScienceNow32.pdf?docID=3721

Ames, B. N. "Micronutrients Prevent Cancer and Prevent Aging." *Toxicology Letters* 102–103 (1988): 5–18.

Cross, A. J., Leitzmann, M.F., Subar, A.F., Thompson, F.E., Hollenbeck,A.R., and A. Schatzkin, A. (2008). "A Prospective Study of Meat and Fat Intake in Relation to Small Intestinal Cancer." *Cancer Research*, 68(22); 9274-9279

Diet and Cancer Report (2007). American Institute of Cancer Research, World Cancer Research Fund (2007). Food, Nutrition, Physical Activit and Prevention of Cancer.

http://www.dietandcancerreport.org/cancer_resource_center/downloads/summary/english.pdf

Gross, L. S., L. Li, E. S. Ford, and S. Liu.(2004). "Increased Consumption of Refined Carbohydrates and the Epidemic of Type 2 Diabetes in the United States: An Ecologic Assessment." *Am Journal Clinical Nutrition* 79 (5): 774–779.

Kurahashi, N., M. Inoue, M. Iwasaki, S. Sasazuki, and A. S. Tsugane. "Dairy Product, Saturated Fatty Acid, and Calcium Intake and Prostate Cancer in a Prospective Cohort of Japanese Men." *Cancer Epidemiology, Biomarkers & Prevention* 17 (April 2008) (4): 930–7.

Lewis, C. J., and E. A Yetley. "Health Claims and Observational Human Data: Relation between Dietary Fat and Cancer." *American Journal of Clinicakl Nutrition* 69 (June 1, 1999) (6): 1357S–1364S.

McNaughton, S. A., G. C. Marks, and A. C. Green.(2005). "Role of Dietary Factors in the Development of Basal Cell Cancer and Squamous Cell Cancer of the Skin." *Cancer Epidemiology. Biomarkers Prevention.* 14 (7): 1596–1607.

Ornish, Dean. *The Spectrum.* New York: Ballantine Books/Random House, 2007.

Sonestedt, E., Ulrika Ericson, Bo Gullberg, Kerstin Skog, Håkan Olsson, and Elisabet Wirfält. "Do Both Heterocyclic Amines and Omega-6 Polyunsaturated Fatty Acids Contribute to the Incidence of Breast Cancer in Postmenopausal Women of the Malmö Diet and Cancer

Cohort?" *The International Journal of Cancer* (UICC International Union against Cancer) 123 (2008) (7): 1637–1643.

Chapter 15: Foods for Long Life

Gavrilov, L. A., and N. S. Gavrilova. *Biology of Life Span: A Quantitative Approach.* City of publication: Harvard Academic Publications, 1991.

Goldman, Robert, with Ronald Klatz and Liza Berger. *Brain Fitness: Anti-Aging Strategies to Fight Alzheimer's Disease, Supercharge Your Memory, Sharpen Your Intelligence, De-Stress Your Mind, Control Mood Swings, and Much More.* City of publication: Publisher, 1995.

Hall, Stephen S. *Merchants of Immortality: Chasing the Dream of Human Life Extension.* City of publication: Houghton Mifflin Company, 2003.

Kurzweil, R., and T. Grossman. *Fantastic Voyage: The Science Behind Radical Life Extension.* City of publication: Rodale, 2004.

Miller, P. L., and Monica Reinagel. *The Life Extension Revolution: The New Science of Growing Older without Aging.* City of publication: Bantam, 2005.

Pauling, Linus. *How to Live Longer and Feel Better.* City of publication: W. H. Freeman and Company, 1986.

Ward, Dean. *Biological Aging Measurement: Clinical Applications.* City of publication: The Center for Bio-Gerontology, 1988.

Ward, Dean, John Morgenthaler, and Steven Wm Fowkes. *Smart Drugs II: The Next Generation: New Drugs and Nutrients to Improve Your Memory and Increase Your Intelligence.* City of publication: Smart Publications, 1993.

Williams, R. J. *Nutrition Against Disease.* City of publication: Pitman Publishing Corporation, 1971.

Chapter 16: Antioxidants, Vitamin D, Enzymes, and Gene Therapy

Antioxidant therapy

Anand, Preetha, Ajaikumar B. Kunnumakkara, Robert A. Newman, and Bharat B. Aggarwal, . "Bioavailability of Curcumin: Problems and Promises." *Molecular Pharmaceutics* 4 (2007) (6): 807–18.

Arts, I. C., and P. C. Hollman. "Polyphenols and Disease Risk in Epidemiologic Studies." *The American Journal of Clinical Nutrition* 81 (2005) (1 Suppl): 317S–325S.

Aviram, M., and M. Rosenblat. "Paraoxonases and Cardiovascular Diseases: Pharmacological and Nutritional Influences." *Current Opinion in Lipidology* 16 (2005) (4): 393–9.

De Vera, Mary, M. Mushfiqur Rahman, James Rankin, Jacek Kopec, Xiang Gao, and Hyon Choi. "Gout and the Risk of Parkinson's Disease: A Cohort Study." *Arthritis & Rheumatism* 59 (2008) (11): 1549–54. doi:10.1002/art.24193. PMID 18975349.

EFSA Panel on Dietetic Products, Nutrition and Allergies. "Scientific Opinion on the Substantiation of Health Claims Related to Various Food(S)/Food Constituent(s) and Protection of Cells from Premature Aging, Antioxidant Activity, Antioxidant Content and Antioxidant Properties, and Protection of DNA, Proteins and Lipids from Oxidative Damage Pursuant to Article 13(1) of Regulation (EC) No 1924/2006." *EFSA Journal* 8 (2010) (2): 1489.

Frei, B.. "Controversy: What Are the True Biological Functions of Superfruit Antioxidants?" (April 1, 2009) Retrieved February 5, 2010., http://ssf.f15ijp.com/wiki/index.php/Polyphenol_antioxidant

U.S. department Human Health Services, Food and Drug Administration, Center for Food Safety and Applied Nutrition (2008). Guidance for Industry, Food Labeling; Nutrient Content Claims; Definition for "High Potency" and Definition for "Antioxidant" for Use in Nutrient Content Claims for Dietary Supplements and Conventional Foods, July 2008.

Kurien, Biji T., Anil Singh, Hiroyuki Matsumoto, and R. Hal Scofield. "Improving the Solubility and Pharmacological Efficacy of Curcumin by Heat Treatment." *ASSAY and Drug Development Technologies* 5 (2007) (4): 567–76.

Lotito, S., and B. Frei. "Consumption of Flavonoid-Rich Foods and Increased Plasma Antioxidant Capacity in Humans: Cause, Consequence, or Epiphenomenon?" *Free Radical Biology and Medicine* 41 (2006) (12): 1727–46.

Nair, Hareesh B., Bokyung Sung, Vivek R. Yadav, Ramaswamy Kannappan, Madan M.Chaturvedi, and Bharat B. Aggarwal. "Delivery of Antiinflammatory Nutraceuticals by Nanoparticles for the Prevention and Treatment of Cancer." *Biochemical Pharmacology* 80 (2010) (12): 1833–1843.

Haytowitz, B. and Bhagwat, Seena (2010). US Department Of Agriculture, Agriculture research service, Beltsville Human Research Center, (2010).

Database for the Oxygen Radical Absorbance Capacity (ORAC) of Selected Foods—2010 Release 2.

http://www.orac-info-portal.de/download/ORAC_R2.pdf

Stauth, David. "Studies Force New View on Biology of Flavonoids." *EurekAlert!* Adapted from a news release issued by Oregon State University.

http://www.eurekalert.org/pub_releases/2007-03/osu-sfn030507.php

Stocker, R., Y. Yamamoto, A. McDonagh, A. Glazer, and B. Ames. "Bilirubin Is an Antioxidant of Possible Physiological Importance." *Science* 235 (1987) (4792): 1043–6. Bibcode:1987Sci … 235.1043S.

Williams, Robert J., Jeremy P.E. Spencer, and Catherine Rice-Evans. "Flavonoids: Antioxidants or Signalling Molecules? *Free Radical Biology and Medicine* 36 (2004) (7): 838–49.

Zawiasa, A., M. Szklarek-Kubicka, J. Fijałkowska-Morawska, D. Nowak, J. Rysz, B. Mamełka,and M. Nowicki.. "Effect of Oral Fructose Load on Serum Uric Acid and Lipids in Kidney Transplant Recipients Treated with Cyclosporine or Tacrolimus." *Transplantation Proceedings* 41 (2009) (1): 188–91. doi:10.1016/j.transproceed.2008.10.038. PMID 19249511.

Vitamin D therapy

Chapuy, M. C., Arlot, M.E., Duboeuf, F., Brune, J., Crouzet, B., and Amond, S. (1997). (1997). "Prevalence of Vitamin D Insufficiency in an Adult Normal Population." *Osteopros Internationalt* 7 : 439–443.

deLuca, H. F.. "Overview of General Physiologic Features and Functions of Vitamin D." *Amer Journal of Clinical Nutrition*, 80 (2004) (Suppl): 1689s–1696s.

Feskanich, D., Ma, J., Fuchs, C.S., Kirkner, G.J., Hankinson, S.E., Holles, B.W., and Giovannucci, E.L. (2004). "Plasma Vitamin D Metabolites and Risk of Colorectal Cancer in Women." *Cancer Epidemiology Biomarkers Preview* 13(9) : 1502–1508.

Garland, C. F., Garland F.C., Gorham, E.D., Lipskin, M., Newmark, H., Mohr, S.B., and Holick, M.F. (2006). . "The Role of Vitamin D in Cancer Prevention." *American Journal of Public Health*, 96(2): : 252–261.

Glovannucci, E., Liu, Y., Rimm, E.B., Hollis, B.W., Fuchs, C.S., Stampfer, M.J. and Willet, W.C. (2006). "Prospective Study of Predictors of Vitamin D Status and Cancer Incidence and Mortality in Men." *Journal of National Cancer Institute*, 98 (7): 451–459.

Gorham, E. D., Garland, C.F., Garland, F.C., Grant, W.B., Mohr, S.P., Lipkin, M., Newmark, M., Gioannucci, E., a]Wei, M., and Holick, M. (2005). "Vitamin D and Prevention of Colorectal Cancer." *Journal of Steroid Biochemistry and Molecular Biology*, 97 : 179–194.

Holick, Michael F. "High Prevalence of Vitamin D Inadequacy and Implications for Health." *Mayo Clinic Proceedings* 81 (2006): 353–373.

Holick, Michael F. "Vitamin D Deficiency." *New England Journal of Medicine* 357 (2007) (3) 266–281.

Holick, Michael F. (2005). "Vitamin D for Health and Chronic Kidney Disease." *Seminars in Dialysis* 18 : 266-275.

Hypponen, E., Larra, E., Reunanen, E., Jarvelin, M.R., and Vitanen, S.M. (2001). "Intake of Vitamin D and Risk of Type 1 Diabetes: A Birth Cohort Study." *Lancet* 358 : 1500-1503.

Li, Y. C. "Vitamin D Regulation of the Renin-Angiotensin System." *Journal of Cell Biochemistry* 88 (2003): 327–331.

VanAmerongen, B. M., Dijkstra, C.D., Lips, P., and Polemen, C.H. (2004). "Multiple Sclerosis and Vitamin D: An Update." *European Journal of Clinical Nutrition* 58 (8): 1095–1099.

Wolpowitz, D., and B. A. Gilchrest. "The Vitamin D Questions: How Much Do You Need and How Should You Get It." *Journal of American Academy of Dermatology* 54 (2006): 301–317.

Zimmerman, A. "Vitamin D and Diseases Prevention with Special Reference to Cardiovascular Disease." *Progress in Biophysics and Molecular Biology* 92 (2006): 39–48.

Zimmerman, A., Schleithoff, S.S., Tendrich, G., Berthold, H.k. (2003). "Low Vitamin D Status: A Contributing Factor in the Pathogenesis of Congestive Heart Failure." *Journal of Ameicanr College of Cardiology* 41 (1): 105–112.

Enzyme therapy

Gonzalez, N. J., and L. L. Isaacs. (1999). "Evaluation of Pancreatic Proteolytic Enzyme Treatment of Adenocarcinoma of the Pancreas, with Nutrition and Detoxification Support." *Nutrients and Cancer* 33 (2): 117–124.

Gonzalez, N. J., and L. L. Isaacs. "The Gonzalez Therapy and Cancer: A Collection of Case Reports." *Alternative Therapeutic Health Medicine* 13 (2007) (1): 46–55.

Gotze, H., and S. S. Rothman. "Enteropancreatic Circulation of Digestive Enzymes as a Conservative Mechanism." *Nature* 257 (1975) (5527): 607–609.

Howell, Dr. Edward. *Enzyme Nutrition: The Food Enzyme Concept.* Wayne NJ: Avery Publishing Group, Inc., 1985.

Howell, Dr. Edward. *Food Enzymes for Health & Longevity.* Twin Lakes WI: Lotus Press, 1994.

Jensen, Bernard (1981). *Tissue Cleansing through Bowel Management. Maximbooks,*Escondido, CA:

Liebow, C., and S. S. Rothman. "Enteropancreatic Circulation of Digestive Enzymes." *Science* 189 (1975) (4201), 472–474.

McClain, M. E. *Scientific Principles in Nursing.* St. Louis: CV Mosby Company, 1950, p. 168.

Moskvichyov, B. V., E. V. Komarov, and G. P. Ivanova. (198 "Study of Trypsin Thermodenaturation Process." *Enzyme Microbiology and Technology* 8 : 498–502.

Santillo, M. H., Humbart, N.D. (1993). . *Food Enzymes: The Missing Link to Radiant Health.* Hohm Press, Prescott AZ

Saruc, M., S. Standop, J. Standop, F. Nozawa, A. Itami, K. K. Pandey, S. K. Batra, N. J. Gonzalez, P. Guesry, and P. M. Pour. "Pancreatic Enzyme Extract Improves Survival in Murine Pancreatic Cancer." *Pancreas* 28 (2004) (4): 401–412.

Wiggins, Shively F. H. "Case of Multiple Fibrosarcoma of the Tongue, with Remarks on the Use of Trypsin and Amylopsin in the Treatment of Malignant Disease." *JAMA* 47, 2003–2008, 1906.

Shively F. H. (1969). *Multiple Proteolytic Enzyme Therapy of Cancer.* Dayton: Johnson-Watson.

Gene therapy

Beard, J. *The Enzyme Treatment of Cancer.* London: Chatto and Windus, 1911.

Cassileth, B. *The Alternative Medicine Handbook.* New York: W. W. Norton & Co, 1998.

Dale, P. S., C. P. Tamhankar, D. George, and G. V. Daftary. (2001). "Co-Medication with Hydrolytic Enzymes in Radiation Therapy of Uterine Cervix: Evidence of the Reduction of Acute Side Effects." *Cancer Chemotherapy and Pharmacology.* 47 (July Suppl):S29–34.

Ernst, E. "Complementary Therapies in Palliative Cancer Care." *Cancer* 91 (June 1, 2001) (11):2181–5.

Gonzalez, N. J., and L. L. Isaacs. "Evaluation of Pancreatic Proteolytic Enzyme Treatment of Adenocarcinoma of the Pancreas, with Nutrition and Detoxification Support. *Nutrients and Cancer* 33 (1999):117–124.

Gonzalez, N. J., and L. L. Isaacs. (2007). "The Gonzalez Therapy and Cancer: A Collection of Case Reports." *Alternative Therapeutic Health Medicine* 13 (1), 46-55.

Gotze, H., and S. S. Rothman. "Enteropancreatic Circulation of Digestive Enzymes as a Conservative Mechanism." *Nature* 257 (1975) (5527): 607–609.

Gujral, M. S., P. M. Patnaik, R. Kaul, H. K. Parikh, C. Conradt, C. P. Tamhankar, and G. V. Daftary. (2001). "Efficacy of Hydrolytic Enzymes in Preventing Radiation Therapy-Induced Side Effects in Patients with Head and Neck Cancers. *Cancer Chemotherapy and Pharmacology* 47 (July Suppl): S23–28.

Liebow, C., and S. S. Rothman. "Enteropancreatic Circulation of Digestive Enzymes." *Science* 189 (1975) (4201): 472–474.

Martin, T., K. Uhder, R. Kurek, S. Roeddiger, L. Schneider, H. G. Vogt, R. Heyd, and N. Zamboglou. (2002). "Does Prophylactic Treatment with Proteolytic Enzymes Reduce Acute Toxicity of Adjuvant Pelvic Irradiation? Results of a Double-Blind Randomized Trial." *Radiotherapeutic Oncology* 65 (1):17–22.

Moskvichyov, B. V., E. V. Komarov, and G. P. Ivanova. (1986). "Study of Trypsin Thermodenaturation Process." *Enzyme Microbial Technology* 8 : 498–502.

Sakalova, A., P. R. Bock, L. Dedik, J. Hanisch, W. Schiess, S. Gazova, I. Chabronova, D. Holomanova, M. Mistrik, and M. Hrubisko. (2001). "Retrolective Cohort Study of an Additive Therapy with an Oral Enzyme Preparation in Patients with Multiple Myeloma." *Cancer Chemotherapeutic Pharmacology.* 47 (JulySuppl): S38–44.

Saruc, M., S. Standop, J. Standop, F. Nozawa, A. Itami, K. K. Pandey, S. K. Batra, N. J. Gonzalez, P. Guesry, and P. M. Pour. "Pancreatic Enzyme Extract Improves Survival in Murine Pancreatic Cancer." *Pancreas* 28 (2004) (4), 401–412.

Vickers, A. (2040). "Alternative Cancer Cures: 'Unproven' or 'Disproven'?" *Cancer Journal for Clinicians.* 54 (2): 110–8.

Chapter 17: Exercise Therapy

Aker, P. D., A. R. Gross, C. H. Goldsmith, and P. Peloso. (1996). "Conservative Management of Mechanical Neck Pain.", British Medical Journal, 313: 1291-1296

Ammer, K. "Physiotherapy in Seronegative Spondylarthropathies: A Systematic Review." *European Journal of Physical Medicine and Rehabilitation* 7 (1997): 114–119.

Augustinus, L., D. Noordzij, K. Verkerk, and D. De Visser D. *Systematische Review Naar Actieve Bewegingsvormen Bij Reumatoide Arthritis.* Rotterdam: Hogeschool Rotterdam (in Dutch), 2000.

Smidt, N., Henrica, C.W. de vet, Boulter, L.M., and Dekker, J. (2005). Effectiveness of exercise therapy: A best-evidence summary of systemic reviews, Australian Journal of Physiotherapy, 51(2): 71-85

Elders, L. A. M., A. J. van der Beek, and A. Burdorf. "Return to Work after Sickness Absence Due to Back Disorders: A Systematic Review of Intervention Strategies." *International Archives of Occupational and Environmental Health* 73 (2000): 339–348.

Van Tudler, M., Malmivaara, A., Esmail, R., and Koses,B. (2000). "Exercise Therapy for Low Back Pain: A Systematic Review within the Framework of the Cochrane Collaboration Back Review Group." *Spine* 25(21): 2784–2796.

Flegal, K. E., Kishiyama,S., Zajdel, D., Hass, M, and Oken, B.S. (2007). "Adherence to Yoga and Exercise in a Six Month Clinical Trial." *MBC Complementary and Alternative Medicine* 7 : 37.

Gordon, N. F., M. Gulanick, F. Costa, G. Fletcher, B. A. Franklin, E. J. Roth, and T. Shephard. "Physical Activity and Exercise Recommendations for Stroke Survivors." *Circulation* 109 (2004): 2031–2041.

Kraag, G., B. Stokes, J. Groh, A. Helewa, and C. Goldsmith. "The Effects of Comprehensive Home Physiotherapy and Supervision on Patients with Ankylosing Spondylitis—A Randomised Controlled Trial." *Journal of Rheumatology* 17 (1990): 228–233.

Peeke, P. M., and G. P. Chrousos.(1995). "Hypercortisolism and Obesity." Annal of New York Academy of Sciences 771 : 665–678.

Pina, I. L., C. S. Apstein, G. J. Balady, R. Belardinelli, B. R. Chaitman, B. D. Duscha, B. J. Fletcher, J. L. Fleg, J. N. Myers, and M. J. Sullivan.: "Exercise and Heart Failure: A Statement from the Pomeroy, V. M., and

R. C. Tallis. "Physical Therapy to Improve Movement Performance and Functional Ability Poststroke. Part 1. Existing Evidence." *Clinical Gerontology* 10 (2000): 261–290.

Rossy, L. A., M. A. Buckelew, S. P. Buckelew, N. Dorr, K. J. Hagglund, J. F. Thayer, M. J. McIntosh, J. E. Hewett, and J. C. Johnson. "A Meta-Analysis of Fibromyalgia Treatment Interventions." *Annals of Behavioral Medicine* 21 (1999): 180–191.

Smidt, N., Henrica, C.W. de vet, Boutler, L.M., and Dekker, J. (2003) "Effectiveness of Exercise Therapy: A Best-Evidence Summary of Systematic Reviews and Prevention." *Circulation* 107: 1210–1225.

Smith, D. S., E. Goldberg, A. Ashburn, G. Kinsella, K. Sheikh, P. J. Brennan, T. W. Meade, D. W. Zutshi, J. D. Perry, and J. S. Reeback. "Remedial Therapy after Stroke: A Randomised Controlled Trial. *British Medical Journal* 282 (1981): 5–9.

Stiller, K., and N. Huff. "Respiratory Muscle Training for Tetraplegic Patients: A Literature Review." *Australian Journal of Physiotherapy* 45 (1999): 291–299.

van den Ende, C. H. M, T. P. M. Vliet Vlieland, M. Munneke, and J. M. W. Hazes. *Dynamic Exercise Therapy for Rheumatoid Arthritis. The Cochrane Library, Issue 1.* Oxford: Update Software, 2002.

van der Lee, J. H., I. A. K. Snels, H. Beckerman, and G. J. Lankhorst. "Exercise Therapy for Arm Function in Stroke Patients: A Systematic Review of Randomized Controlled Trials. *Clinical Rehabilitation* 15 (2001): 20–31.

van Tulder, M. W., B. W. Koes, and L. M. Bouter. "Conservative Treatment of Acute and Chronic Nonspecific Low Back Pain: A Systematic Review of Randomized Controlled Trials of the Most Common Interventions." *Spine* 22 (1997): 2128–2156.

van Tulder, M. W., A. Malmivaara, R. Esmail, and B. W. Koes. *Exercise Therapy for Low Back Pain. The Cochrane Library, Issue 1.* Oxford: Update Software, 2002b.

van Tulder, M. W., R. Ostelo, J. W. Vlaeyen, S. J. Linton, S. J. Morley, and W. J. Assendelft. "Behavioral Treatment for Chronic Low Back Pain: A Systematic Review within the Framework of the Cochrane Back Review Group. *Spine* 25 (2000b): 2688–2699.

Werner, R. A., and S. Kessler. "Effectiveness of an Intensive Outpatient Rehabilitation Program for Postacute Stroke Patients." *American Journal of Physical Medicine and Rehabilitation* 75 (1996): 114–120.

Chapter 18: Health Values of Meditation and Yoga

Baptiste, Sherri, and Megan Scott. *Yoga with Weights for Dummies*. City of publication: Betterworldbooks, an online book seller, 2012.

Birdee, Gurjeet S. et al"Characteristics of Yoga Users: Results of a National Survey." *Journal of General Internal Medicine* Vol. 23 (October 2008) (10)1653–1658.

Burley, Michael. *Hatha Yoga: Its Context, Theory and Practice*. Delhi, India: Motilal and Banarasidas, 2000.

Carmody, Denise Lardner, and John Carmody. *Serene Compassion*. City of publication: Oxford University Press US, 1996, p. 68.

Coulter, H. David. *Anatomy of Hatha Yoga: A Manual for Students, Teachers, and Practitioners*. Honesdale PA: Body and Breath, Inc., 2001.

Radhakrishnan, Sarvepalli. *Indian Philosophy*, Vol. II. London: George Allen & Unwin Ltd., 1971pp. 19–20.

Sarbacker, Stuart Ray. *Samādhi: The Numinous and Cessative in Indo-Tibetan Yoga*. City of publication: SUNY Press, 2005, pp. 1–2.

Yogani. *Advanced Yoga Practices—Easy Lessons for Ecstatic Living*. City of publication: Publisher, 2010.

Chapter 19: Stress Control by Foods

Benson, H., and M. Z. Kilppur. *The Relaxation Response*. New York: Harper Torch, 2000.

Goleman, Tara Bennet. *Emotional Alchemy: How the Mind Can Heal the Heart*. Nevada City CA: Harmony Books, 2001.

Muesse, Mark W. *Practicing Mindfullness: An Introduction to Meditation*. City of publication: Harvard University, 2010.

Chapter 20: Sleep Therapy

Aton, S. J., Seibt, J., Dumoulin, M., Jha, S.K., Steinmetz, N., Coleman,T., Naidoo, N., andv Frank, M.G. (2009). "Mechanism of Sleep Dependent Consolidation of Cortical Plasticity." *Neuron* 61(3): 454–466.

Ayas, N. T., White, D.P., Al Delamy, W.K, Mason, J.E., Stampfer, M.J., Spiezer, F.E., Patel, S., and Hu F.B. (2003). "A Perspective Study of Self-Reported

Sleep Duration and Incident Diabetes in Women." *Diabetes Care* 26(2): 380–384.

Danguir, J., and S. Nicholaidis. "Dependence of Sleep on Nutrients' Ability." *Physiology of Behavior* 22 (1979): 735–740.

Gorman, Christine. "Why We Sleep." *Time*, December 20, 2004, pp. 46–59.

Huber, R., Ghiraldi, M.F., Massumini, M., and Tononi, G. (2004). "Local Sleep and Learning." *Nature* 430 : 78–81.

Spiegel, K. "Metabolism and Appetite Regulation: A Scientific Workshop on Sleep Loss and Obesity." Interacting Epidemics, March 27–28, 2006, Washington DC.

Spiegel, K., R. Leproult, and E. Van Cauter. "Impact of Sleep Debt on Metabolic and Endocrine Function." *Lancet* 354 (1999): 1435–1439.

Spiegel, K., Tesali, E., Penev, P., and van Cauter, E.(2004). "Sleep Curtailment in Healthy Young Men Is Associated with Decreased Leptin Levels, Elevated Ghrelin Levels, and Increased Hunger and Appetite." *Annals of Internal Medicine* 141 : 846–850.

Vyazovsky, V. V., Cirelli, C., Pfister-Genskov, M., faraguna, U., and Tononi,G. (2008). "Molecular and Electrophysiological Evidence for Net Synaptic Potentiation in Wake and Depression in Sleep." *Nature. Neuroscience* 11 :200–208.

Chapter 21: Belief and Faith Therapy

Gowari, A. and Hight, Ellen (2001). "Spirituality and Medical Practice." *American. Family Physician* 63 (2001) (1): 81–89.

Begley, Sharon. *Train Your Mind, Change Your Genes.* New York: Ballantine Books/Random House, 2007.

Benor, Daniel J. *Spiritual Healing: Scientific Validation of Healing Revolution.* Southfield MI: Vision Publications, 2000.

Church, Dawson. *The Gene in Your Genes.* Fulton CA: Elite Books, 2007.

Kluger, Jeff. "The Biology of Belief." *Time*, February 12, 2009, pp. 3–7.

Kluger, Jeffrey. "How Faith Can Heal." *Time* February 23, 2009, pp 61-72.

Lipton, Bruce H. *The Biology of Belief: Unleashing the Power of Consciousness, Matter, and Miracles.* Carlsbad CA: Hay House, Inc., 2008.

Lipton, Bruce H., and S. Bhaerman. *Spontaneous Evolution.* Carlsbad, CA: Hay House Inc., 2009.

Mathews, Dale A., McCullough, M.E., Larson, D.B., Korig, H.G., Swyers, J.P., and Milano, M.G. (1998). . "Religious Commitment and Health Status." *Archives of Family Medicine* 7(2): 118–124.

Chapter 22: The Methylation Diet for DNA Stability and Gene Expression

Jump, B. P., and S. D. Clarke. (1999). "Regulation of Gene Expression by Dietary Fat." *Annual Review in Nutrition* 19 : 63–90.

Milagro, E. I., Campion, J., Garcia-Diaz, D.F., Goyenechea, E., Patemain, L,, and Martinez, J.A. (2009). . "High Fat-Diet-Induced Obesity Modifies the Methylation Pattern of Leptin Promoter in Rats." *Journal of Physiological Biochemistry* 65 (1): 1-9

Ross, S. A. "Diet and DNA Methylation Interactions in Cancer Prevention." 983 (2003): 197–207.

Slattery, M. L., Benson, J., Ma, K.N., Schaffer, D., and Potter, J.D. (1997). "Are Dietary Factors Involved in DNA Methylation Associated with Colon Cancer." *Nutrition and Cancer* 28 (1): 52–62.

Lim U. and Song M.A. (2012). "Dietary and Lifestyle Factors in DNA Methylation." *Methods in Molecular. Biology* 76: 876–883.

Stafford, P., Abdelwahab, M.G., Kim do Y., Pressel, M.C., Rho, J.M., and Scheck, A.C. (2010). . "The Ketogenic Diet Reverses Gene Expression Pattern and Reduces ROS Level When Used as an Adjuvant Therapy for Glio" *Nutrition and Metabolism* 7 : 74.

Switzeny, O. J, Mulliner, E., Wagner, Karl-Heinz, Bath, H., Aumuller, E., and Hasslburger, A.G. (2012). "Vitamin and Antioxidant-Rich Diet Increases MLH1 Promoter DNA Methylation in DMT2 Subjects." *Clinical Epigenetics* 4 : 19.

van den Veyer, I. B. "Genetic Effects of Methylation Diets." *Anual. Review in Nutrition* 22 (2002): 255–282.

Chapter 23: Summation

Anderson, J. W. "Structured Weight Loss: Metaanalysis of Weight Loss at 24 Weeks and Assessment of Effects of Intervention Intensity." American Journal of Clinical Nutrition, 21 (2004) (2) 67–69.

James W Anderson, Elizabeth C Konz, Robert C Frederich, and Constance L Wood (2014). Long-term weight-loss maintenance: a meta-analysis of US studies, http://ajcn.nutrition.org/content/74/5/579.full.pdf

Carrano, Andreas C., Liu, Z., Dillin, A., and Hunter, T. (2009). "A Conserved Ubiquitination Pathway Determines Longevity in Response to Diet Restriction." *Nature* 460 : 396–399.

Commander, David, and Michael Rape. "The Ubiquitin Code." *Annual Review in Biochemistry* 81 (2012): 203–229.

Coulter, H. David. *Anatomy of Hatha Yoga: A Manual for Students, Teachers, and Practitioners.* Honesdale PA: Body and Breath, Inc., 2001.

Franz, M. J., Vanwarmer, J.J., Crain, A.L., Boucher, J.L., Histon, T, Caplan, W., Bowman, J.L., and Pronk, N.P. (2007). "Weightloss Outcomes: A Systematic Review and Metaanalysis of Weight Loss Trials with 1 Year Follow Up." *Journal of American. Dietetic Association.* 107 (2007) (10) 1755–1767.

Kaati, G., L. O. Bygren, M. Pembrey, and M. Sjostrom. "Transgenerational Response to Nutrition, Early Life Circumstances and Longevity." *Journal of Human Genetics* 15 (2007): 784–790.

Kurzweil, Raymond, and Terry Grossman. *Fantastic Voyage: Live Long Enough to Live Forever.* New York: Rodale, Inc., 2004.

Lipton, Bruce H. *The Biology of Belief: Unleashing the Power of Consciousness, Matter, and Miracles.* Carlsbad CA: Hay House, Inc., 2008.

McGowan, P. O., M. J. Meaney, and M. Szyf.. "A Diet and the Epigenetic Programming of Phenotypic Differences in Behavior." *Brain Research* 1237 (2008): 12–24.

Reth, Michael. "Matching Cellular Dimensions with Molecular Sizes." 14 Nature Immunology, (2013): 765–767.

Ryu, K. Y. Fujuki, N., Kazantzis,M., Garza, J.C., Bouley, D.M., Stahl, A., Lu X-Y, Nishino, S., and Kopito, R.R. (2010). "Loss of Polyubiquitin Gene Ubb Leads to Metabolic and Sleep Abnormalities in Mice." *Neuropathology and Applied Neurobiology* 36 (4): 285–99. Siri-Tarino, P. W., Q. Sun, F. B. Hu, and R. M. Krauss. "Meta-Analysis of Prospective Cohort Studies Evaluating the Association of Saturated Fat with Cardiovascular Disease." *The American Journal of Clinical Nutrition* 91 (2010) (3): 535–46.

Welchman, Rebeca L., Gordon,C., and Mayer, R.J. (2005). "Ubiquitin and Ubiquitin Like Protein as Multifunctional Signals." *Nature Reviews Molecular Cell Biology* 6(8): 599–609.

Subject Index

W

Y

Z

About the Author

Triveni P. Shukla is a food technologist of international repute by way of his global engagements in food product design, food process deployment, and food marketing. All along his 40 year long professional carrier he has been critical of the impacts of foods and beverages in retail commerce on public health for a very simple reason of disproportional consumption of breakfast, lunch, snack, and dinner foods. He has been very critical of misappropriation and misplacement of calories and nutrients available in foods common to retail stores and away from home eating establishments. He made his prime food industry carrier during the period when all major US food laws on public education as to nutrition and nutrients were in making including later laws on health claims and food supplements during nineties.

He sensed in pain that food marketing overrules public health concerns and that the "health maintenance and education" intent of our laws is never reduced to practice in terms of public health promotion neither in the United States of America where a majority of processed food developed and evolved nor in other nineteen countries he practiced his art and technology of food during a period of over 25 years.

This book springs from his concern about the impact of daily food molecules we consume on (1) gene expression as they connect to and cure chronic diseases, (2) maintenance of our second genome of gut bacteria, and (3) upkeep and health of our second brain- the enteric nervous system. His analysis of works from centers of repute and his critical examination of commercial diet programs convinced him of deception and misrepresentation of foods by which our cells in the body live and without which they die. This book is a sincere effort in proper communication of food values to DNA stability and gene expression, a food dependent process by which we live and maintain our health. Even more important, he believes, is the fact that what, how much, and when we eat has routine impact on our second brain and second genome associated with our digestive system. He expounds in this book that the key to long lasting good health is "Methylation Diet" designed without additives and food substitutes and a well managed lifestyle that stabilizes and in combination trigger our genes for a good metabolism, physiology, and health.

Dr. Shukla holds a Ph.D. in food technology from University of Illinois, Urbana-Champaign, IL. He is corner stone member of Council of Agricultural Science and Technology. He sees through the technologies of food processing and food products with a keen and curious mind of fundamental biochemist.